I0488432

CHANGE

DESCRIBED-EXPLAINED-PREDICTED

The Galactic Force Driving Solar System,
Planetary and Human Change

BY Paul R. Drewfs

Published by Paul Drewfs
2010

Published by Paul Drewfs
Address: 6801 Baker Avenue N.E.
Albuquerque, New Mexico 87109
E-mail: pdrewfs@aol.com

Copyright © 2007 by Paul R. Drewfs

All rights reserved including the right of reproduction in whole or in part. No part of this publication may be reproduced or transmitted in any form or by any means, electronic or mechanical, including photocopying, recording, or by any information storage and retrieval system, without permission from Paul R. Drewfs. Reviewers, researchers, and educators may quote passages using accepted standards of attribution, so long as the excerpts do not exceed ten percent of the total content of this publication.

First paperback edition 2010

Printed in the United States of America

International Standard Book Number
ISBN 978-0-0557-62063-0

Dedicated to Rudolf Clausius' 1862 statement of the second law of thermodynamics – *"The algebraic sum of all the transformations occurring in a cyclical process can only be positive, or, as an extreme case, equal to nothing."*

PREFACE

Imagine an invisible tsunami nearly 4 million times larger than our entire solar system rushing through you at the speed of light. An indomitable self-propagating gravitational upsurge irrevocably transforming everything it passes through. Now picture that monster-breaker twisting, stretching and squeezing every particle in our solar system for over ten thousand years. Every foot-pound of pressure of that spiraling swell leveraging sufficient torque to turn dust to stars, tilt the vertical axis of our entire solar system, and suck most of the angular momentum from our Sun.

Sound like the end of the World? Well, it isn't. You are living on the leading edge of one of those massive waves this very minute. What's more, that great tidal-wave of torque is steadily advancing our species, evoking cyclic Earth changes, and synchronizing the motions of our entire solar neighborhood. Why if not for hundreds of thousands of waves just like it, our solar system wouldn't exist. Take away the good work of a half million more and our Galaxy would wind up in a tight little knot at the Galactic Center – and there'd be no more Milky Way.

For the last two decades scientists have been observing, measuring, modeling, and simulating these massive gravitational waves of torque. Yet, unless you subscribe to an astrophysics journal for the mentally-math-mad you probably haven't heard about them. How could anyone fail to notice such massive waves? Easy, they are totally invisible. Their products, however, are not, and include no less than our galaxy, solar system, this planet, and our fundamental human sense of a physical self.

What is genuinely spooky about all of this is that our ancient ancestors discovered the effects of these gravitational torque waves at least 3,000 years ago. Back then they called the effects of those waves *"World Ages"* and *"Cycles of Change."* They termed the stellar rush brought on by each new wave, *"The Quickening Time,"* and set their calendars by it.

This book weaves together dates, numbers, and assertions bequeathed to us by our long dead predecessors with those of fairly recent discoveries. Why; to put a few of these prescient primeval assertions to the test. But the results reveal much more than the validity of a few primordial beliefs. The findings reported in these pages extend our knowledge and understanding of this solar system, our planet, and our most fundamental sense of self. Consequently, this book brings together basic perspectives too long torn asunder and resurrects frighteningly accurate ancient insights too long overlooked by the fragment-fixated minds of modern men. In their perpetual haste to compulsively ferret out multitudinous details, our beloved scientists have failed to integrate the greatest hits of our long dead progenitors. A situation now made ironically correctable by the cutting edge discoveries mated-up in this book.

Contemporary researchers have rigorously demonstrated that galactic waves of gravitational torque are produced by gargantuan bars of elder stars fast rotating about galactic centers. They have demonstrated those waves universally responsible for the evolution, rotational speed, and distribution of all the stellar matter comprising all the visible galaxies. More recently those same scientists have learned that those tugging waves of torque generate Herculean ripples in the very fabric of space and time. Great stirrings that compress energy and matter, extrude long narrow streams of dust and gas, facilitate new star formation, and feed the massive central star that comprises every galactic nucleus.

Putting that new found knowledge to work, still other investigators have modeled the birth, maturation, death, and resurrection of all of the known types of visible galaxies – including our own. These remarkably detailed simulations now spin the entirety of the 13.75 billion years since the birth of our universe. From such models leading cosmologists have discovered that in the absence of such stirring-stellar-torques, galaxies collapse. Yet, given a strong rotating stellar bar, galactic structures like the spiral arms of our Milky Way, emerge, mature, stabilize, and persist for tens of billions of years.

A few emboldened science-bent souls have modeled the periodic effects of these gravitational torque waves on the Galactic orbit of our solar neighborhood. It was those results that led us to revisit the outrageously keen insights of our ancient forefathers. An effort richly rewarded by the discovery of uncanny matches between the long derided perspectives of the ancients and our most recent cosmological breakthroughs.

The idea that *World age cycles* count out periodic *waves of change* is one of man's oldest and most widespread beliefs. That primordial assertion all but vanished in the last millennia, and for an acutely dark reason. About a thousand years ago the belief in rhythmic waves of change shot straight to number one on the heresy charts. The good news; simply repeating the concept is no longer considered adequate justification for the public burning, beheading, quartering, or exiling of those folks who dare extol it. The bad news; a valid, reliable, practical, and economic predictor of fundamental change has been lost to us for a very long time.

Well, heretical or not, the objective evidence inscribed in these pages clearly vindicates that periodic wave-change hypothesis. What is more, the resulting knowledge condemns a fair few sacred-cows straight to the slaughterhouse. Such is the price of objectively validating what has been subjectively torn asunder by the minutia-mad-minds of modern men.

CONTENTS

1 INTRODUCTION 1

2 WAVES OF HUMAN CHANGE 9

Competing Views of Human Change 9
What is a Human? 14
Why start 282,000 years ago? 16
Maya Asserted Change Intervals 18
Human Changes at issue 19
What is social diffusion? 24
Testing Extreme Gradualism 31
Argument for Punctuated Equilibrium 33
Testing the Maya Theory of Human
Evolution 35

3 WAVES OF WORLD AGE EPOCHS 47

Testing for World Age Epochs 47
Human Change in the Current Period 54
Waves of Human Change 56

4 WAVES OF HUMAN EVOLUTION 61

Emergence and Extinction 61
Bottlenecks, Mutations, and Migrations 66
Human Change Summary 69

**5   WAVES OF CELESTIAL
      MOTION   73**

Precession   75
Solar System Z-Axis Gain   87
Solar System Galactic Orbital Speed   90
Obliquity   92
Inclination   93
Earth's Orbital Eccentricity   96
Description, Explanation, and Prediction   99
Interpretation   102

**6   WAVES OF CLIMATE
      CHANGE   105**

Paradoxical Views   105
800,000 Years of Global Climatic Data   106
Weather Born Change Illusion   109
Historic Long Lived Greenhouse Gas
Emissions   114
Current Epoch Motion, Temperature, and
CO_2 120
Global Warming   125
The Unanswered Question   127

**7   WAVES OF GRAVITATIONAL
      TORQUE   129**

An Ancient Clue   129
Criteria   131
An Invisible Wave-like Force; Gravity   133
Sufficient Force; Torque   134
Waves   136
Gravity Waves   137
Gravitational Torque Waves   139

Originating at the Galactic Center   142
Wavelength, Amplitude, and Pitch Angle 145
Poisson distributed   157

8   WAVES OF WORLD AGES   161

Sourcing and Forcing Trains   161
Testing Gravitational Torque Waves   163
     From 285,012 BC to 1500 AD   163
     From 1880 to 2007   166
Confirmation from the Edge   171

**9   WAVES OF ECONOMIC
CHANGE   177**

Crashing Clues   177
Economic History Lessons   178
Building Perspective   180
Stochastic Mysticism   181
Testing Gravitational Torque Wave
Effects   195
Prediction of Per Capita Consumption   201
Prediction of Real US Gross National
Product   202
Prediction of Global Wages   203
Prediction of Real US Consumption   207
Prediction of US Gross Domestic
Product   208
Prediction of US Money Supply   209
Prediction of the Velocity of Money   211
Prediction of the US Consumer Price
Index   212
Purging Perplexity   214

10 WAVELETS OF SYNCHRONIZATION   **217**

Maya Sub-cycles   217
Unique Maya Day   219
The Maya Long Count Day   222
Maya Long Count Sub-cycles   225
Sub-cycles in Perspective   228
Indexing the *K'ul* Response and Resonance   229
Testing the *K'ul* Resonance   233
Validating the *K'ul* Resonance   238
Compression and Resonance Distributions   239
K'ul Improved Predictions   240
Scary Smart Maya   244

11 WAVES OF HUMMING EARTH   **247**

Measuring Earth's Gravity   248
Human Sensing of Earth's Hum   258
Sensitivity of the Vestibular System   261

12 WAVES OF HUMMING IDENTITY   **271**

The Default Mode Network   272
Source of the Default Mode Net Rhythm   282
Definitely Neuronal   282
Case for an External Stimulus   284
Locator Candidates   288
The Most Likely Candidate Source   292

**13   WAVES OF HUMAN
        INFLUENCE   297**

    Question   297
    Data   299
    Measures   300
    Hypotheses   300
    Results   303
    Interpretation   317
    Theoretic Implications   318

**14   TSUNAMI TRAINS OF
        TRANSITION   323**

    Maya Prophecy of Transformation   323
    Wave Train and Caboose Quandary   326
    Resonant Swing Solution   327
    Transformation Precedence   331
    Tiny Echoes of the 2.5 Wave-Anomaly   336

**15   WORLD AGE RESONANT PHASE
        SYNCHRONIZATION   343**

    A Composite Theory; WARPS   343
    Whomp of WARPS   349
    Wave Warp Drive   353
    A Falsifiable Theory   358

**16   WAVES OF CASCADING
        TRANSFORMATION   363**

    Non-Destructive   363
    Precedence   364

Generalization 368
A Downside 372

17   WAVES OF EARTH COMPRESSION   381

Earth's Tummy Tuck Trend 381
Earth's Long Term Waistline Reduction 387
Predicting Earth's Waist Warps 390
Predicting Earth's Waistline Anomalies 394

18   WAVE TRAIN SIGNALS, SWITCHES, AND SIDINGS   399

Accelerating Rate of Human Genetic
Change 402
Population and the Rate of
Genetic change 404
Meaning is Usage; Not Possession 409
Evolving Switch Settings 412
Accelerating Rate of Stressed
Populations 414
GTW Cosmic Ray Forcing 419

19   CONCLUSIONS   429

Counting the Unknown Cause 429
Measuring & Predicting Effects 430
Revealing the Cause 430
Conjoining Cause and Effect 432
Drilling Deeper 433
A Mind Warping Prediction 434
Value Added 436

20   HYPOTHESES SUMMARY   437

Waves of Human Change   437
Waves of Celestial Motion   440
Waves of Climate Change   440
Waves of Gravitational Torque   442
Waves of Economic Change   444
Wavelets of Synchronization   446
Waves of Humming Earth   448
Waves of Humming Identity   449
Waves of Human Influence   450
Tsunami Waves of Transition   452
World Age Resonant Phase
Synchronization (WARPS)   453
Waves of Earth Compression   453
Wave Train Signals, Switches, &
Sidings   456

END NOTES   459

TABLE 8 ENDNOTES FOR ESTIMATES OF
DISTANCE FROM SUN TO GALACTIC
CENTER   497

TABLE 8 PRECESSION ESTIMATE
ENDNOTES   499

SUBJECT INDEX   501

AUTHOR PAGE   513

1

INTRODUCTION

Human change occurs in waves; cyclic pulses that explain the variations in the last 280,000 years of our oft sordid history and predict our precarious future. The timing and distribution of our most dramatic changes are recorded in our cumulative fossils, genome, and history. Those records clearly reflect the wave-like nature of *human* change. What's more, those waves correlate in the extreme with the fundamental motions of this planet, our solar system, and at least 800 millennia of Earth's climatic variation. Yes, even climate change, as evidenced by accumulating Antarctic ice core data, paints the rhythmic shifts in Earth's global temperature and those infamous long lived greenhouse gases.

Strange as it may seem, the long trains of historic waves of change are a dead-on match to *World Age* cycles once diligently tracked by long departed civilizations. Civilizations separated not just in time, but by continents and oceans. Civilizations that included, among others, the: ancient Mesoamericans, Vedic East Indians, Sumerians, Assyrians, Egyptians, Babylonians, Chinese, and peoples of the later Hellenic and Islamic cultures.

No known past civilization more precisely tracked the great cycles of change than the Mesoamerican Maya. It was the ancient Maya who most effectively used their World Age *Epoch* counting Precession cycles to accurately describe, explain, and predict human, planetary, and solar system changes. With their insights forbidden by their Spanish conquers, the Maya left a major mystery unsolved; the physical source of the monumental cycles of synchronizing change that they so diligently tracked and calculated.

Adding to the confusion, people today falsely acknowledge such long lived cycles as the *precession of the equinoxes*. A physical phenomenon that modern scientists have gotten patently wrong for 465 years, and horribly underestimated in its role and importance. Worse, most scientists still ascribe their version of precession to an impossible wobble in the vertical axis of Earth's rotation. The early Maya, by contrast, correctly localized the Precession of World Ages to the tightly repeating cyclic wobble in the vertical rotational axis of our entire solar system.

The ancient Maya ardently insisted that World Age defining Precession cycles are driven by forces emanating from the Galactic Center. An assertion snickered and scoffed at since the days of Isaac Newton. Modern astronomers argue that the Galactic Center is much too far away and much too weak a gravitational source to influence the motions of objects in our solar system. At the same time and in direct contradiction, their astrophysicist contemporaries successfully demonstrated the Galaxy-wide effects of Gravitational Torque Waves. Waves that emanate from the Galactic Center, just as asserted by the Maya at least 2,100 years ago. These massive wave-warps of tugging gravitational torque propagate at the speed of light from the great bar of stars fast rotating about our evolving Galactic Center. From the edges of that stellar bar spread waves that impose a twisting force on the very fabric of space and time (space-time). That force stirs and synchronizes the stellar medium from the Galactic Center to well beyond the outermost reaches of the great Galactic plane.

This book relates prior research to the results of a series of analytic tests and shares results interpretations. The tests reported were performed to confirm or deny the repeated occurrence of the World Age cycles of change. Those very cycles once counted out in fixed length *Epochs* by the ancient Maya.

That done, World Age cycles are quantitatively related to recent discoveries about Gravitational Torque Waves to identify and describe the astronomical source of those confirmed cycles of change. To cut right to it, the density compression effects of passing Gravitational Torque Waves are used to describe, explain, and predict the history of man, Global climate change, the major motions of our planet and solar system, human behavior, and the physical dynamics of our autonomic sense of self.

These pages scrutinize and integrate a number of fundamental assertions concerning the basic workings of our Galaxy, solar system, and our selves. Presented analytic test results negate and supplant a fair number of tightly clung to views of our oft over-sanctified science. The tendency to overstate and over generalize is constrained, by formally stating each *investigative* hypothesis put to the test. An objective perspective is further forced on the work by stating prevailing views directly contradicting those hypotheses. Prevailing- and opposing-views take the form of what are labeled *null* and *alternative* hypotheses.

An atypically conservative numeric limit (i.e., statistical significance level) is set for rejection or acceptance of each hypothesis, be it null, alternative, or investigative. That limit is .001, meaning there can be no more than one chance in a thousand of rejecting a hypothesis when it is true, or of accepting it when it is false. Fortunately, the overwhelming majority of the analysis results reported in this book have yielded a statistical significance level with less than one chance in a million or less of falsely accepting or rejecting a hypothesis. In the chapter titled 'Waves of Human Influence' near the end of the book, that confidence limit is reset to the .05 level, owing to the nature of the tests and the data.

The last chapter restates all of the hypotheses and the declarations of rejection or acceptance addressed in the book. That listing is provided solely for purposes of summative reference and review. Worry not; you forfeit nothing by not reading those hypotheses and declarations up front. Each hypothesis is stated in the chapter and section where it is tested, accepted, or rejected.

The original objective was not to define the driving source of solar system, planetary, and human change. The initial intent was to test the ancient Maya concepts of *World Age* change cycles and partitioning *Epochs*. It was reasoned that if the Maya had gotten it right, their declared cycles ought to predict human history and global climatic change. Nailing that down a bit tighter, it was hypothesized that the "Great and Grand cycles" indexed by Degrees (day-count to the Maya) of Maya Precession should accurately predict the historic grouping of 280,000 years of human adaptation, invention, and innovation. A further original objective was to see if those same cycles could predict 800,000 years of recorded changes in global temperature and the concentration of long lived greenhouse gases. Things changed, when the Maya cycles proved excellent predictors of those very things.

Given those early discoveries, it was posited that those same cyclic Ages and Epochs should account for key aspects of celestial motion. Tests were performed to see if the Maya cycles could predict: true Precession (the wobble in the vertical axis of our solar system); Earth's Obliquity (the cyclic tilt in the vertical axis of our planet); Inclination (the tilt of our planet's solar orbit); and Earth's orbital Eccentricity (the deviation of our planet's solar orbit from a perfect circle). When the Maya cycles did precisely that, it was realized that the true scope of the challenge still lay ahead. It was abundantly clear that all those significant predictive relations had to be the result of some positively enormous clockwork-like wave force.

Consequently, a meticulous search of the scientific literature was performed. Sought was evidence of gargantuan invisible waves of energy capable of modifying literally anything and everything. Well, everything from the most microscopic constituents of space-time to the largest macroscopic objects in our Galaxy. The hunt was on for an outrageously large Poisson or normally distributed wave force that self-propagates at the speed of light. Daunting as all that was it soon became clear that the mysterious wave force had to have a wavelength of 10,250.7 light years and a wave height of 5,125.36 light years (where one light year is the distance light travels in a year; 9,460,730,472,580.8 kilometers).

Of course any genuinely viable candidate wave force would also have to constitute a viable source of Solar System: Precession, Z-Axis vertical gain, Galactic orbital oscillation and velocity variation. Then too, that wave force would also have to be the source of Earth's celestial motions and variations in equatorial and polar diameters. Oh yes, and the arrivals and departures of truly candidate waves had to correlate with Earth's global climate history, predict the diffusion of human innovation for the last 280 millennia, and count out seven million years of human-like species emergences and extinctions.

It hardly seemed likely that evidence of such an all encompassing wave force lay languishing in the existing scientific literature. Strangely enough, the long gone Maya pointed the way. The ancient Maya had staunchly insisted that the unimaginably massive wave force driving their World Ages and Epochs arises from the Galactic Center. So, the search was begun there.

Imagine the surprise at the discovery that for well over two decades astrophysicists have been studying invisible Galactic Gravitational Torque Waves (GTW) arising from galactic centers. As it turned out, those GTW were the very waves sought.

Literature described density waves (also known as GTW) produce outrageously large gravitational compression effects on literally everything at every level of energy and matter. Compression effects that correlate perfectly with Maya World Age Precession, the dates of human change, and solar system motion. What's more, those wave born compression effects predict Earth's history of global temperature change and variations in atmospheric greenhouse gases. Even more astounding, the accelerating arrival of the first GTW in each World Age correlates perfectly with: hominid (human like) species emergences (so called *speciation* events), genetic diversity reductions (so called genetic *bottlenecks*), major human migrations (as indexed by regionally specific population *genetic markers*), and all of the hominid species extinctions known to have occurred in the last seven million years.

All that constituted a major set of discoveries, to be sure, but like most Earth shaking breakthroughs, there's an inevitable price to be paid. In the current case, that price is the seriously disconcerting news that we are presently living in the leading edge of just such a change forcing GTW. Yep, that GTW now moving through our tiny solar neighborhood is of the Earth transforming kind.

And, what's more, it could well become one of those historically associated with human genetic bottlenecks, speciation events, great migrations, and extinctions.

One could not help but wonder; could the *coming* rate of human and Earth changes possibly exceed the rate of change from 1500 AD to now? The shocking answer detailed in this book is a resounding *YES!* The rate of human innovation and change from the mid-Renaissance to the present represents less than 1.5 percent of the magnitude of accelerating change that awaits us in the remainder of the current wave – and that maybe a monumental underestimate.

If the Maya had it right – and the evidence strongly suggests they did – we are about to become as fundamentally different from our immediate ancestors as they were from our long extinct Neanderthal cousins. A point that suggests we might be well advised to attend to the knowledge and insights bequeathed us by our long-expired elders. In the prophetic words of Winston Churchill, "The further backward you look the further forward you can see."[1]

2

WAVES OF HUMAN CHANGE

No one contests the fact that human beings change. The evidence is abundant and overwhelming. Throughout human history human beings have adapted physically, varied in individual and collective behavior, and invented and socially diffused a veritable ocean of novel ideas, practices, and objects. Such time indexed shifts in man and his artifacts are prominently evidenced by the unearthed and interpreted fossil record, the sequenced genetic blueprint known as *the expressed* human genome, and surviving historical records.

Competing Views

What is hotly contested is how much human beings change, how fast, and how often. These overheated battle-lines have been drawn, challenged, and re-drawn since the early 1870s. Opposing camps fling edgy word daggers back and forth at one another through an ever evolving literature. Jargon riddled journals continue to host emboldened attempts to strengthen the latest position and weaken prior ones. Engaging factions christen their staunchly defended positions with a litany of exotic names. Yet the only real differences between these notions congeal down to this: Arguments of how often and how fast humans transition from a stable state of *static equilibrium* (periods of little to no discernable change), to a state of *dynamic equilibrium* (periods of clearly evident dynamic change).

The competing theories of the rate of human change are:

1. <u>Extreme Gradualism</u>: The idea that Human change is continuously distributed and equally probable over time (i.e., exhibiting as a decidedly flat rectangular distribution).

2. <u>Extreme Punctuated Equilibrium</u>: The notion that significant Human change is time concentrated in discrete irregularly spaced evolutionary jumps. Bursts of change occurring when a population undergoes substantial stress or threat of extinction, and changes occur solely in one or more suddenly geographically separated subpopulations.

3. <u>Gradualism with Punctuated Equilibrium</u>: The view that the rate of Human change sporadically fluctuates from very fast to very slow, with the static equilibrium condition representing nothing more than periods of ultra slow change.

Over 2000 years ago our wizened ancestors held a much more definitive view of human change. Laid alongside the three leading contemporary theories, the newer notions seem downright subjective, noncommittal, and ambiguous.

A Maya shaman of 2,100 years ago, for example, would posit that human *speciation* (the emergence of a new human species) when enabled, is forced into existence at World Age Cycle change points; intervals spaced 25,626.8 years apart (or 360 degrees on the Maya World Age circle, which the Maya counted in days). The Maya shaman would insist that any new species emerges solely in the first even numbered Epoch (i.e., each spaced 5,125.36 years or 72 degrees apart) on the World Age cycle that triggered speciation.

Maya shaman comfortably endorsed the equivalent of our Darwinian notion of new species adaptation through natural and sexual selection. They had no problem with the idea that any new form of man continues to evolve and drift physiologically. Nor did they doubt that man proceeds to invent and socially absorb genuinely novel ideas, practices and objects. However, they would sternly add that the pace of such changes varies in accordance with the wavelike accelerations and decelerations of alternating World Age cycle Epochs, and that Epoch peaks and valleys are spaced 5,125.36 years apart.

Listening to an ancient Maya shaman describe this serpentine variation in the frequency of human invention and innovation, a modern scientist would interpret him to be describing Poisson or normally distributed sinusoidal

waveforms of change. Waves that concentrate the
frequency of human change at and about the boundaries
of Epoch divided waves 10,250.7 years long. Agreeing
with that characterization, the ancient shaman would
point out that at the end of each World Age Cycle – and
hence the beginning of the next – different species take
different courses. Some spawn a new species. Others
experience a burst of genetic changes within species,
while still others go extinct. Figure 1 illustrates these
Maya World Age Cycle implied waves and the
partitioning Epochs of change that punctuate them.

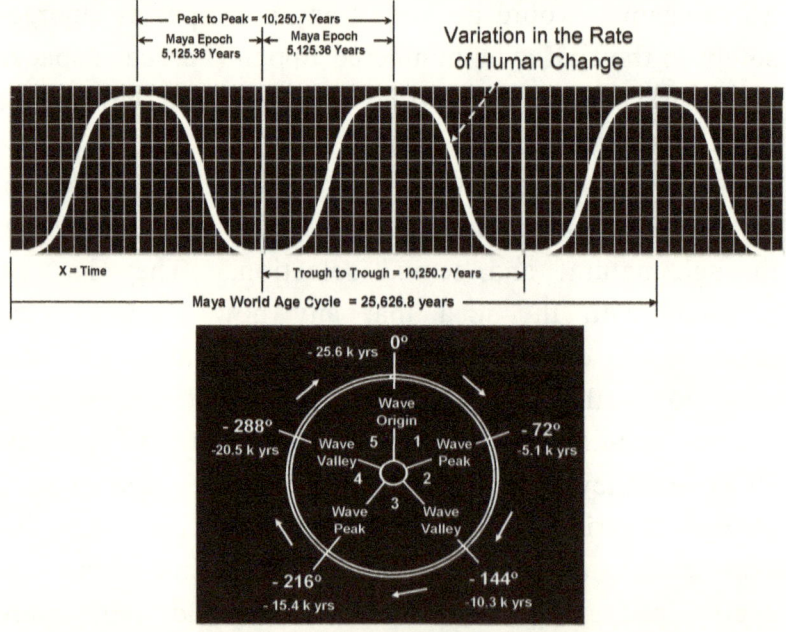

Figure 1 the Mayan View of Change

The Maya considered their view of change more
than mere theory. To them it was the quintessential

natural force that describes, explains, and predicts change. Not just human change, but all change.

It was seriously disconcerting to find that the Maya theory of Human Change subsumes and supersedes our best modern theories of human evolution. What shocks one's socks off, however, is that the Maya theory better lends itself to scientific test and evaluation. What's more, the Maya theory renders our competing contemporary theories fundamentally falsifiable. If evaluated against the objective *demonstration* criteria of modern science, it is the Maya theory of human evolution that deserves high marks.

Here's why; only two things were needed to confirm or deny the cited ancient Mayan contentions: 1) a representative database of solidly dated human changes spanning the definitive history of man; and 2) a verified means of mapping the 25,626.8 year long Maya World Age cycles and 5,125.36 year long *Epochs* onto that historical database. As if anticipating every future need, the long gone Maya left behind the very calendar dates needed. What's more, multiple scholars had long ago mapped those dates onto our error ridden Gregorian calendar.

After converting the Maya theory to investigative hypotheses, a survey of the paleontology, archeology, genetic, and anthropological literature was conducted. Using the fossil, genetic, and historical evidence afforded by that literature, a representative database of the dates of Human Change spanning the last 282,000 years was constructed. Before discussing the analysis and results, a few foundational terms and concepts need to be shared.

13

Key ideas used by scientists to qualify and characterize what a human is, and what is meant by human innovation and change.

What is a Human?

The term "Homo" is defined as the genus of the family of *Hominidae* [i.e., *the Hominids*] whose sole surviving species is *Homo sapiens* (us), but which also includes all extinct humans.[2] *Hominids* constitute a long lived family of highly varied modern human ancestor aspirants. The prevailing school of thought dominating modern evolutionary theory defines a highly theoretical and oft revised hominid family tree. Currently, that tree begins with the species "*Sahelanthropus tchandensis,* who is believed to have lived six to seven million years ago."[3] According to the accepted fossil record, as described by Carl Zimmer (2005) in the "Smithsonian Guide to Human Origins,"[4] *Sahelanthropus tchandensis* tops the tree of all of the Hominid species cataloged in Table 1.

In the main, no attempt is made to imply that the majority of tabled hominid species are our literal ancestral relatives. There is a single exception; the more genetically diverse version of *Homo sapiens* that lived from 200,000 years ago to 90,000 years ago. That species is unquestionably the genetic progenitor of surviving *Homo sapiens*. The rest of the controversial job of lineage-linkage is left to contemporary paleontologists, archeologists, anthropologists, and geneticists. Fortunately, such speculative and frequently revised leaps of ancestry are not at issue here. Tests of the hypotheses addressed in this investigation require only accepted fossils, and the available genetic markers that help discriminate each species.

Table 1 Hominid Species Emergence, Extinction, and Longevity

Species	Hominid Species	Years Ago		Years
		Emergence	Extinction	Longevity
1	Sahelanthropus tchadensis	7,000,000	6,000,000	1,000,000
2	Orrorin tugenis	6,400,000	5,600,000	800,000
3	Ardipithecus	5,600,000	4,500,000	1,100,000
4	Australopithecus anamensis	4,200,000	3,900,000	300,000
5	Australopithecus afarensis	3,900,000	3,000,000	900,000
6	Australopithecus africanus	3,500,000	2,400,000	1,100,000
7	Australopithecus garhi	2,700,000	2,400,000	300,000
8	Paranthropus boisei	2,400,000	1,200,000	1,200,000
9	Homo habilis	2,400,000	1,550,000	850,000
10	Homo ergaster	1,850,000	1,500,000	350,000
11	Homo erectus	1,800,000	30,000	1,770,000
12	Homo heidelbergensis	600,000	100,000	500,000
13	Homo sapiens (earliest)	200,000	90,000	110,000
14	Homo neanderthalensis	150,000	30,000	120,000
15	Homo sapiens (current)	150,000		150,000
16	Homo floresiensis	98,000	18,000	80,000

Reference: Carl Zimmer, 2005, Smithsonian Intimate Guide to Human Origins, Madison Press Ltd., Toronto, Canada, p.6-7

Why start 282,000 years ago?

The *fossil* search was halted at 282,000 years ago for several reasons. First, the further back in time one goes, the fewer *accepted* fossils there are in the unearthed and interpreted fossil record. What's more, the further back in time a fossil is dated, the greater the error of measurement associated with that dating. The term 'accepted fossil' simply implies that a given fossil dating and interpretation fits reasonably well into some prevailing faction's favored model of hominid evolution. As such, an *accepted* fossil is only mildly controversial, while all others remain the targets of open academic warfare.

Paleontology, archeology, and anthropology are *fuzzy* sorts of sciences. These fields of inquiry depend a lot on *pure dumb luck* in their search for valid and reliable fossil evidence. Evidence that must survive the continually shifting views of controlling collegial associations. As pointed out by Carl Zimmer (2005), these disciplines face a far more fundamental challenge as well.

> Only a tiny fraction of all hominids that ever lived have been memorialized in rock. What determined whether a hominid turned into a fossil or not? Its fate was largely dependent on how it died. If a hominid was killed by a leopard, most of its bones would be digested, gnawed, or otherwise destroyed. The remnants would be worked over by insects and bacteria, and finally washed away into a river and then the sea.

A better candidate for fossilization would have died by drowning in a still, muddy pool. Its body might then be gradually covered in sediment. As its flesh was devoured by mud-dwelling bacteria, its skeleton would remain joined together. Over the course of millions of years, the bones would be transformed to stone.

...

A single fossil can take years to pull from the rock, but the payoff is enormous – a glimpse at our ancestry.[5]

Another reason the fossil hunt was halted at 282,000 years ago, was the convergence of key events around that time. Fossil evidence of *Homo Heidelbergensis* (Heidelberg man) and the much older *Homo erectus* continue to be found and dated to that time. The genetic founding of *Homo neanderthalensis* (Neanderthal man) is evidenced in the fossil record around then as well. Equally important, 282,000 years ago is 82,000 years before the suspected emergence of the now nearly extinct version of the earliest *Homo sapiens*. It is also 87,000 years before the first accepted fossil evidence of our own (truly modern) man emerged.

A starting point of 282,000 years also encompasses a series of genetic "bottlenecks" – species threatening genetic diversity narrowing events – that needed to be included in this study. In addition, a starting point of 282,000 year ago is 132,000 years before the first fossil and genetic evidence of currently predominating *Homo sapiens* (i.e., the estimated birthdays of our literal genetic "Adam" and "Eve"). It was thought important to establish a substantial baseline of changes preceding our literal ancestors. For it is against that baseline that we later distinguished ourselves.

Maya Asserted Change Intervals

The end-date of the Human Change database was tightened up in another way as well. Since the plan was to test the descriptive and predictive accuracy of the ancient Maya 25,626.8 year long World Age cycles of change, that interval was multiplied by 11 to derive a more specific starting point of 281,894.8 years ago. Hence, the fossil and historical data records are grouped into 11 periods of 25,627 years each, numbered -1 to -11. The minus sign in front of the Maya World Age Precession Period numbers simply indicates that the age numbers increase in the negative direction moving backward in time from zero.

The calendar date of zero was set at the Gregorian calendar date of 1500 AD around the middle of the Renaissance, some 508 years ago, as of 2008. This was done for multiple reasons that are detailed as they come up in the course of this book. Positive Period 1 in the database thus represents the last 508 years of Human change from 1500 AD to 2008.

Readers familiar with Maya cosmology may view the calendar placement of zero as a mistake. After all, the Maya Long Count Calendar end-date has been translated and verified as coinciding with the Northern Winter Solstice of Friday December 21, 2012 AD.[6] In the course of this investigation something important was discovered; the beginning of Period 1 and the end-date of the Maya Long Count calendar represent two distinctly different dates. It turns out that these two dates coincide with two different planetary, solar system, and Galactic conditions.

While the zero point date of 1500 AD was independently derived, the Maya may well have done the same. In his 1998 book, "Maya Cosmogensis 2012," John Major Jenkins provides evidence supporting that assumption. According to Jenkins, the astronomical events predicted by the Maya to culminate on the Winter solstice of 2012, actually begin over 450 years earlier.[7] The Maya may well have set what equates to a 0 point at 512 years prior to 2012 AD, which equals 1500 AD. 512 years is one tenth of one Maya *Epoch* of 5,125.36 years, an Epoch sub-unit-scaling value known to be employed by the Maya. Further, the date of "1507 AD" coincides with the long observed sacred Mesoamerican "New Fire Ceremony."[8] Regardless, the database zero point was set to 1500 AD.

Human Changes at issue

Human Changes of interest in this investigation include any novel idea, practice, object or physiological adaptation diffused within a population of hominids that lived from 281,895 years ago to the present. All that was required is evidence that any such candidate innovation fit within the accepted interpretations of the fossil, genetic, or historical records. In the case of fossil and genetic evidence, it was also required that changes be scientifically dated by multiple sources using multiple established dating methods.

As indicated earlier, the older the fossils used to define and date a human change, the less accurate the reported dating of it is likely to be. Carolyn Barry mercifully puts the carbon dating version of this problem this way:

> "Because most archeological remains contain carbon, the method of choice for determining age is carbon dating in which scientists compare the relative amounts of stable carbon isotope to one that radioactively decays. Radiocarbon ages are reliable – to a point. Corroborating data from ice cores, corals, and tree rings have pushed the dependable carbon dates back to about 26,000 years, but in preceding millennia the dates became increasingly less certain. By 50,000 to 60,000 years ago, a radiocarbon date [as opposed to the many alternative kinds of dating that are used] might be off as much as 2,000 years from the true date [± 1,000 years].
>
> ...
>
> Advances in reducing sample contamination, improved techniques to extract specific compounds out of samples, and a new source for ancient tree ring data have offered hope that the 60,000-year [radiocarbon] benchmark is within reach."[9]

The error of measurement for included fossil based Human Change dates was set to 2,500 (± 1,250) years. For dates more recent than 75,000 years ago, this error term is overly conservative. For the more distant fossil dates reaching back from 75 to 280 millennia, this error term becomes increasingly cavalier. In the main, though, fossil dates tend to be severely "underestimated" as opposed to overestimated.[10]

From 281,895 years ago to around 25,627 years ago, such fuzzy-dated fossilized remains and artifacts dominate the evidence of human change. Starting around 51,000 years ago, Neanderthals and modern *Homo sapiens* began leaving our scientists a bit more to work with. At that point advanced hearths, tools, adornments, and artwork (like rock and cave painting) substantially augment the scant fossil record. Beginning around 12,800 years ago, fossil evidence to include genetic material teased from petrified bones and human feces begin to gradually be supported by more direct historical evidence of what is termed *human civilization*. Around 10,000 years ago widely spaced remnants of rare innovations begin to give way to great hordes of man-made leavings.

Early on a niggling issue arose; what to do about simultaneously reoccurring variant discoveries. All sorts of intriguing little variant knock-offs and reinventions surround genuinely unique innovations. For example, the first: hunter-gather religious center, city, walled city, city with running water, city with flushing toilets, sewers, and so on, are all unique breakthroughs. In more recent history, every ancient city on every continent is dated and counted as a unique innovation in the literature. Within such subsets primordial cities vie for the title of *first and oldest*, not only locally, but regionally and globally. As a result, the number of human change events multiplies with a bursting vengeance as time decreases toward the present.

All the dated inventions and innovations found were faithfully collected. All the while it was worried that the great explosion of nearer term historic innovations would mask the effects of any cycles of change that lay hidden in the data. The concern was unwarranted, though initially that fact was unknown. So, at first the data were cut and analyzed in two different ways. One cut focused on the subset containing what was foolishly segregated out as definitive *firsts*. The other used the whole grand garbage-can of overlapping human changes reported in the literature. It was a great relief to discover that the trends in the history of human change are so pronounced that not even man's vainglorious attempts at geo-local self-deification could suppress them. In the end, all the variant cases and their multitudinous echoes were analyzed together.

The trying struggle to define definitive firsts led to an important discovery. The innovation of mining affords a partial case in point. The fossil record indicates that *Homo sapiens* first began mining the Earth about 100,000 years ago. And yet our ancestors didn't literally go for gold until around 8,000 years ago; brittle and annealed copper until 6,200 years ago; silver until 6,000 years ago; and lead until 5,500 years ago. This staggering inventory of man's ore targets is repeated multiple times on multiple continents, and in multiple geographical areas and locals. It was suspected there was more to this story than first caught the eye, and indeed there was.

There was something hidden in this twisted tangle of repeating firsts. The example of steel-making points out another aspect of it. While the ancient Chinese first made steel in 200 BC, the Syrians did it in Damascus – using what amounts to an accidental application of nanotechnology – in 900 AD. The diffusion of Bessemer Steel, however, did not occur until 1856 AD. Each steely instance is a genuine first, even though the global diffusion of this enabling technology did not really get going until the middle of the nineteenth century.

A series of key insights and cautions are to be found here. One is that at least some social network diffusion of steel-making had to have occurred in each of these cases. Otherwise, we'd never have heard of them. Another key clue is that just because a couple of inventive *Homo sapiens* get together and make a breakthrough that is no guarantee that their grand success will be accepted by the greater population. Clearly, *invention* does not equal the social diffusion of an *innovation*. What's more, evidence of the diffusion of an innovation is far more likely to survive than any record of its invention.

But wait! There's still more to it. It's the really big deal called *simultaneity*. It has recently been found that the domestication of plants and animals happened near simultaneously in the Middle East, South America, and East Asia. The parallel emergence of cities and city states also spans oceans and continents. This sort of simultaneity in the human change data was trying to tell us something – something hugely important.

Cycles of reinvention and parallel diffusion point like a hunting dog to a reoccurring global source of human change. The data are clear. Whenever human beings diffuse an innovation in one place, they're most likely doing it at another geographically separated location at approximately the same time. In short, *simultaneity* is the rule, not the exception. What's more, just because a novel idea, practice, or object fails to diffuse within a population when first invented – worry not – it'll get multiple chances in multiple places later on. Something periodically goes to work on human beings; something that spans Hominid species, eras, oceans, and continents.

What is social diffusion?

The most complete review of the research on the acceptance, adoption, and diffusion of modern human innovations was done by the late Everett M. Rogers from 1962 to 2003. Five editions of his book "Diffusion of Innovations" document in excess of forty-six years of research into this subject. We read the first edition of his book in graduate school. A few years later, we chanced to meet its renowned author during a plane ride from Chicago to Los Angeles. While developing the database of human changes, we were reminded of that pleasant airborne conversation. We decided to find Rodgers and arrange a meeting to discuss the work on this book. Alas, it was soon discovered from the World Wide Web that he had recently passed away. So, we picked up the fifth edition of his book and reread all 471 pages. While a tough read, the weighty compendium is recommended to anyone interested in understanding the diffusion of innovations through human social networks.

Rogers' book offers many valuable research-driven concepts relating to human change. The few shared here are absolutely foundational to this discussion. They are:

> The main elements in the diffusion of new ideas are: (1) *an innovation* (2) that is *communicated* through certain *channels* (3) *over time* (4) among members of a *social system*.
>
> *1. Innovation*
>
> An *innovation* is an idea, practice, or object perceived as new by an individual or other unit of adoption [in other words a social network of people].
> ...
> The characteristics of an innovation, as perceived by the members of a social system, determine its rate of adoption. [These] Five attributes are: (1) relative advantage, (2) compatibility, (3) complexity, (4) trialability, and (5) observability.
>
> *Re-invention* is the degree to which an innovation is changed or modified by a user in the process of adoption and implementation.
>
> *2. Communication Channels*
>
> A *communication channel* is the means by which messages get from one individual to another. [Today] Mass media channels

are the most effective in creating knowledge of innovations, whereas interpersonal channels are [and have always been] more effective in forming and changing attitudes toward a new idea, thus in influencing the decision to adopt or reject a new idea.

...

3. Time

Time is involved in diffusion in (1) the innovation-diffusion process, (2) innovativeness [in the population], and (3) an innovations rate of adoption. The *innovation-decision process* is the process through which an individual (...) passes from first knowledge of an innovation to forming an attitude toward the innovation, to a decision to adopt, reject, to implementation of the new idea, and to confirmation of this decision. We conceptualize five steps in the process: (1) knowledge, (2) persuasion, (3) decision, (4) implementation, and (5) confirmation. ... The decision stage leads to (1) *adoption*, a decision to make full use of the innovation as the best course of action available, or (2) to *rejection*, a decision not to adopt the innovation.

4. Social System

A *social system* is a set of interrelated units [people] that are engaged in joint problem solving to accomplish a common goal. A system has *structure*, defined as the patterned arrangement of the units of a system [individuals and small and large groups], which gives stability and regularity to individual behavior in a system. The social and communication structure of a system facilitates or impedes the diffusion of innovation in the system. One aspect of social structure is *norms,* the established behavior patterns for the members of a social system.[11]

In his book Rogers relates the concept of *social equilibrium,* and how that equilibrium is affected by the diffusion of innovations within established social systems. He points out that, "The undesirable, indirect, and unanticipated consequences of an innovation usually go together, as do the desirable and anticipated consequences (Generalization 11-2)."[12] Rogers defines three states within social systems associated with an innovation and change. These are:

> *Stable equilibrium* occurs when almost no change is occurring in the structure or functioning of a social system. *Dynamic equilibrium* occurs when the rate of change in a social system is commensurate with the systems ability to cope with it. *Disequilibrium* occurs when the rate of change is too rapid to permit the [social] system to adjust.[13]

A core notion in the present research is that when the individuals comprising a social system first confront an innovation, they are naturally predisposed to resist its adoption and diffusion. On that long ago plane ride with Rogers, we suggested to him that instinct-born and experience-reinforced fears of "disequilibrium" create an innate human resistance to change. We asserted that some external force must move the members of a social system to overcome their resistance and actively embrace an innovation. We asked him if we had somehow misinterpreted his book.

He thought about that for a minute, pivoted in his seat, and said, "No, no, you're right. A single factor I call '*innovativeness*,'[14] describes, explains, and predicts the majority of variation in the rate of adoption and diffusion of innovations." Encouraged, we asked him if the state of development of the population under study made any difference in their innovativeness.

"No," He replied, "It makes no difference whatever. The same simple model describes all social groups equally well, regardless of their state of development."

The implications of Rogers' single "*innovativeness*" variable model would not trigger the notion of concentrated waves of Human Change in us for three decades. Back then it was his work on the sub-categorization of populations into classes of what he called *adopters* that had captured everyone's attention.

When we met Rogers he was already famous for his "Adopter Categorization on the basis of Innovativeness." He'd developed these categories years before while still a doctoral student at Iowa State University.[15] He defined "innovativeness" as "the degree to which an individual ... is relatively earlier in adopting new ideas than other members of a social system."[16] Rogers had long since demonstrated that the adopter distribution closely approached normality (approximated the normal distribution with its classic bell shaped curve). Figure 2 depicts Rogers final "Adopter Categorization on the Basis of Innovativeness," meaning willingness to accept and adopt an innovation.

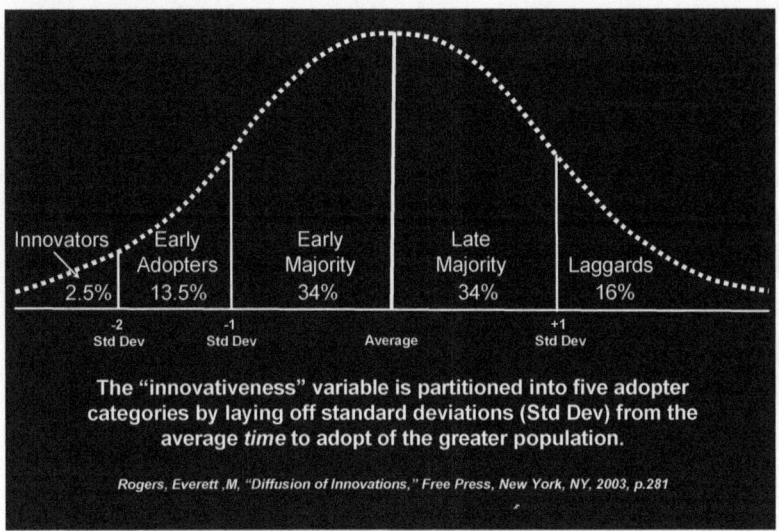

The "innovativeness" variable is partitioned into five adopter categories by laying off standard deviations (Std Dev) from the average *time* to adopt of the greater population.

Rogers, Everett ,M, "Diffusion of Innovations," Free Press, New York, NY, 2003, p.281

Figure 2 Rogers' Adopter Categorization
Source: Rogers, Everett M., 2003, Diffusion of Innovations, Free Press, New York, NY, p.281

Rogers' categories are still frequently cited and used by sociologists, journalists, and marketing types all over the World. In the figure the innovativeness dimension is measured by the time at which an individual adopts a particular innovation. The *innovativeness* variable is partitioned into five adopter categories by laying-off standard deviations from the average time of adoption for a given innovation.

As Rogers pointed out in his book, the adopter category into which an individual falls is not fixed. An individual's role is determined by the individual's situation and his predisposition toward *each particular* innovation. People shift roles (adopter categories), depending upon the particular idea, practice, or object they confront. While the distribution of innovativeness for a given innovation is percentage-fixed across any population of ample number, the time it takes for a particular innovation to diffuse varies, often substantially. That is of course assuming the novelty being introduced is not rejected outright at some point in the diffusion process.

It is a bit of a paradox, but *inventors* are rarely the first people to fully adopt an innovation. The social diffusion of an innovation does not truly begin until a few venturesome *"innovators"* (as opposed to *inventors*) decide to adopt it. Only then does an innovation stand a reasonable chance of gaining widespread acceptance and showing up in the historical record.

Rodgers spent his entire adult life studying the diffusion of innovations in contexts ranging from remote primitive tribes to those folks entangled in the World Wide Web. Yet all these studies were limited to people who lived within the last 100 years. Until now, no one has attempted to see if the rate of the adoption and diffusion of innovations has varied over the greater course of human history. Rogers had no way of knowing that the people he'd been studying were exhibiting the effects of a historically rare and definitively cyclic willingness to embrace change and risk social disequilibrium.

Testing Extreme Gradualism

Extreme Gradualism is that **Alternative Hypothesis** which asserts that the dates of changes exhibited in the behavior, artifacts, and genetics of hominids are equally probable and flatly distributed throughout time. In this test, time includes the eleven periods of 25,626.8 years comprising the hominid history of interest.

To test this hypothesis a Chi Square (χ^2) analysis was performed. The analysis compared the Observed frequencies (O) of the dates of Human Changes with the Expected (E) flat (rectangular distribution) of theoretic change frequencies implied by Extreme Gradualism. The Observed frequencies of human innovation consist of those independently obtained fossil, genetic marker, and historical record changes reported by paleontologists, archeologists, anthropologists, geneticists and historians from around the world.

The Expected theoretical frequencies were arithmetically generated using the Extreme Gradualism hypothesis of equally probable and distributed changes. The Expected frequency of changes for each of the 11 periods was set to a constant value, a Probability of Change = 16.73, which is the total number of changes divided by eleven, the number of historic Periods of interest. The Observed and Expected frequencies spanned the eleven periods from 281,894.8 years ago (279,882.8 BC) to the start of the current period (the Period spanning the 508 years from 1500 AD to the year 2008). The analysis answered the question, is the difference between Observed (O) and Expected (E) Frequencies of human innovation statistically significant? If the Expected Frequencies are not significantly different from the Observed Frequencies, the theory of *extreme gradualism* is in evidence. Otherwise, it is not.

The analysis revealed that the Observed frequencies are significantly different from the flat Expected theoretical frequencies of Extreme Gradualism. Given this less than surprising result, we rejected the Extreme Gradualism hypothesis. The $\chi 2$ (pronounced Chi Square) value calculated is 944.9348. This value is far greater than the critical value for significance required at the .001 level of confidence. The accepted $\chi 2$ criterion for hypothesis rejection at the .001 level of confidence with 10 degrees of freedom is 35.5572. The difference between Observed and Expected frequencies is significant at the 1.352×10^{-196} level (i.e., slightly less than the estimated probability of our universe springing forth from literally nothing by pure random chance.[17])

The test result cannot reasonably be explained by sample fluctuation. Clearly the dates of Human Changes, as indexed by the unearthed and interpreted hominid fossil record, genetic analyses, and historical records are not equally distributed and equally probable over the last 282,000 years. In short, the rate of human change varies over time, and Extreme Gradualism is just plain wrong.

Argument for Punctuated Equilibrium

Further support for rejecting the notion of Extreme Gradualism, lends support for the viability of the theory known as "Punctuated Equilibrium" and it comes from the work of geneticists. For more than thirty two years there had been intense debate over the hypothesis that most morphological evolution occurs during relatively brief sporadic episodes of rapid change that punctuate much longer periods of stasis (i.e., static equilibrium).[18] Wikipedia, the free encyclopedia, defines this "*Punctuated equilibrium*" as:

> Punctuated equilibrium is a theory in evolutionary biology. It states that most sexually reproducing populations will show little change for most of their geological history, and that when phenotypic evolution [changes in observable characteristics and traits] does occur, it is localized in rare, rapid events of branching speciation (called cladogenesis).
> Punctuated Equilibrium is commonly contrasted against the theory of phyletic [Extreme] gradualism, which states that evolution generally occurs uniformly and by the steady and gradual transformation of whole lineages (anagenesis). In this view, evolution is seen as generally smooth and continuous. In 1972 paleontologists Niles Eldredge and Stephen Jay Gould published a landmark paper developing this idea [of punctuated equilibrium]. Their paper was built upon Ernst Mayr's theory of geographic

speciation [emergence of a new species], Michael Lerner's theories of developmental and genetic homeostasis [equilibrium], as well as their own empirical research. Eldredge and Gould proposed that the degree of gradualism championed by Charles Darwin was virtually nonexistent in the fossil record, and that stasis [static equilibrium] dominates the history of most fossil species.[19]

In 1954 Ernst Mayr published a paper emphasizing the homogenizing effects of gene flow and the stabilizing influence of large interbreeding populations.[20] Gould summarized Mayr's theory and its consequences for punctuated equilibrium, in a 1977 essay for Natural History magazine[21]:

A new species can arise when a small segment of the ancestral population is isolated at the periphery of the ancestral range. Large, stable central populations exert a strong homogenizing influence. New and favorable mutations are diluted by the sheer bulk of the population through which they must spread. ... Small peripheral isolates are a laboratory of evolutionary Change.[22]

Gould goes on to identify how the fossil record masks speciation (the emergence of genuinely new species from existing ones):

What should the fossil record include if most evolution occurs by speciation in peripheral isolates? Species should be static through their range [history] because our fossils are the remains of large central populations. In any local area inhabited by ancestors, a descendant species should appear suddenly by migration from the peripheral region in which it evolved. In the peripheral region itself, we might find direct evidence of speciation [the emergence of a new species], but such good fortune would be rare indeed because the event occurs so rapidly in such a small population."[23]

Gould, like nearly all geneticists, paleontologists, archeologists, and anthropologists, is quick to ascribe the source of human change to global climatic conditions. That seemingly incontrovertible fact is seriously tempered in a later chapter. Here and now, we need to let the ancient Maya inspired Hypothesis H_1 have its way with Alternate Hypotheses 2 (Extreme Punctuated Equilibrium) and 3 (Gradualism with Punctuated Equilibrium).

Testing the Maya Theory of Human Evolution

Maya inspired **Hypothesis H_1** asserts that a significant portion of the variation in the literature reported dates of human change occurring over the last 282,000 years, is described, explained, and predicted by Degrees of Maya Precession (e.g. World Age Cycles); *where one World Age Cycle equals 25,626.8 years and 360 degrees.* **Hypothesis Null$_1$** posits that there is no significant relation between the variation in reported dates of human change and Degrees of Maya Precession (e.g. Maya World Age Cycles).

Extreme Punctuated Equilibrium, Alternate Hypothesis 2, states that significant Human changes are concentrated in discrete evolutionary jumps (i.e., exhibiting as spikes of dynamic change) that occur when one or more small sub-populations become geographically separated from a greater population undergoing substantial stress or the threat of extinction.

In contrast, **Gradualism with Punctuated Equilibrium, Alternate Hypothesis 3**, posits that the rate of Human change continuously fluctuates from very fast to very slow with the static equilibrium condition representing nothing more than a period of ultra slow change.

Please note that our less-than-bold modern theorists – in sharp contrast to the Maya – heavily hedged their bets. They set no quantitative event or time interval criteria whatsoever for the occurrence of substantial human change. They also reject any notion of a reoccurring cyclic rhythm in the human change process.

In order to test Maya inspired Hypothesis H_1, each of the Human Change dates was coded using three continuous interval measures: (1) Raw Years Ago; (2) Period Years (which equals raw Years Ago divided by the applicable Maya World Age Precession Cycle Period Number, with each period scaled from 0 to 25,626.8 years); and (3) Degrees of Maya World Age Precession (scaled from 0 to 360 degrees per Precession Cycle, calculated at a fixed constant rate of 71.1857 years per degree of Maya Precession).

As described in detail in Chapter 5, Maya World Age Precession is based on the vertical wobble in the Z-axis of the Solar System, and each cycle is of a constant duration of 25,626.8 Gregorian calendar years. In contrast, the modern Copernican view of precession is based on a wobble in the Z-axis of the Earth and varies dramatically in cyclic duration. In Chapter 5, Maya fixed constant Precession is demonstrated astronomically valid and reliable, while Copernican precession (the Lunisolar model) is proven invalid and unreliable.

The Period Year transformation of database event dates simply concentrates the dates (Raw Years Ago) of specific Human Change events into a single time span (Precession Cycle) of 25,626.8 years. It soon became apparent that the Period Year transformation is needed to compensate for the disproportionate sparseness of acceptable fossils in the more distant dates of the database. Figure 3 depicts the effects of the Period Year transformation for the oldest 184 Human Change dates in the database. Figure 3 contrasts the frequencies of human changes in raw Years Ago with their concentration into Period Years.

The two graphs of Figure 3 make a key point abundantly clear. In the absence of the Period Year transformation, the paltry frequencies of the Human Change records from the most distant past seriously under-represent any trends in the dates of Human Change. The concentrating transformation to Period Years recovers the representative Human Change distribution using available data.

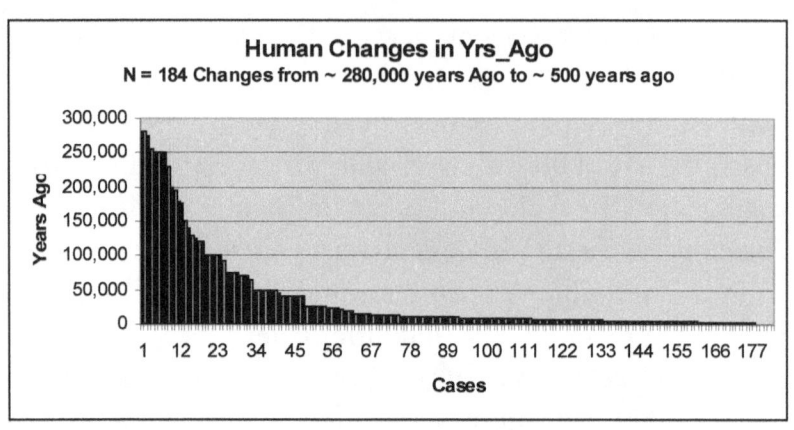

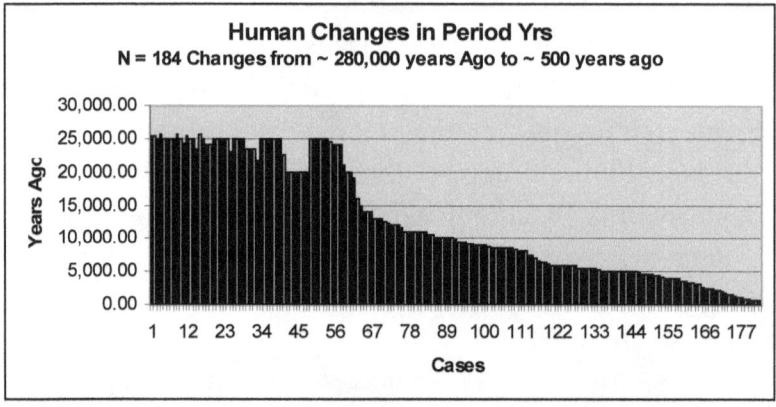

Figure 3 Effects of the Transformation of Human Change Dates from Years Ago to Period Years

To test Maya Hypothesis H_1, both linear correlation and regression analyses were performed. These analyses relate the dates of Human Change to Degrees of Maya World Age Cycles (Maya Precession). Both statistical procedures were first run using the Human Changes measured in raw Years Ago. Then, the procedures were repeated using the change data transformed into Period Years.

As hypothesized, a significant positive relation of .74 (a correlation coefficient = 0.737) was found between the dates of Human Change expressed in raw Years Ago and Degrees of Maya Precession. The sample analyzed consisted of 719 cases of Human Change, with the number of cases minus 2 equaling 717 degrees of freedom. This correlation is significant at the .001 level of confidence. Hence the null hypothesis of no relationship between Maya World Ages and the dates of Human Change is rejected. Clearly we can say with a high degree of confidence that Degrees of Maya Precession is significantly related to the historic untransformed (raw) dates of Human Change.

A regression analysis of the raw (meaning untransformed) dates of Human Change was performed. Maya Degrees of Precession describe, explain, and predict 55 percent of the variation (an Adjusted R Square of 0.55) in Human Change measured in raw Years Ago. This predictive relation is significant at the .001 level of confidence. A relation explaining a little over half the variation in the dates of Human Change may seem weak, but rest assured it is a respectable value in the world of regression analysis.

The counting relation between Maya Degrees of Precession and the dates of Human Change transformed into Period Years was tested next. The correlation was expected to improve, but the *perfect* relation (i.e., a correlation coefficient of 1.00) that resulted was unanticipated. A perfect correlation means that the dates of Human Change count precisely with the corresponding changes in Maya Degrees of Precession. For all intents and purposes, Degrees of Maya Precession equal the dates of Human Change expressed in Period Years at a confidence level boarding on certainty. To cut straight to it, in the world of correlation analysis, correlation coefficients simply cannot get any better than this.

Next run was the regression analysis of the change dates transformed into Period Years. Not surprisingly, the regression analysis echoed the correlation and yielded a perfect predictive relationship between Maya Degrees of Precession and the dates of Human Change in Period Years (i.e., an Adjusted R Square of 1.00). Figure 4 depicts the line fit of the perfect prediction of the dates of Human Change in Period Years using Degrees of Maya World Age Precession.

The prediction is significant in the extreme with a confidence level to match. In essence, the analysis indicates that Degrees of Maya Precession describe, explain, and predict 100 percent of the variation in the dates of Human Change expressed in Period Years. Here is clear evidence of *World Age Cycles of Change* exhibited in the accepted human fossil, genetic, and historical records.

Scientists are immediately suspicious of any statistical relation approaching perfection; and for good reason. It is exceptionally rare to find a genuinely perfect correlation and predictive relationship. When one does show up, scientists are immediately afraid that it is just another *fire trucks cause fires* sort of *spurious* correlation result. There were two possible reasons for the perfect correlation found. Either the transformation to Period Years had produced a spurious correlation, or the transformation had truly corrected for the lack of distant past event data and the correlation is real. To find out which, it was necessary to replicate the results with a data set that didn't require the transformation to Period Years.

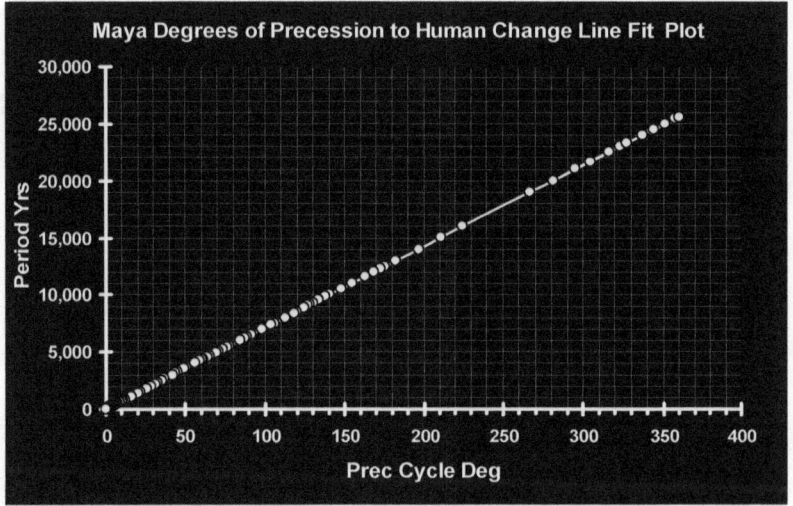

Figure 4 Maya Degrees of Precession (World Ages) Perfect Correlation and Prediction of the Dates of Human Changes in Period Years

The analyses were replicated using only the near term change data from 1500 to 2008 AD in raw years. Once again Maya Degrees of Precession perfectly predicted (the adjusted R-Square = 0.9999) the dates of human innovation, only this time all the data were in raw *untransformed* Years Ago. Thus, the predictive relation discovered for the entirety of human history in Period Years was replicated for the last 508 years (536 cases and

535 degrees of freedom), with no need to transform the data. The perfect correlation is real.

Figure 5 depicts the percentile ranks of the Normal Probability plot from the full Period Year regression analysis using all 282,000 years of data transformed into Period Years. Figure 5 reveals two key points of interest.

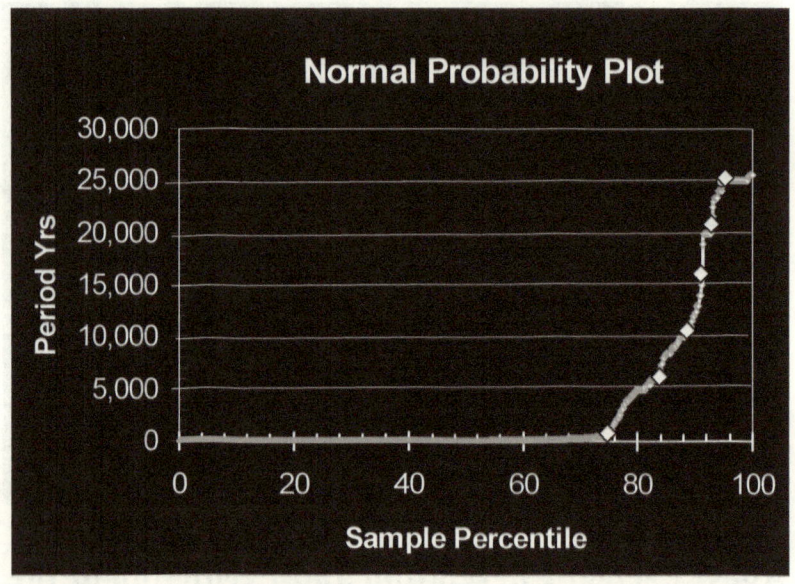

Figure 5 Percentage of the Sample of 719 Human Changes by Date of Change in Period Years

First, it is obvious from the figure that 80 percent of the sample of the 719 dates of Human Change transform to the interval extending from less than or equal to 5,000 years ago. The use of the transformation to Period Years is justified in order to compensate for the dramatic rarity of fossils and historical data in the majority of human history (i.e., from 5,000 to 282,000 Years Ago). The transformation is of course further vindicated by the replication of the analysis results using the untransformed current period data.

Another insight is afforded by Figure 5. It is revealed within the dramatic rise in the upper 75th to the 100th percentile of changes. That portion of the plot is punctuated by distinct *inflection points* (e.g., sudden abrupt upward shifts in the direction of the curve of changes). Even where the data is thinnest (between Period Years 15,000 and 20,000 years ago), these inflections are spaced just over 5,000 years apart. These abrupt break points in the frequency of Human Change were definitively predicted by the ancient Maya.

The Maya asserted that Human Changes are arrayed about a World Age Precession Cycle, a circle 25,626.8 years in circumference. They further insisted that each such *Cycle* is subdivided – like a sliced pie – into 5,125.36 year long *Epochs* spaced precisely 72 degrees apart on that circle. The Maya claimed that each of these five Epochs punctuate each World Age cycle. These punctuating spike-points are strongly indicated in the historical data of human change plotted in the figure. It seems that Degrees of Maya World Age Precession really do predict the cyclic rhythms of human change and innovation.

Acceptance of Maya inspired Hypothesis H_1 leads to the rejection of both *Alternative Hypotheses* 2 and 3. In truth, it renders those modern theories incomplete, ambiguous, and incorrect. Where Maya Hypothesis H_1 specifies fixed interval cycles of change, neither contemporary alternative hypothesis does. Alternative Hypothesis 2, Extreme Punctuated Equilibrium, suggests irregularly spaced event driven human evolutionary jumps, where Maya inspired Hypothesis H_1 posits the *regularly* spaced cyclic surges of Human innovation in evidence.

Alternative hypothesis 3, Gradualism with Punctuated Equilibrium, demands rapid *irregularly* spaced local or regional event driven pulsing variations in the rate of human change. Such irregularly spaced groups of change are simply not in evidence.

Acceptance of Hypothesis H_1 and rejection of alternatives 2 and 3 has serious implications. It means that while there is a wave-like pulsing in the rate of human change, those pulses are *regularly* spaced and globally determined, not irregularly spaced and local event driven.

Figure 6 depicts a few of the major fossil evidenced innovation events (genetic and cultural) associated with the eleven 26,626.8 periods of human history. In truth, there are 184 innovation dates in the archeological portion of the database, and 525 more innovation dates in the historical portion of the database collected and analyzed.

The available literature indicates that the innovation descriptive fossils and artifacts cited in the figure were dated by multiple methods. Such methods include various geologic strata depth and soil analyses, radiometric, and accelerated mass spectroscopy (AMS) dating, extinct animal and plant species date support, and other forensic techniques. Further, all of the reported fossil dates in the database have been reviewed and consensus accepted by qualified professional paleontologists, archeologists, or anthropologists in multiple published peer reviews.

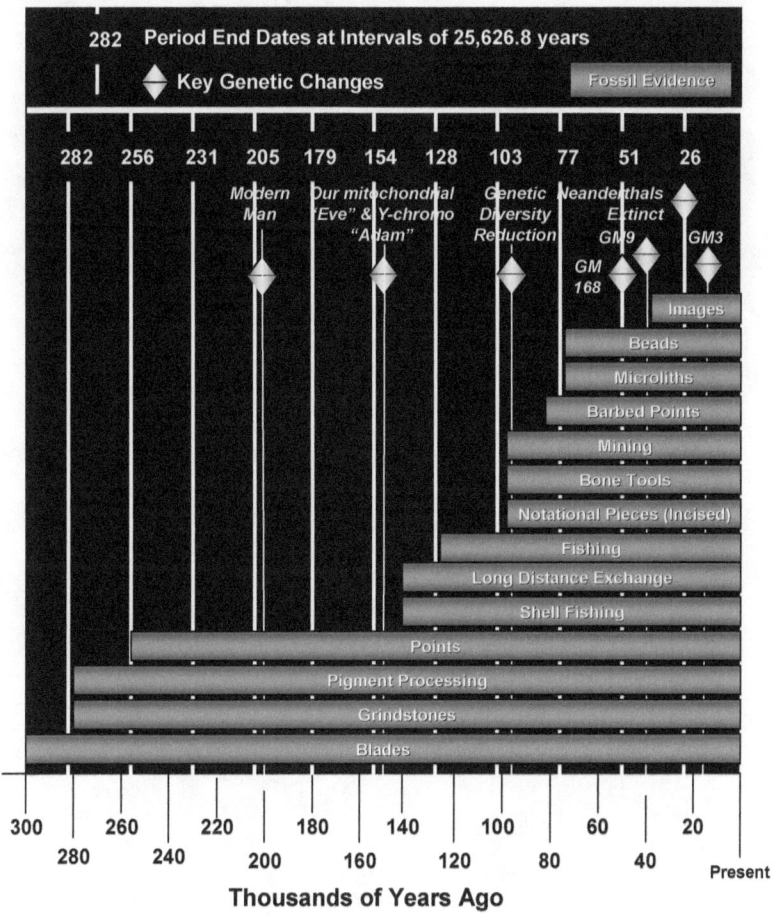

Figure 6 Fossil Evidenced Change Events

The strong suggestion of Maya predicted Epoch (pie-slice point) spikes in the Human change data were simply too intriguing to ignore. The next chapter relates and tests that ancient assertion.

3

WAVES OF WORLD AGE EPOCHS

Testing for World Age Epochs

So far three major discoveries have been demonstrated. They are: 1. the dates of Human change are concentrated in specific time intervals correlated with and predicted by Maya World Age cycles. 2. Human change is not equally distributed or erratically spiked across the historic record. 3. Inflection points in the normal probability curve of the dates of Human Change suggest that Maya *Epochs* of change are real.

Indeed, just as the Maya insisted over two millennia ago, there appear to be five 5,125.36 year long Epochs punctuating each Maya Precession Cycle of 25,626.8 years. These Epochs seem to form peaks and valleys in the rate of the diffusion of Human innovation at and about the Epoch boundaries.

To explore the emergent issue of *World Age Epochs*, the Human Change records in the database were coded with their associated Epoch numbers. This was done by simply numbering each Epoch interval of 5,125.36 years (each 72 Degrees of Maya Precession) beginning at 1500 AD, the designated zero point. This resulted in the grouping of database records in even numbered and odd numbered sets of Epochs. The Epochs were further stratified into 1,023 year segments, as exemplified for each wave of 10,250.7 years in Table 2.

Table 2 Stratification of Even-Odd Epoch Pairs

	Interval ID	Range in Yrs	From	To
Even Epochs	1	1024	1	1025
	2	1024	1026	2050
	3	1024	2051	3075
	4	1024	3076	4100
	5	1024	4101	5125
Odd Epochs	6	1024	5126	6150
	7	1024	6151	7175
	8	1024	7176	8200
	9	1024	8201	9225
	10	1024	9226	10250

Each pair of odd and even Epochs forms a hypothetical '*wave,*' having a wavelength (measured peak to peak) of 10,250.7 years. According to the ancient Maya, during even numbered Epochs the rate of change *accelerates,* while during odd numbered Epochs, the rate of change *decelerates*.

The Epochs thus progress backward through time within each World Age Precession Period from 1500 AD to 277,988 BC. For reasons described earlier, Maya Epochs were explored using the dates of Human Change transformed into Period Years. Further, since interest was focused solely in pre-Renaissance Epoch effects, positive Period 1 (the current Epoch from 1500 AD to the year 2008) was excluded from the initial analyses.

Figure 7 depicts the resulting frequencies of Human Change for these Even and Odd Epochs. The frequencies are further consolidated for purposes of illustration into a single wave of 10,250.7 years.

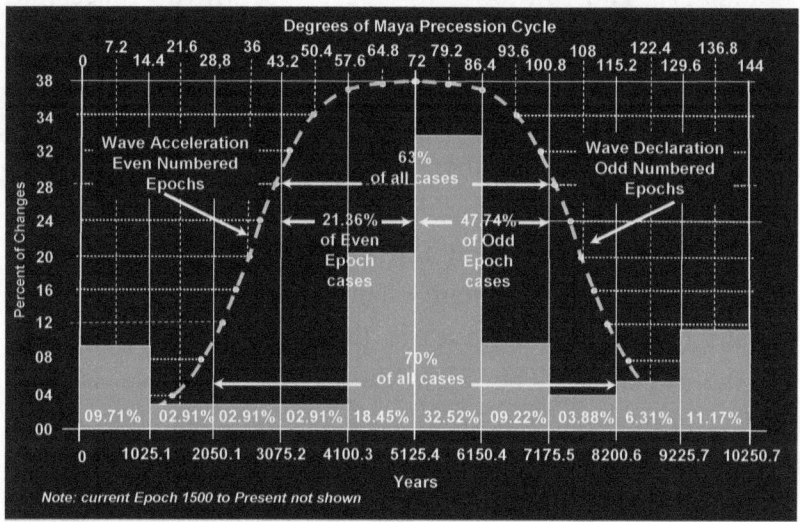

Figure 7 the Distribution of Human Changes within Odd and Even Numbered Epoch Pairs

Percentages of Human Change in each 1,025 year division of the wave paired Epochs are shown just above the year scale values at the bottom of the figure. The percentages exhibit a rapid rise in the number of changes on the even side of the Epoch boundary, and a rapid fall in the number of changes on the odd Epoch side of the wave, just as asserted by the ancient Maya. The spike in the changes is strongly off-center (skewed) to the odd side of the figure between 72 and 79.2 degrees (i.e., between 5,125.36 and 5,637.86 years). Some 70 percent of all the cases of Human Change in the database are concentrated within ± 3,075 years of the Epoch boundary that divides the Even and Odd numbered Epochs in the figure. A majority (63 percent) of the Human Changes are concentrated within ± 2,050.16 years of the Epoch boundary. Thus, in terms of Degrees of Maya Precession, 63 percent of all the historic dates of Human Change fall within ± 28.8 degrees of the 72 degree boundary separating the even from the odd numbered Epochs.

The mean number of Human Changes is 20.6 (equating to a rate of .02 changes per year) with one standard deviation about the mean of ± 19.21 changes. The highly skewed human change distribution leans and breaks (like a water-wave) to the odd Epoch (right) side of the mean. The distribution of the wave is also seriously peaked.

A mean nearly equal to the standard deviation forming a distribution that is forward skewed are the trademark characteristics of energy and surface waves. Such waves are conventionally modeled using the Poisson distribution. Consequently, Hypothesis H_2 was formulated.

Hypothesis H_2 asserts that the number of Human Changes per 5,125.36 year long Maya Epoch of 72 Degrees of Maya Precession, describe a distribution not significantly different from the continuity transformed Poisson distribution. **Hypothesis Null$_2$** asserts that the frequency of Human Change is neither Poisson nor normally distributed across the 5,125.36 year long (72 degree) partitions characterizing the five Epochs of Maya Precession defined World Ages.

The Poisson distribution was selected for reasons in addition to those already cited. Foremost among them is the fact that "the number of inventions of an inventor over his or her career" and "the mutations in a given stretch of DNA after a certain amount of radiation" are "modeled as Poisson distributions."[24]

"The Poisson distribution was discovered by Denis Poisson (1781-1840) and published, together with his probability theory in 1838 in his work ... *Research on the Probability of Judgments of in Criminal and Civil Matters*."[25] The rate of invention by an individual is obviously not the same as the rate of the diffusion of human innovation throughout an entire social network. Still, it seemed likely that both phenomena would have much in common – human inventors and adopters. Hence, it was thought that the Poisson and perhaps the normal distribution might best describe the human change data.

In the test Hypothesis H_2, the Observed frequencies of Human Change were those occurring in each 1,024 year interval. Changes were grouped by simply dividing the number of changes in each interval by the total number of sampled changes; which is 206. Next, the Expected (E) probabilities corresponding to each of the Observed (O) rates of Human Change were calculated. This was done using an automated *Poisson* distribution function generator. In accordance with the literature on the Poisson distribution, a "continuity transformation" was performed on the Expected probabilities. The "continuity transformation adds a constant of 0.5 to each Poisson equation calculated probability, and is required whenever the observed mean is greater than 10."[26] Since the calculated mean of the Observed Human Changes was 20.6, the continuity transformation was mandatory.

A Chi Square test was conducted comparing the Observed and Expected values. The resulting Chi Square value was 2.12. In any Chi Square analysis, the higher the value of the Chi Square calculated, the greater the number of discrepancies between the Observed and Expected Probabilities. In order to reject Hypothesis H_2 at the .001 level of confidence, a Chi Square value greater or equal to 27.88 is required at the applicable 9 degrees of freedom. Hence, Hypothesis H_2 is retained. There was no significant difference between the Observed rates of Human Change and the Expected values of the Poisson probability distribution. Figure 8 plots both the Observed and Expected Poisson Probabilities in terms of the consolidated Even and Odd Epoch pairs.

The Chi Square test was repeated. This time *Normal Probability Distribution* values were generated for the Expected Frequencies. The Chi Square value returned was 3.74. This is substantially less than the required value of 27.88 needed to retrain the Null Hypothesis at the .001 level of confidence with 9 degrees of freedom. Despite the sombrero shaped and skewed character of the Observed distribution, the Observed rates of human change are not significantly different from the Expected normal distribution frequencies. The Null Hypothesis assertion of significant differences is thus rejected. Clearly, the rate of human change can be described by either the normal or the Poisson distribution, but cannot possibly be approximated by a flat distribution of equally probable and distributed Changes.

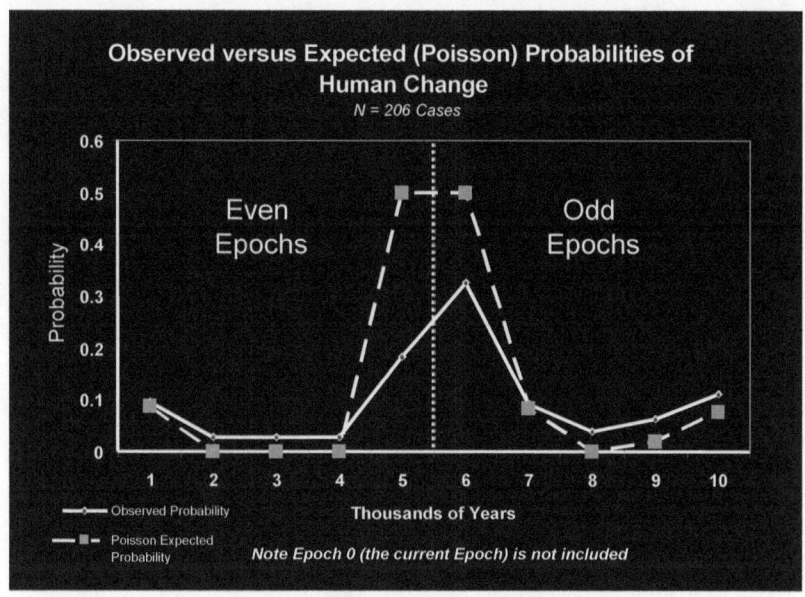

Figure 8 the Distribution of Observed and Expected Rates of Human Change

For the periods prior to 1500 AD, the dates of Human Change are undeniably concentrated about Maya World Age Epoch boundaries. These concentrations occur in a way approximating the bell-shapes of the Poisson and Normal probability distributions. The Maya posited *Epochs of Change* do indeed punctuate the history of Human Change. Further, Degrees of Mayan Precession (*World Ages cycles*) solidly predict the tsunami-like change surges (even numbered Epochs) and change recessions (odd numbered Epochs) in evidence. Once again, the early Mesoamericans had it right. Though when these analyses were first performed there was little indication of just *how right* they had actually gotten it.

Human Change in the Current Period

The Chi-square test of Hypothesis H_2 had revealed no significant differences between the Period -1 to -11 observed frequencies of Human Change and the Expected values of the Poisson and normal distributions. Still, the observed distribution was strongly sombrero shaped and highly skewed. There was concern that the sparse fossil record, exceedingly rare genetic evidence, and the stuttering start of the historical record might be misleading. In short, the nearer term evidence needed to be replicated in order to confirm or deny the reliability of the results.

From 1500 AD to the present mankind has experienced the *Renaissance, Pre-industrial Age, Industrial Revolution, Atomic Age, Space Age, Digital Revolution and Computer Age.* That dramatic history is frequently cited by scientists and historians as an unprecedented increase in the rate of human invention and innovation. Given the majestic upsurges of change just witnessed in Periods -1 to -11, the asserted uniqueness of the last 508 years seemed questionable. What does make the current Epoch unique is simply that we know so much more about it. The changes since 1500 AD are much greater in observable number and far better documented. Consequently, the human changes occurring from 1500 AD to the present better represent any upsurge in the frequency of human change. That fact led to formulation of Hypothesis H_3.

Hypothesis H_3 declares that the frequency of Human Change from 1500 AD to the present, describes a distribution not significantly different from the continuity transformed Poisson distribution and the untransformed normal distribution. **Hypothesis Null$_3$** counters that the frequency distribution of Human Change within the current Epoch (from 1500 AD to the present) is significantly different from both the Poisson and normal distributions.

In computing the rate of change needed to test Hypothesis H_3, the complete distribution of Human Change must be described in terms of two opposing Maya Epochs. That is, a total of 10,250.7 years, one complete wave of change. That meant that the Human change rate numerator must be divided by the full duration of 10,250.7 years.

As before, a Chi-square analysis was conducted to evaluate Hypothesis H_3. The analysis compared the Observed (O) frequencies (probabilities) of Human Change in the current Epoch, with the Expected (E) frequencies of the Poisson function. The normal distribution Expected frequencies did not have to be calculated. The reason for this is that during the last 508 years of the current Epoch (the tail of the complete distribution) the Expected probabilities produced by the Poisson and Normal distribution functions are identical.

Thus, only a single test of Hypothesis H_3 was needed to address both distributions.

The calculated Chi Square value equals 0.04. As stated, in any Chi Square analysis, the higher the computed value of the Chi Square the greater the number of discrepancies between the Observed and Expected Probabilities. The Chi-Square value was nearly zero. Clearly the Null Hypothesis assertion that there are significant differences between the observed and expected frequencies is false and rejected at the .001 level of confidence. To reject Hypothesis H_3 a Chi Square value greater or equal to 18.46 is required. Thus, Hypothesis H_3 is retained. There are no significant differences between Expected and Observed change frequencies in the current Period, solidly replicating the results from the test of Hypothesis H_2.

Figure 9 plots the Expected Poisson probabilities of Human Change for the last 508 years of history. The plotted curve tightly approximates the Observed rates of Human Change.

Waves of Human Change

One could not help but be intrigued. Each pair of opposing 5,125.36 year long Maya Epochs describe a Poisson distributed 10,250.7 year long wave of human change. Two of these waves make up 20,501.44 years of a 25,626.8 year long Maya World Age, which equals one complete Maya World Age Precession cycle.

Figure 10 depicts the Expected (Predicted) rate of change per 1,000 years for a full wavelength (one complete pair of Maya Epochs, or 10,250.7 years). The Poisson probabilities plotted in the figure represent the rise and fall of the history of man's changes, just as posited by the ancient Maya.

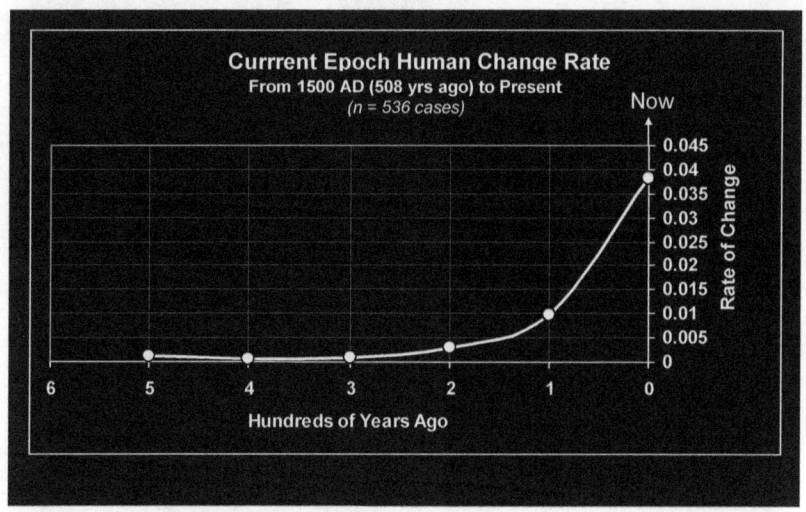

Figure 9 Current Epoch Poisson Distributed Rate of Human Change

According to John Major Jenkins (1998; 2002) the Maya believed that the first 5,125.36 years (even numbered Epoch) of each new World Age brought about major changes in Hominids. Jenkins states:

> While the scenario may seem bleak, the Maya doctrine of World Ages extends back over four previous epochs [4 times 5,125.36 years = 20,501.44 years], each of which ended in cataclysm and the transformation of humanity into something completely new, a

new being better suited for life in the new world.[27]

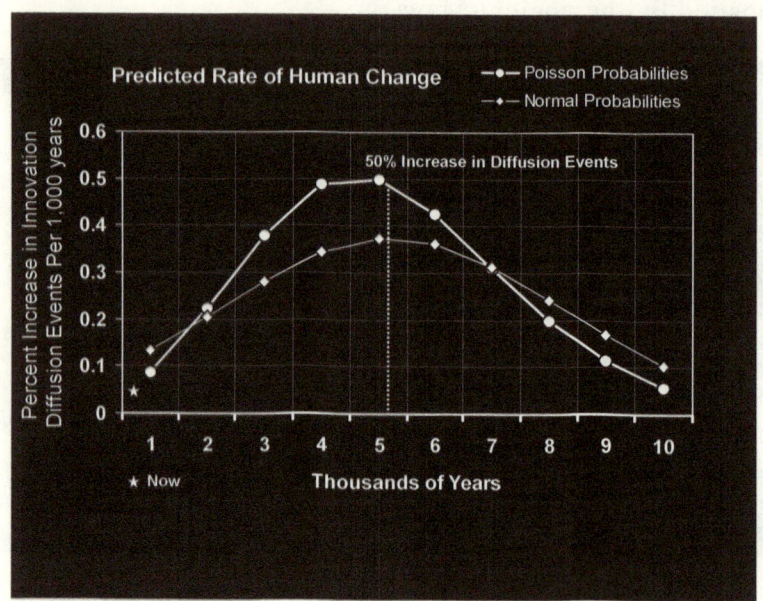

Figure 10 Predicted Poisson and Normal Human Change Rate Distributions for 10,250.7 Years

Jenkins describes the specific prediction that the ancient Maya and Toltec shaman (as well as surviving Maya) believe await us following the Winter Solstice of Friday December 21, 2012.

At *Chichén Itzá* [a 2,000 year old Mayan city in the Yucatan Peninsula between the modern cities *Cancún* and *Mérida*, Mexico] ... the Toltecs and the Maya began to act as one [around 850 AD]. Both cultural traditions joined in the powerful proclamation that what we call the twenty-first century, measured with two different cosmological alignments in the Great Year of precession [25,626.8 years], would be a great era of transformation for humanity, one in which the old world is destroyed and a new world rises from the ashes. According to both the Maya and Toltec myth, humanity saves itself at these critical nexus points only by transforming, by mutating into something totally unrecognizable, a new being altogether.[28]

Upon re-reading this prediction it was recalled that the hypotheses of Maya World Ages and Epochs had been retained. What's more, the hypotheses of Extreme Gradualism, Extreme Punctuated Equilibrium, and the popular combination of the two had been rejected. The ancient Maya had clearly stated a viable replacement for the three failed modern theories of human evolution. Namely, the Maya had provided a testable World Age Epoch theory of cyclic speciation, adaptation, and extinction. The next chapter puts those particular aspects of the Maya theory of evolution to the test.

4

WAVES OF HUMAN EVOLUTION

Emergence and Extinction

T he Maya theory of human evolution led to formulation of **Hypothesis H₄.** It states that: a significant portion of the variation in the fossil record driven dates of Hominid species emergence (speciation) and extinction events is described, explained, and predicted by the range of dates comprising the first even numbered Epoch in each applicable Maya World Age. **Hypothesis Null₄** counters that Hominid speciation and extinction events exhibit no counting predictive relation with the dates of Maya World Age Cycle first even numbered Epochs.

Table 3 lists accepted dates for each Hominid species emergence and extinction. These dates are those published by Carl Zimmer (2005) in his, "Smithsonian Intimate Guide to Human Origins," Madison Press Ltd., Toronto, Canada, pages 6-7.[29] In the table, the Zimmer provided dates of speciation and extinction are contrasted with the corresponding Maya World Age first even numbered Epoch dates.

Table 3 Observed and Predicted Species Emergence and Extinction Dates

	Hominid Species Events	World Age Epoch Date	Zimmer Date	Diff In Yrs	Percent Diff
1	Sahelanthropus tchadensis Emergence	7,000,000	7,000,000	0	0.00%
2	Sahelanthropus tchadensis Extinction	6,000,000	6,001,796	1,796	0.03%
3	Orrorin tugenis Emergence	6,400,000	6,401,574	1,574	0.03%
4	Orrorin tugenis Extinction	5,600,000	5,596,893	3,107	0.06%
5	Ardipithecus Emergence	5,600,000	5,596,893	3,107	0.06%
6	Ardipithecus Extinction	4,500,000	4,489,815	10,185	0.23%
7	Australopithecus anamensis Emergence	4,200,000	4,200,000	0	0.00%
8	Australopithecus anamensis Extinction	3,900,000	3,900,398	-398	-0.01%
9	Australopithecus afarensis Emergence	3,900,000	3,900,398	-398	-0.01%
10	Australopithecus afarensis Extinction	3,000,000	3,000,000	0	0.00%
11	Australopithecus africanus Emergence	3,500,000	3,495,496	4,504	0.13%
12	Australopithecus africanus Extinction	2,400,000	2,388,418	11,582	0.48%
13	Australopithecus garhi Emergence	2,700,000	2,695,939	4,061	0.15%
14	Australopithecus garhi Extinction	2,400,000	2,388,418	11,582	0.48%
15	Paranthropus boisei Emergence	2,400,000	2,388,418	11,582	0.48%
16	Paranthropus boisei Extinction	1,200,000	1,199,334	666	0.06%
17	Homo habilis Emergence	2,400,000	2,388,418	11,582	0.48%
18	Homo habilis Extinction	1,550,000	1,547,859	2,141	0.14%
19	Homo ergaster Emergence	1,850,000	1,850,254	-254	-0.01%
20	Homo ergaster Extinction	1,500,000	1,496,605	3,395	0.23%
21	Homo erectus Emergence	1,800,000	1,800,000	0	0.00%

22	Homo erectus Extinction	30,000	30,000	0	0.00%
23	Homo heidelbergensis Emergence	600,000	594,542	5,458	0.91%
24	Homo heidelbergensis Extinction	100,000	100,000	0	0.00%
25	Homo sapiens (earliest) Emergence	200,000	200,000	0	0.00%
26	Homo neanderthalensis Emergence	150,000	150,000	0	0.00%
27	Homo neanderthalensis Extinction	30,000	30,000	0	0.00%
28	Homo sapiens (current) Emergence	150,000	150,000	0	0.00%
29	Homo floresiensis Emergence	98,000	98,000	0	0.00%
30	Homo floresiensis Extinction	18,000	18,000	0	0.00%
	= Dates Different in General Literature				

In all, three such tables were constructed. Each table held alternative speciation and extinction event dates taken from the existing paleontology, archeology, anthropology, and genetics literature. These alternate dates are not included in table. All of the analysis results were the same, regardless of the differences in the various published dates of speciation and extinction analyzed. The fiercely argued for differences in dates are simply not large enough to make any difference.

An initial test of Hypothesis H_4 was limited the data of the last 1.85 million years. The data set was shortened in response to fears that the fossil record for the distant past was too sparse to enable a valid test of the hypothesis. That assumption proved false. The Maya theory of World Age Epoch driven human evolution is sufficiently robust to overcome all such limitations. Moreover, it soon became apparent that the accepted Hominid fossil record of the last 7 million years most likely represents less than ten percent of the total number

Hominid speciation, mutation, and extinction events that likely occurred in that time.

The correlation analysis performed for the accepted events over the full 7 million years were definitive. The correlation resulted in the unequivocal rejection of Hypothesis Null$_4$. That null hypothesis states that there is no counting relation between the observed (literature cited) and Maya predicted dates of speciation and extinction. The Literature observed and Maya predicted dates are near perfectly correlated. The correlation coefficient equals .9999 (approximately 1.00), and is significant well beyond the .001 level of confidence.

The subsequent regression analysis echoed the correlation result. The range of dates defining the first Even-numbered Epoch in Each World Age Cycle predicts 99.98 percent of the variation in the dates of species emergence and extinction events. The results were significant to 77 decimal places at a level of confidence approaching certainty. Figure 11 depicts this exceptionally strong predictive relation.

Hypothesis H$_4$ is retained. A significant portion of the variation in the fossil record driven dates of Hominid species emergence (speciation) and extinction events is described, explained, and predicted by the range of dates comprising the first even numbered Epoch in the applicable Maya World Age.

The implications of the finding were nothing short of staggering. The results suggest that each first even numbered Epoch in each new Maya World Age cycle is near perfectly associated with the major changes in all known Hominid species. Over the course of 7 million years, the total expected number of such Epochs equates to no less than 273 *major* Hominid genetic changes. The overwhelming majority these species changes have yet to be discovered. While not every World Age cycle would necessarily produce a Hominid species extinction or emergence, the implication is clear. There ought to be many more than 30 such species related events in evidence. If *Homo sapiens* and our cousins the Neanderthals emerged in parallel in less than 200,000 years, there should be approximately 35 to 70 Hominid species in the fossil record – not 30.

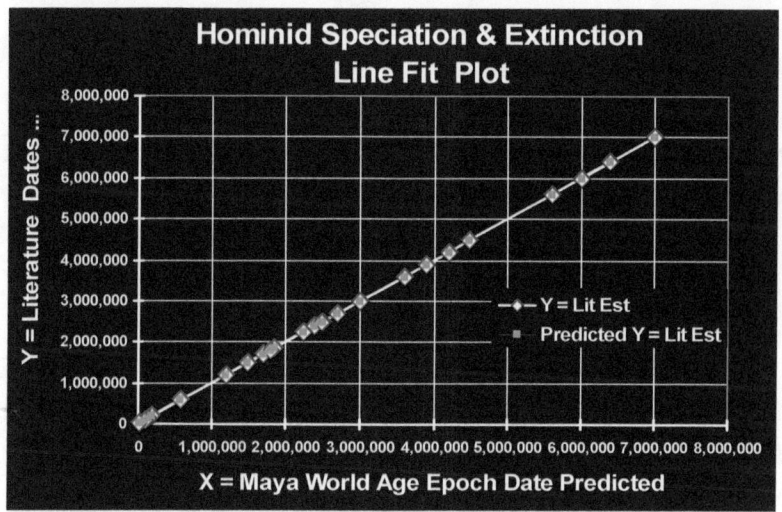

Figure 11 the Near-Perfect Prediction of Hominid Speciation and Extinction Events

In fact, the number of missing Hominid species could be much larger. *Homo floresiensis* (also known as the "hobbit" species) is believed to have emerged and gone extinct in a span of about 80,000 years.[30] If we generalize from that example, some 88 species or more events appear to be missing from the known fossil record. Even if nature skipped a beat now and then, the spanned Maya World Ages should have produced many more species than have so far been discovered.

It was at once apparent that further evidence of such changes was likely languishing in the database. That evidence takes the form of: *genetic bottlenecks* (the narrowing of genetic diversity in a population due to severe population reductions), *mutations* exhibited as genetic markers, and *major migrations* spanning multiple continents. Table 4 lists the major bottleneck, mutation, and migration events described in the prevailing literature. Alongside each literature cited event date is the corresponding first Even-numbered Epoch Date from the applicable Maya World Age cycle. Based on the data in Table 4, Hypothesis H_5 was formulated.

Bottlenecks, Mutations, and Migrations

Hypothesis H_5 asserts that a significant portion of the variation in the fossil record driven dates of Hominid species genetic bottlenecks, mutations, and major migrations is described, explained, and predicted by the range of dates of the first even numbered Epoch in applicable Maya World Ages.   **Hypothesis Null$_5$** counters that the dates of Hominid genetic bottlenecks, mutations, and major migrations exhibit no counting relation with Maya World Age Cycle first even numbered Epoch dates.

Table 4 Hominid Genetic Bottlenecks, Mutations, and Major Migrations

Case	Event Description	Years Ago		
		Literature Dates	World Age Epoch Date	Diff
1	Neanderthal Species Arrives in Europe	128,000	123,008	-4,992
2	Population Bottleneck No 1.	100,500	100,500	0
3	Homo Sapiens Genetic Diversity Reduction	98,000	98,000	0
4	Australasian & N. Eurasian populations divide	71,000	71,000	0
5	Population Reducing Bottleneck No. 2 in Africa	71,000	71,000	0
6	H. sapiens migrate out of Africa	70,000	70,000	0
7	Population Reducing Bottleneck No. 3 in Africa	67,000	67,142	142
8	Genetic Marker 168 Emerges	50,000	50,000	0
9	H. sapiens Austrailian Population Estab.	50,000	50,000	0
10	Population Reducing Bottleneck No. 4	50,000	50,000	0
11	Neanderthals Begin to be Displaced by H. sapiens	45,000	46640	1,640
12	Genetic Marker 9 Emerges	40,000	46,640	6,640
13	H. sapien W. Asian & Europe Popultations Established.	40,000	46,640	6,640
14	Genetics; Microcephalin Allele Selection	40,000	46,640	6,640
15	H. sapien E. Asian Popultions Established	30,000	30,000	0
16	Genetic Marker 3 Emerges; New World	20,000	20,000	0
17	H. Sapiens New World Populations Established.	20,000	20,000	0
18	M3 genetic marker Emerges	10,000	15,888	5,888

The correlation analysis revealed a near perfect counting relationship. The literature cited event dates in Table 4 and World Age first even numbered Epoch dates are near perfectly correlated. The correlation coefficient equals 0.996 (1.00) with a level of confidence well beyond the .001 level (in fact, the confidence level extends to eighteen decimal places). With Hypothesis Null$_5$ summarily rejected, a subsequent regression analysis yielded an adjusted R square of 0.9919 (1.00). That means that nearly 100 percent of the variation in the literature reported dates of Human genetic bottlenecks, mutations, and migrations is predicted by Maya World Age first even numbered Epoch dates. Figure 12 depicts this strong predictive relation.

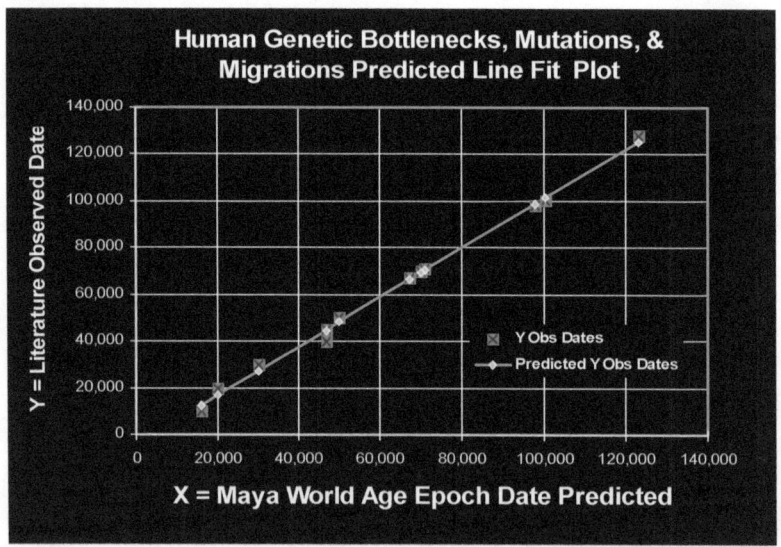

Figure 12 Maya World Age First Even Numbered Epoch Prediction of Human Change Events

Human Change Summary

The dates of recorded Human Changes over the last 281,895 years are concentrated in Maya World Age Precession Cycles. Each World Age Cycle describes a perfect circle with a circumference of 25,627 years. Within each Maya defined World Age Cycle, the dates of human change are further concentrated in two and one-half (2.5) Poisson distributed waves. Each wave has a wavelength of 10,251 years, and spans 144 degrees of each World Age Cycle. Each wave is subdivided into two Epochs of 5,126 years each spanning 72 degrees of the World Age Cycle. As a result, Degrees of Maya World Age Precession describe, explain and predict the overwhelming majority of the variation in the dates of human change over the definitive history of modern man.

Prevailing theories of human evolution fail to describe, explain, and predict the dates of the emergence and extinction of Hominid species. These theories of extreme gradualism, punctuated equilibrium, and the combination of the two, further fail to reliably account for the dates of genetic bottlenecks, genetic mutations, and major migrations in evidence for known Hominid species.

The Maya World Age Epoch theory and model of human evolution describes, explains, and predicts 99 percent of the variation in hominid speciation and extinction events known to have occurred in the last 7 million years. Further the first even Epoch theory and model describe, explain, and predict 99 percent of the variation in the dates of Hominid genetic bottlenecks, genetic mutations, and major migrations that have occurred over the last 130,000 years.

Results indicate massive shortfalls in the currently accepted history of Hominids. Existing fossils and genetic evidence appear to substantially under-estimate the total number of Hominid speciation, extinction, and genetic mutation events. Worse, there is no reason to assume that the force driving Maya World Age Precession Cycles, waves, and Epochs suddenly emerged in the last 7 million years. There is also no reason to assume that the natural forces driving Maya World Age Precession Cycles, waves and Epochs predict only Hominid speciation, extinction, and mutation events. It would seem that our paleontologists, archeologists, anthropologists, and geneticists have but nicked the surface of evolutionary history.

Without question, Hominid history must be vastly richer than is presently believed. What's more, the totality of Hominid history could well have been repeated five times in 50 million years, instead of the once in 7 million years as is currently asserted.

Ancient Maya World Age Precession Cycles, waves, and Epochs have proven a powerful tool for the description, explanation, and prediction of variations in the rate of Human Change and innovation. This arcane cyclic Maya time keeping system is, however, only a man-made tool. What is missing is a definition of the natural force that drives Maya Precession Cycles.

What is the rhythmic force that imposes Poisson and normally distributed waves of human change. This great deficit takes center stage today. True, we are currently living in the *last* even Epoch of a World Age. However, due to a genuinely rare event that the Maya called *the Transformation,* we may soon find ourselves abruptly cast into the first even Epoch of a New World Age. One of those Epochs associated with new species emergences, extinctions, genetic bottlenecks, major adaptations, and mass migrations. The quest for the driving force behind such surges of change could not be more relevant. Hence, the characterization and discovery of that World Age cycle driving force is the central focus of the next three chapters.

5

WAVES OF CELESTIAL MOTION

Maya World Age Precession cycles describe, explain, and predict *much more* than the dates of Human Change. Pairs of Maya Epochs describe waves that drive the fundamental motions of our planet and solar system. This chapter reveals that:

1) The ancient definition of Maya Solar System Precession is true. In contrast, the widely held Lunisolar Model of the precession of the equinoxes defined by Nicolaus Copernicus, Sir Isaac Newton, and adapted by Jean-le-Rond D' Alembert et al, is invalid, unreliable, and misleading;

2) Degrees of Maya *Precession* meter out our solar systems' Z-Axis Tilt, Galactic Orbital oscillation, and Galactic Orbital Speed;

3) Degrees of Maya Precession describe, explain, and predict the variation in the cyclic tilt of the Earth's vertical axis of rotation (*Obliquity*) – one Earth Obliquity cycle equals 8 Maya Epochs;

4) Degrees of Maya Precession describe, explain, and predict Earth's cyclic variations in the angle struck between the plane of the Earth's solar orbit, and either the Earth's or the Sun's equator (*Inclination*) – one short term Inclination cycle equals 14 Maya Epochs, and one long term Inclination cycle equals 20 Maya Epochs.

5) Degrees of Maya Precession describe, explain, and predict the short and long term variations of Earth's solar orbit from a perfect circle (*Eccentricity*) – one short term Eccentricity cycle equals 20 Maya Epochs, and one long term Eccentricity cycle equals 81 Maya Epochs.

The cited declarations constitute a testable theory of celestial motion. That theory asserts that Maya World Age Precession Cycles, Earth's cyclic Obliquity and Earth's solar orbital Inclination and Eccentricity, are all products of periodic wave forces acting on our solar system. Specifically, trains of waves having a wavelength of 10,250.7 light years and amplitude (wave height) of 5,125.36 light years. These energy – as opposed to ordinary matter – waves travel (propagate) at the speed of light. Hence the wavelength of 10,250.7 light years takes 10,250.7 calendar years to pass through our solar system. The theory asserts that 2.5 waves of the unnamed force produce the Maya World Age Precession cycle, which equals 25,626.8 years divided into 5 Maya Epochs.

At the end of this chapter those contentions are put to the test. Specifically, tested is **Hypothesis H$_6$** which states that: for any given Maya World Age Precession cycle Epoch of 5,125.36 years, the elapsed Maya Degrees of Precession describe, explain, and predict all (100 percent) of the variation in Earth's Obliquity, Inclination, and solar orbital Eccentricity. **Hypothesis Null$_6$** counters that Maya World Age Cycle Degrees of Precession are not significantly related to Earth Obliquity, Inclination, or Eccentricity. The null hypothesis represents the view that Precession, Obliquity, Inclination, and Eccentricity are all independent (uncorrelated) phenomena resulting from independent sources.

Precession

The currently accepted view of precession is *not* the Maya World Age Precession constant of 25,626.8 years neatly inscribed about a circle of 360 degrees. Instead, the accepted view of precession is that encapsulated by the formulas comprising the Earth centric "Lunisolar Model."[31]

The Lunisolar Theory and Model of the precession of the equinoxes was originated by Nicolaus Copernicus in 1543 in his *De revolutions*.[32] That model was subsequently adapted by Sir Isaac Newton in 1687, repaired by Jean-le-Rond D' Alembert in 1754, and has been heavily tuned by countless others ever since.[33] The Lunisolar Model is also that form of precession touted (as recently as 2003) by the National Aeronautics and Space Administration (NASA).[34]

So just what does Lunisolar Theory say? Walter Cruttenden (2003) describes it this way:

> Lunisolar theory ... states that the Earth's changing orientation to inertial space is principally due to the gravitational forces of the Moon and the Sun acting on the *oblate* Earth [meaning a spheroid flattened at the poles]. The lunisolar forces [i.e., that of the Sun and the Moon] are thought to produce enough torque to slowly move the Earth's spin axis in a clockwise motion so that after a period of approximately 25,770 years (one of many different currently calculated estimates in the rate) the Earth would have completed one retrograde motion relative to the Sun and the fixed stars. In this theory the Earth is thought to act like a wobbling top.[35 & 36]

Alas, all is not well in Lunisolar land. One glaring problem is that Lunisolar Theory predicts highly variable values for the duration of precession in the distant past and far future. This anomalous variation is ascribed to the torques exerted on the Earth by the nearly *constant* masses of the Sun and the Moon, and more recently by a host of equally all but *constant* lesser sources.

Figure 13 illustrates the prevailing view and Lunisolar version of the precession of the equinoxes.

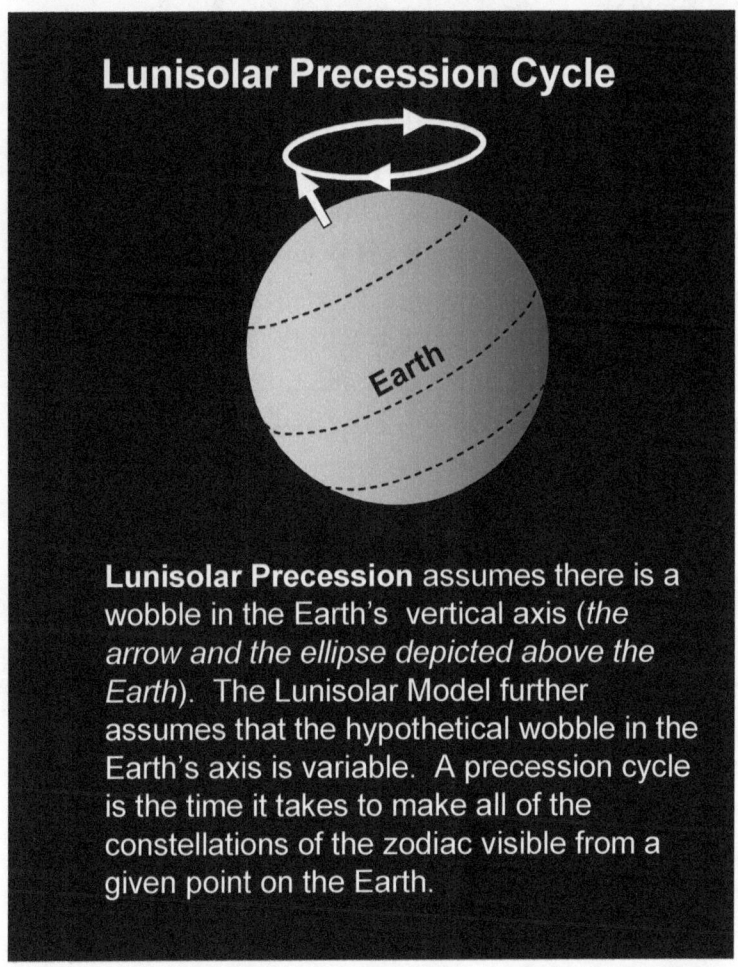

Figure 13 the Lunisolar Concept of Precession
Source: http://en.wikipedia.org/wiki/Precession, revisited March 3, 2008

In point of fact, the Sun and Moon and other cited torque perpetrators do vary in mass. They *loose* exceedingly small amounts of mass over time. That fact stands in direct opposition to the Lunisolar Model which currently predicts *increasing* precession values.

Another genuinely dreadful problem with the Lunisolar Model is that given the forward looking trend in Lunisolar predictions, the Earth's axial wobble will eventually violently disrupt the stability of the planet. Most astronomers prefer to simply ignore this dubious and highly unlikely event, as it does not adversely effect their current observations.

Equally indicting is the fact that if you extend the Lunisolar Model trend back in time a few hundred thousand years, the model produces precession values approaching zero[37]; an assertion paleo-climatologists ardently insist could have never happened.

The principal motive that drove Copernicus to put forth the idea of a wobbling Earth spin axis was that he needed to explain the well-known phenomena of the precession of the equinox. That is, he needed to account for the fact that the equinoctial point preccesses [wobbles] backward through the stars of the Zodiac at the rate of about one degree per 72 years to explain the motions of the Earth. He said the axis must *wobble or experience liberation,* and dubbed this the third motion of the Earth. But, "Copernicus never said this wobble was due to local forces."[38] "It was Newton who named the Sun and the Moon as the sources of this notional wobble in the Earth's vertical axis of rotation."[39]

Therefore, the originating motive for the Lunisolar Model was to account for the observed difference between the solar year (i.e., one complete Earth orbit around the Sun) and the sidereal year – the slight offset of the observed position of fixed stars lying outside the solar system, as seen by an Earth-bound observer. To adjust for this distant star declination angle difference twixt solar tropical and sidereal years, precession came to be defined in the following mind-numbing way:

> Precession is the angular difference or rotation time difference between Earth's 360-degree rotation on its axis (tropical day) and Earth's complete rotation relative to the fixed stars (sidereal day). After one complete 360° revolution of the Earth around the Sun (tropical year), this daily rotation difference accumulates to approximately 50.26" (arc-seconds). It takes Earth about 3.34 time seconds, and not roughly 1223 seconds, to rotate through this angle. The 1223 seconds represent an angle of approx. 50.26" [arch seconds] in a 360° orbit period of 31,556,925.97 seconds (a tropical year).[40]

In addition to making a person's head hurt really badly, there are serious problems with this hastily accepted and overlong held to notion of precession. One in particular renders Lunisolar Theory patently false and unreliable for long term predictions. It was strongly expressed in a clear fit of frustration by Walter Cruttenden in 2003:

> The Earth does not change orientation to the Perseid meteor shower, or to the Moon, or to eclipses, or to any points of planetary

occultation or to anything within the solar system, because local wobbling of the Earth does not cause precession. What we call precession only occurs relative to fixed stars and objects 'outside our solar system,' because precession is actually due to the motion of the solar system itself.

The solar system ... moves as a single reference frame at the rate of about 50 arc seconds annually relative to inertial space. All bodies within that reference frame maintain their relative gravitational relationships, the Earth does not experience precession within that frame, and therefore only the tropical frame applies. All bodies [observed from the Earth] outside that reference frame must be adjusted for precession and the sidereal frame applies.[41]

Figure 14 illustrates the Maya Solar System Precession Model and Theory, which Cuttenden and other modern investigators have rediscovered for themselves.

In 2003, Walter Cruttenden, Vince Days, and the folks at the Binary Research Institute (BRI) published several papers attacking and invalidating the Lunisolar Model of precession. They did this in a concerted effort to show that our Solar System is a "binary star system." A binary star system is one in which two stars (like our Sun) are gravitationally bound to one another resulting in their mutual orbit of a common center of gravity in-between them. Cruttenden et al reasoned that if "80 percent of the stars observed are binary star systems,"[42] there is a very good chance that ours is as well, and further, that that is the source of *solar system* precession.

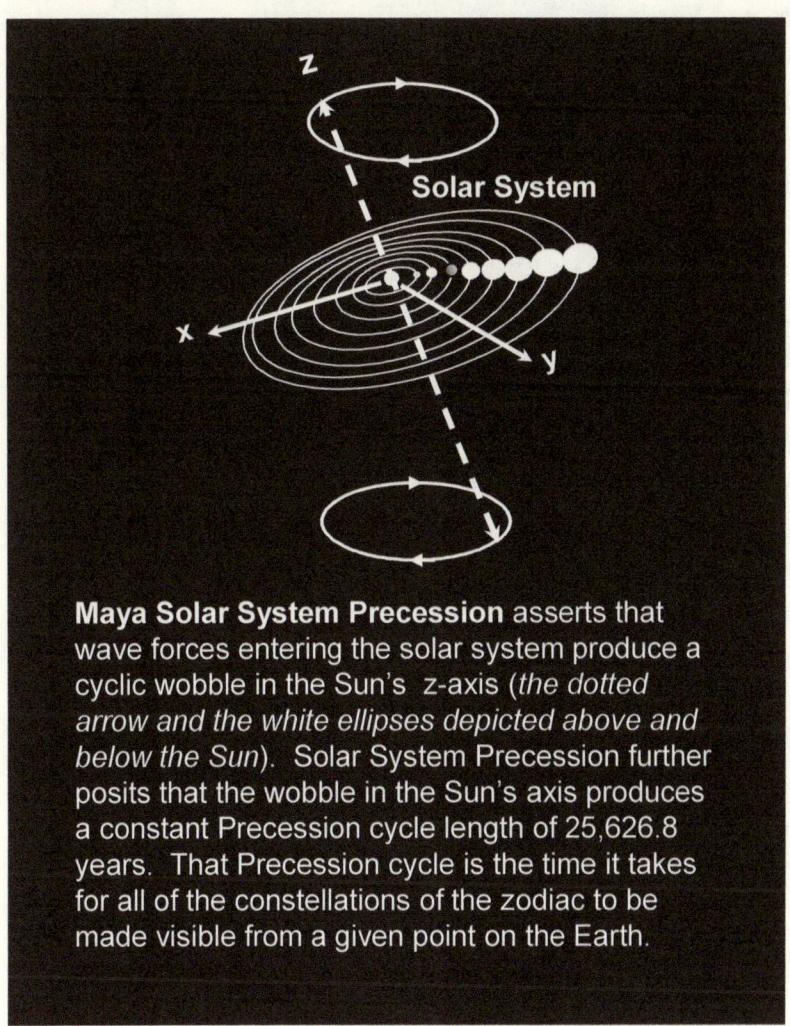

Maya Solar System Precession asserts that wave forces entering the solar system produce a cyclic wobble in the Sun's z-axis (*the dotted arrow and the white ellipses depicted above and below the Sun*). Solar System Precession further posits that the wobble in the Sun's axis produces a constant Precession cycle length of 25,626.8 years. That Precession cycle is the time it takes for all of the constellations of the zodiac to be made visible from a given point on the Earth.

Figure 14 Maya Solar System Precession

Recent research puts the proportion of observable binary star systems at closer to 50 percent.[43] Definitive evidence our solar system is a binary star system has failed to mature in recent astronomical surveys. What's more, Earth and space based observations and theoretic modeling of the solar neighborhood have diminished the likelihood that our solar system is a binary system.

Suffice it to say that no fault is found with BRI's rock solid criticisms of the Lunisolar Model, and BRI's insistence that "precession is due to the motion of the solar system itself."[44]

It is important to note that the Maya World Age Precession Cycle retains all the physics of the Lunisolar Model. The Maya World Age Precession Cycle simply re-assigns the *axial wobble to the center of our solar system* (i.e., the Sun), and turns the rate of precession into a *constant*. There is no significant difference between the two models in their near term daily, monthly, and yearly time differences ascribed to tropical and sidereal measures. There is, however, a dramatic difference in the causative factors driving the two precession models and in their long term predictions.

Maya World Age Precession posits that the vertical (Z) axis of the plane of our solar system: tilts, oscillates, and accelerates relative to the Galactic Center over fixed intervals of 25,626.8 years. The Maya correctly ascribed these effects to forces originating outside our solar system. Forces thus far characterized in this book as Poisson distributed waves. Force waves that are named and detailed in later chapters. The Lunisolar Model, in contrast, falsely ascribes the source of *variable* precession to local within solar system forces; forces that vary too little and in the wrong direction to support current Lunisolar Model predictions.

The Lunisolar Model has been disproved by arguments put forth by A.L. Berger (1977), Carlos Santagata (2002), Walter Cruttenden (2003), Vince Dayes (2003), and others. Those sources provide more than sufficient evidence to warrant rejection of the Lunisolar Model. There is no reason to belabor that litany of indicting Lunisolar Model faults here, although they are legion. Summarizing his 2003 condemnation of the Lunisolar Model, Cruttenden reports,

> Based on several years of study this institution [BRI] has found that precession occurs relative to objects outside our solar system (fixed stars, quasars, other galaxies, etc) but does not occur relative to objects within the solar system (the Moon, eclipses, and other planetary occultation, comet debris, etc)."[45]

In January of 2005, the Sirius Research Group put it this way:

> There is extremely clear scientific evidence that the rotational axis of the Earth does NOT precess (wobble) relative to the Sun and other objects within the solar system as the Earth makes a complete 360° orbit around the Sun each tropical year. However, Earth's celestial pole slowly changes its orientation relative to the stars. Thus, for an observer on Earth the stars change their position in the sky (declination) by approximately 47° during half a precession cycle (i.e. from winter to summer solstice) -- but not the Sun, it repeats the same yearly path in the sky![46]

How well do the quantities of conventional precession map onto Maya Precession Cycles and Epochs? The answer is near *perfectly*. Moving the Copernican variable wobble from the Earth to a constant wobble in the vertical axis of our solar system, keeps all the observed values characterizing conventional precession, but eliminates all the nasty contradictions and horribly wrong distant past and future predictions produced by the Lunisolar Model.

We start where Copernicus started. That is with the need to "explain the well known phenomena of the 'precession of the equinox' whereby the equinoctial point wobbles backward through the [stars of] the Zodiac at a rate of about one degree per 72 years [more precisely one degree every 71.185556 years]."[47] That wobble-rate works out to 72 degrees of precession for every 5,125.36 year long Maya Epoch, which is a perfect fit to the Maya model. That solar system wobble-rate further results in 144 degrees of precession for each Epoch Pair (*wave*) of 10,250.7 years, which is another dead-on match to the Maya Precession Cycle. What's more, that accepted wobble rate results in a total of 360 degrees of precession for every 25,626.8 year long Maya World Age Precession Cycle. In short, the ancient Maya score a perfect 10. Copernicus and Newton rate a big fat zero, however, because their variable Lunisolar precession rate flatly fails over the long haul, while the Maya's constant Solar System Precession rate does not.

Ah, but how well does the slow Z-axis tilting of our solar system result in the difference observed between the solar and sidereal years as seen from Earth?

The Rand McNally New Concise Atlas of the Universe states:

> "The Earth's Axis is not constant in direction, but describes a circle 47 degrees in *diameter* in a period of 25,800 years [one of *many* different Lunisolar variable rate estimates in the published literature]. This means that the position of the celestial pole alters [shifts], which in turn affects the position of the celestial equator and of the equinoxes."[48]

By the elementary formula, Pi times the diameter, the Rand McNally Earth relative precession circle circumference = 3.141592653590 * 47 degrees = 147.65486 degrees. 147.66 degrees divided by 25,800 years equals a Rand McNally precession rate of 0.005723056 degrees per year. Substituting the Maya Solar System Precession period, but retaining the Earth relative precession circle circumference, the Maya rate of precession per year becomes 147.654854718720 degrees divided by 25,626.8 years = 0.005761736 degrees per year. Thus, the Maya annual Earth relative rate of precession is just 0.0000387 degree per year greater than the Rand McNally rate. Over a complete Precession cycle of 25,626.8 years that difference accumulates to just under 1 degree, a difference of 0.67 percent.

Another source, the United States Navy, phrased the definition of the Earth relative precession in a way that enables us to roughly validate the estimated difference. That source states:

"We usually think of the Earth's axis as being fixed in direction - after all, it always seems to point toward Polaris, the North Star [the current pole star]. But the direction is not quite constant: the axis does move, at a rate of a little more than a half-degree per century."[49]

Both prior estimates of precession are in solid agreement with the Navy's guidance. The two estimated degrees of precession per year multiplied times 100 years equal 0.572305638 of a degree per century for the Rand McNally estimate, and 0.576173595 of a degree per century for the Maya estimate. The Maya estimate is only 0.003867956 of a degree greater per century.

Maya Solar System Earth relative z-axis tilt (wobble) works out to:

- 0.0057617 degree per year.

- 0.57617 degree per 100 years.

- 2.9 degrees per 512 years.

- 29.53097 degrees per 5,125.36 year Maya Epoch.

- 59.06182477 degrees per 10,250.7 year Epoch pair (force wave).

- 73.827425 degrees per 12,813.4 years *(half Maya Precession Cycle).*

- 147.65486 degrees per 25,626.8 year long Maya Precession Cycle.

The Lunisolar *variable* Earth Z-axis wobble thus translates near perfectly into the *constant* Maya solar system Z-Axis wobble of Precession. The Maya Solar System Precession Cycle just does not fall flat on its face in the prediction of seriously distant past and future Precession values.

Solar System Z-Axis Gain

Our solar system exhibits a sort of oscillation in its trajectory around the Milky Way Galaxy. In addition to following an elliptical path about the Galactic Center, our solar system slowly moves up and down on that path. Actually, the word *slowly* seriously under-represents the lethargy of our Sun's Galactic orbital ups and downs. Our solar system visits to the Galactic northern (top side) of the plane of the Milky Way Galaxy over a period of approximately "31.5 million years," one half of a 62 million year cycle.[50] Figure 15 depicts the vertical oscillation of our solar system relative to the Galactic Plane.

Our solar system is relatively close to the Galactic Plane at the moment, and it will be climbing in the Z direction for tens of millions of years to come. Thus, it was thought reasonable to limit the analysis to vertical gain. Suffice it to say that millions of years from now our solar system will finally reach a maximum altitude (some estimate at around Z = "230 Light years"), and the masses comprising the Galactic plane will then slowly pull us back toward the plane again. [51, 52, 53, 54 & 55]

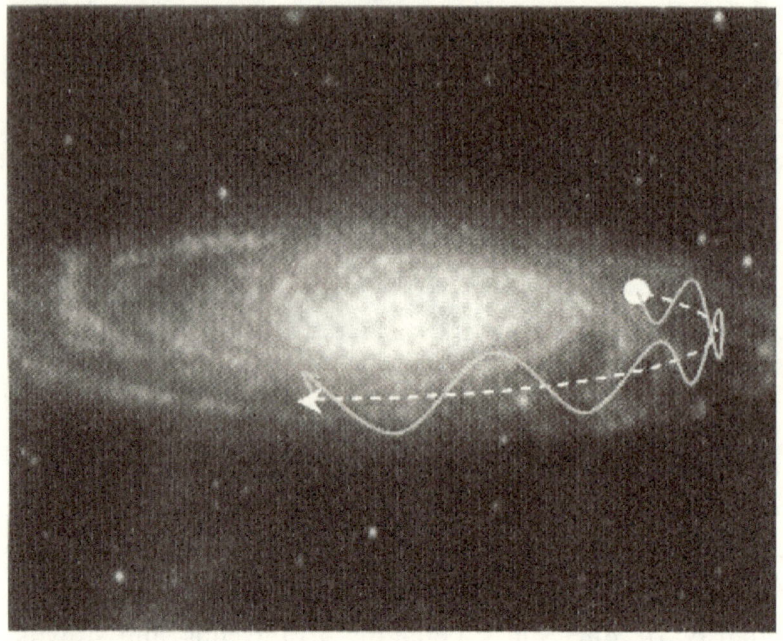

Figure 15 Vertical Oscillations of the solar System about the Galactic Plane
Source: Castelvecchi, D., 2007, Northern Exposure; the Inhospitable Side of the Galaxy, Science News, April 21, 2007, Vol. 171, p.244

The problem with our solar system's ups and downs is not the risk that we will drift away from the Galactic Plane and wonder off into space. Rather, it is radioactive "cosmic ray bombardment." Apparently, the high Galactic North is a sort of unshielded radiation-hell that our solar system visits over extremely long periods of time. Prior visits have been associated by Robert Rhode, Richar Muller, and Mikhail Medvedev et al with some of Earth's most "catastrophic mass extinctions."[56] That's a genuinely grim tale for a time far-far away. At issue here is the rate of solar system Z-Axis gain above the Galactic Plane.

The rate of our solar system's Z-Axis Gain has recently been measured at 7 kilometers/second (7 Km/s).[57] Given the Epoch and Precession Cycle durations, the work reported here was performed in units of light years, as opposed to kilometers. One light year equals 9,460,730,472,580.8 kilometers, or 5,878,625,373,183.608 miles, or about 0.307 Parsecs. Now parsecs are nice and all, but most folks have trouble picturing a *Parsec*. Somehow the brain finds it easier to visualize is the distance that light travels in a year. Astronomical Units (1 AU equals the average distance from the Earth to the Sun) may be easier for folks to picture, but it is way too small a unit of measure for this application. Regardless, the 7 Km/s estimate of Z-Axis Gain previously cited, works out to:

- 0.0000232 light years per year of Maya Precession

- 0.0118784 light years per 512 years of Maya Precession

- 0.118784 light years per 5,125.36 year long Maya Epoch of Precession

- 0.237568 light years per 10,250.7 year Epoch pair Wave of Precession

- 0.2972708 light years per 12,813.4 year long one-half Maya Precession Cycle

- 0.59454176 light years per 25,626.8 year long World Age Maya Precession Cycle

Solar System Galactic Orbital Speed

As stated earlier, our solar system rotates clockwise around the Galactic Center. The scientific literature on our Sun and solar system invariably states that our Sun's Galactic Orbital rate (speed along the Galactic Y-axis, as opposed to angular velocity) as if it were constant. Literature reported estimates of our Sun's orbital speed vary from 210 km/s (kilometers per second) to 280 km/s. Saner sources actually confess our Sun's Galactic Orbital Speed varies, but fail to name the source of that variation.

The ancient Maya were much more courageous. As stated earlier, they insisted that as we approach an even numbered Epoch boundary, everything speeds up (accelerates). Mayan oral tradition calls this acceleration, "The Time of the Quickening."[58] The Maya also claimed that as we move through an odd numbered Epoch everything slows down (decelerates). There is now a veritable mountain of evidence validating these Maya assertions. For now, it is simply asserted that our Sun starts each wave defining Epoch pair at a rate of approximately 210 km/s and gradually accelerates to a maximum speed of approximately 280 km/s by the time our Sun reaches the Epoch boundary 5,125 years later.

Then, our Sun slowly decelerates in the odd numbered Epoch from 280 km/s back to approximately 210 km/s over the ensuing 5,125 years. Hence, each Galactic orbit acceleration and deceleration cycle of our solar system spans a range of ± 70 km/s distributed over 10,250.7 years.

Figure 16 plots our Solar System's Z-Axis Tilt, Z-Axis Gain, and Sun Speed changes onto the Maya Epoch pair waveform.

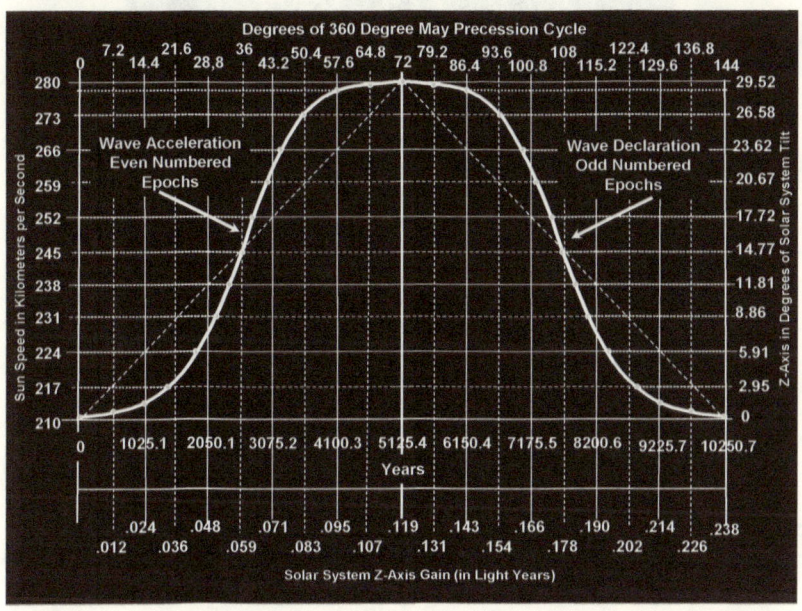

Figure 16 Mapping of Solar System Z-Axis Gain, Z-Axis Tilt, and Galactic Orbital Speed onto One Epoch-Pair Wave

Obliquity

Figure 17 illustrates the concept of Earth's Obliquity cycles.

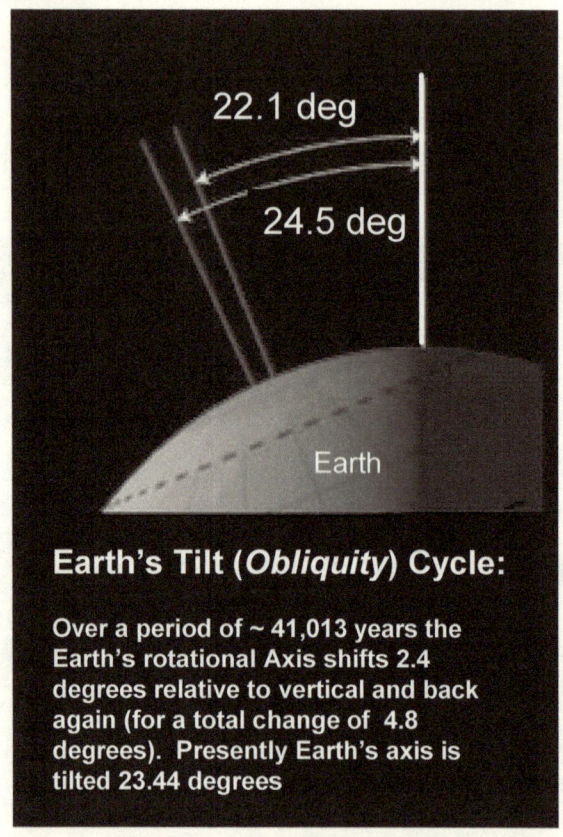

Figure 17 Earth's Obliquity Motion
Source: http://en.wikipedia.org/wiki/Obliquity; Revisited April 26, 2008

Obliquity is the cyclic tilting (as opposed to wobbling) of the Earth's vertical rotational axis. Over "a period of approximately 41,013 years," the Earth's axis of rotation varies from a minimum value of "22.1 degrees to a maximum of 24.5 degrees (a range of 2.4 degrees)," and then returns again to the minimum. The total angular tilt per cycle thus constitutes an axial movement of 4.8 degrees over an estimated period of 41,013 years. That equates to an approximate rate of 0.00017 degrees per year. "The current obliquity value is 23.44°, which is near the middle of the [one half cycle] range of [motion]."[59]

The last maximum obliquity is thought to have occurred in 8,700 BC (*10,708 years ago as of 2008*), the last mean obliquity occurred in 1550 AD, and the next maximum is expected to occur in 11,800 AD.[60]

The last mean Obliquity date of 1550 AD, coupled with the rapid rise in the frequency of Human Changes since the mid-Renaissance (around 1500 AD) are two of the reasons for setting the zero date of the database of Human Changes to 1500 AD.

Multiplying the value 8 times the Maya Epoch duration of 5,125.36 years equals 41,002.88 years, which is just 10.12 years less than the conventional estimated duration of the published Obliquity cycle.[61]

Inclination

The "**Inclination**" angle of the Earth is described in several different ways in the prevailing scientific literature. It is most often characterized as the angle between the plane of the Earth's orbit around the Sun (e.g., the ecliptic) and either the Earth's or the Sun's equator. "An Inclination of 0 degrees means that the orbiting body [the Earth] orbits the Sun in its [the Earth's or the Sun's] equatorial plane, in the same direction as the planet rotates about the Sun [which is counterclockwise]."[62]

Figure 18 illustrates the concept of Earth's Orbital Inclination Cycles.

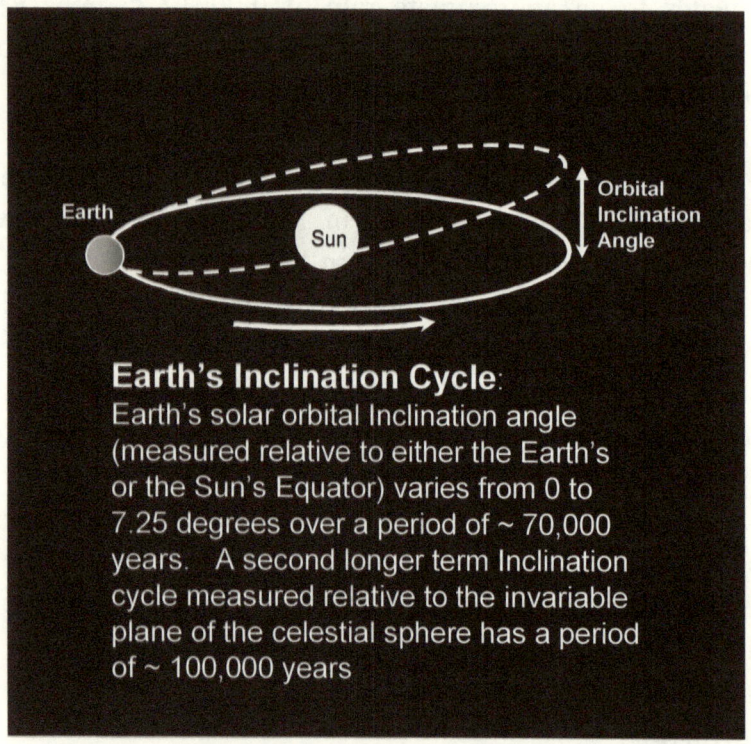

Figure 18 the Earth's Inclination Cycle
Source: Wikipedia, the free encyclopedia,
http://en.wikipedia.org/wiki/Inclination, Revisited January 24, 2008

The Earth's orbital Inclination angle ranges from 0 to 7.25 degrees, measured as the difference in the angle of the *ecliptic* (Earths plane of rotation around the Sun) and the Sun's Equator. The inclination of Earth's solar orbit drifts up and down relative to its present orbit with each cycle having a period of about 71,000 years."[63]

Fourteen times the Maya Epoch duration of 5,125.36 years equals 71,755.04, which is just 755.04 years longer than the frequently published duration of the Earth's Inclination Cycle.

It is interesting to note that available inclination angle data derived from simulations published by F. Varadi in collaboration with M. Ghil and B. Runnegar in 2003)[64] indicates a somewhat longer Inclination cyclic. Those simulations run for the total time of the last million years produced a period of Inclination of 76,880.4 years. That Inclination Cycle interval equals exactly 15 Mayan Epochs.

"More recent researchers have noted a drift in Inclination, and found that the Earth's orbit also moves relative to the orbits of the other planets. The *invariable plane*, the plane that represents the angular momentum of the solar system, is approximately the orbital plane of Jupiter. That Inclination of the Earth's orbit has a 100,000 year cycle relative to the *invariable plane*. It is interesting to note that 'this 100,000-year Inclination cycle closely matches the 100,000-year pattern of ice ages.'"[65] Of more importance here, twenty times the Maya Epoch duration of 5,125.36 years equals 102,507.2 years, which is just 2,507.2 years longer than the approximate 100,000 year invariable plane referencing Inclination Cycle duration. Yet another relatively close match lending support to Hypothesis H_6.

Earth's Orbital Eccentricity

Eccentricity is defined as the variation of the Earth's orbit about the Sun from a perfect circle, and is measured in dimensionless units. Figure 19 depicts Earths' 100,000 year Eccentricity Cycle.

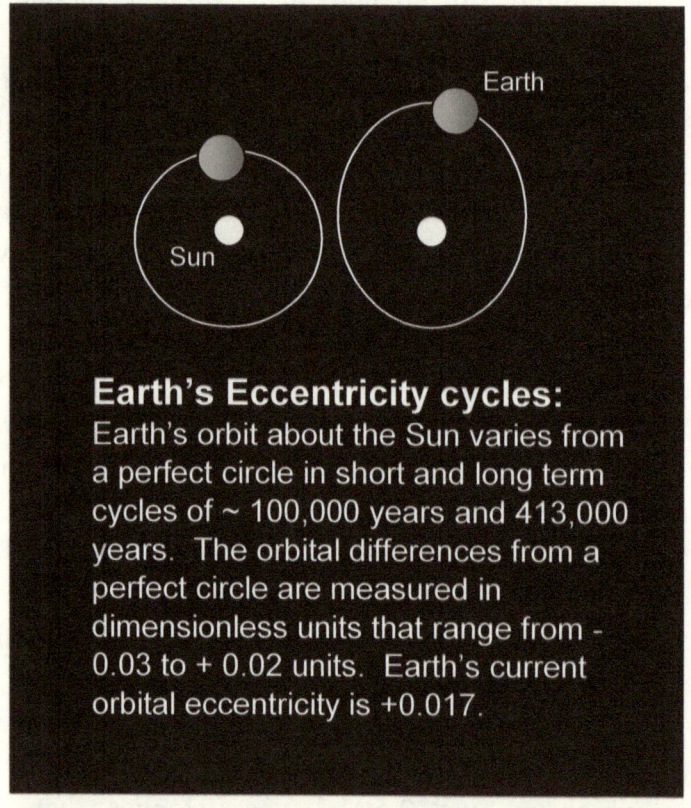

Earth's Eccentricity cycles:
Earth's orbit about the Sun varies from a perfect circle in short and long term cycles of ~ 100,000 years and 413,000 years. The orbital differences from a perfect circle are measured in dimensionless units that range from - 0.03 to + 0.02 units. Earth's current orbital eccentricity is +0.017.

Figure 19 the Earth's Orbital Eccentricity
Source: http://en.wikipedia.org/wiki/Orbital_eccentricity, Revisited March 3, 2008

According to the folks at the University of Washington:

>   The path of the earth around the sun is slightly elliptical, rather than [perfectly] circular. The eccentricity of the ellipse changes periodically with time, with components having periods of approximately 100,000 years and approximately 413,000 years.
>   The eccentricity is defined as the ratio of the focal length of the ellipse (the distance between the foci [points of convergence and divergence] to the length of its major axis [the diameter at the widest part of the ellipse]; the eccentricity of the earth's orbit about the sun has ranged from less than 0.01 to more than 0.05 over the past 600,000 years.[66]

Twenty times the Maya Epoch duration of 5,125.36 years equals 102,507.2 years. That's just 2,507.2 years longer than the grossly estimated 100,000 year duration of Earth's Eccentricity Cycle. Eighty-one times the Maya Epoch duration of 5,125.36 years equals 415,154.16 years, which is just 2,154.16 years longer than the approximate 413,000 year duration of Earth's Eccentricity Cycle.

No one has ever lived long enough to actually witness and measure a solar system Precession cycle or any of the Sun and the Earth's geometric cycle changes. Published estimates are little more than generalizations derived from exceedingly short-term astronomical observations, measurements, and theoretical calculations. The assumed error of measurement for the literature reported estimates of Earth's geometric *Obliquity, Inclination, and Eccentricity* cycles is ± 2.5 percent (a total of 5 percent).

Given that error term and the results related in Table 5, it is asserted that the force waves of Maya World Age Precession measured in Maya Epochs constitute the common denominator of solar system and Earth centric geometric cycles of periodic motion.

Table 5 Relation of Mayan Epochs to Earth Motion Geometry

Cycles	Literature Duration Estimates (In Years)	Number of Mayan Epochs (In Epochs)	Multiplied Maya Epoch Equivalent (In Years)	Difference (In Years)	Percent Difference
Obliquity	41,013	8	41,002.88	10.12	0.025%
Inclination$_1$	71,000	14	71,755.04	755.04	1.063%
Inclination$_2$	76,000	15	76,880.40	880.40	1.158%
Inclination$_3$	100,000	20	102,507.20	2,507.20	2.507%
Eccentricity$_1$	100,000	20	102,507.20	2,507.20	2.507%
Eccentricity$_2$	413,000	81	415,154.16	2,154.16	0.522%

Note: 1 Mayan Epoch = 5,125.36 years

As stated at the beginning of this chapter, **Hypothesis H$_6$** states that: for any given Maya World Age Precession cycle Epoch of 5,125.36 years, elapsed Maya Degrees of Precession describe, explain, and predict all (100 percent) of the variation in Earth's Obliquity, Inclination, and solar orbital Eccentricity. **Hypothesis Null₆** counters that Maya World Age Cycle Degrees of Precession are not significantly related to Earth Obliquity, Inclination, or Eccentricity.

Description, Explanation, and Prediction

In Hypothesis H_6, the words 'for any given Maya World Age Precession cycle Epoch' are of analytic importance. Within each Epoch of 5,125.36 years, any possible relationships between Precession, Obliquity, Inclination, and Eccentricity remain linear. That is, plot as straight lines. The cyclic (oscillating wave like) nature of Obliquity, Inclination, and Eccentricity data over two or more Epochs are non-linear. That means that in any multi-Epoch analysis, the variable data plot curves that violate the assumptions of linear correlation and regression analysis. Since nonlinear (curve fitting) analysis exceeded the scope of this study, Hypothesis H_6 was intentionally limited to within Epoch linear statistical tests. That means that multi-Epoch analyses were done piecewise, Epoch by Epoch.

To test Hypothesis H_6 a dataset extending back in time from time zero (1500 AD) to 287,020 years ago, and forward in time from zero to the future end of the current Epoch in 6,620 AD, was used. That dataset was divided into 13 Maya World Age Precession Periods. 12 negatively signed Maya World Age Periods of 25,626.8 years, and one positively numbered Maya Epoch of 5,125.36 years.

Degrees of Maya Precession (the single predictor in Hypothesis H_6) were calculated for each ± 512.54 year database record interval, within each Maya World Age Period, Wave, and Epoch.

Starting at zero (1500AD) and 288.0000058 degrees of Precession, Maya Precession was calculated at a constant rate of ± 0.014047794 degrees per year. This resulted in a 2008 Maya Precession value 295.1362852 degrees.

Degrees of Earth Obliquity for each ± 512.54 year database record in each Maya World Age Period, wave, and Epoch were calculated using a 0 point (1500AD) value of 23.30 degrees and a constant Obliquity rate of ± 0.000117065 degrees per year. That made the 2008 Obliquity value 23.44 degrees, which equals the currently published estimate.

Degrees of Earth Inclination for each ± 512.54 year database record of each Maya World Age Period, wave, and Epoch were calculated using a 0 point (1500AD) value of 0.00 degrees, and a constant Inclination angle change rate of ± 0.000101038 degrees per year. That made the 2008 Inclination angle 0.0513 degrees. That value equals the currently published estimate.

Earth orbital Eccentricity for each ± 512.54 year database record of each Maya World Age Period, wave and Epoch was calculated using a 0 point (1500AD) value of 0.0164 units, and a constant Eccentricity change rate of ± 0.000001 units per year. That made the 2008 orbital Eccentricity 0.0167079 units, which equals the currently published estimate.

Correlation analyses were performed for each Epoch in the database relating Degrees of Maya Precession to Obliquity, Inclination, and Eccentricity. For all of the 57 Epochs in the database the correlation results were identical to those summarized in Table 6. Signs indicating the direction of correlation have been omitted. The cyclic values of Obliquity, Inclination, and Eccentricity change sign relative to the always positive change in Degrees of Maya Precession. The correlation coefficients, however, remain the same.

Table 6 Correlation of Maya Degrees of Precession with Obliquity, Inclination, and Eccentricity

	Years	Precession	Obliquity	Inclination	Eccentricity
Years	1.00				
Precession	1.00	1.00			
Obliquity	1.00	1.00	1.00		
Inclination	1.00	1.00	1.00	1.00	
Eccentricity	1.00	1.00	1.00	1.00	1.00

Hypothesis $Null_6$ is rejected. As hypothesized, all of the relationships are significant well beyond the .001 level at a probability approaching certainty. Clearly Maya World Age Cycle Degrees of Precession are not only significant, but perfectly correlated with Earth's Obliquity, Inclination, or Eccentricity. The prevailing view that precession, obliquity, Inclination, and Eccentricity are independent (uncorrelated) phenomena that result from independent sources is apparently false.

Hypothesis H_6 is retained. Regression analyses performed for each Maya World Age Precession cycle Epoch of 5,125.36 years confirmed that elapsed Maya

Degrees of Precession describe, explain, and predict 100 percent of the variation in Earth's Obliquity, Inclination, and solar orbital Eccentricity.

Interpretation

Results reported in the prior chapter combine with those just related to indicate the following. Maya solar system Precession is reliably consistent with existing astronomical observations and eliminates the invalidating predictive errors of Lunisolar precession.

Degrees of Maya Solar System Precession describe, explain, and predict 100 percent of the variation in Earth's Obliquity, Inclination, and Eccentricity for at least the last 287,020.16 years of Earth history. These cycles appear to be the correlated manifestations of a single reoccurring periodic wave force emanating from an extra-solar-system source. That yet to be named Poisson distributed wave force must have a wavelength of 10,250.7 years (144 degrees of Maya Precession). Each such wave must form a bell shape subdivided into two Maya Epochs of 5,125.36 years each (72 degrees of Maya Precession). Moreover, there is every reason to suppose that the extra-solar wave force and its effects have persisted for the majority of the history of this planet.

Degrees of Maya Precession describe, explain, and predict the variation in the rate of human change in evidence for the last 282,407 years of human history. The variation in Human change is Poisson or normally distributed in waves having a wavelength of 10,250.7 years (144 degrees of Maya Precession), with each such wave subdivided into two Epochs of 5,125.36 years and 72 degrees of Maya Precession.

The following Hominid species events are perfectly correlated with and are predicted by the first even numbered Epoch of each associated Maya World Age:

- Speciation
- Extinction
- Genetic bottlenecks
- Major mutations
- Major migrations

Modern scientists cite global climate change as the primary historical determinant of human evolution and adaptation. United Nations assembled climatologists now claim that man has become the primary determinant of global climate change. The next chapter puts these seemingly contradictory notions to the test and pits both against an alternate candidate explanation – Maya Degrees of Precession.

6

WAVES OF CLIMATE CHANGE

Paradoxical Views

There is a universally embraced assumption in paleontology, archeology, and anthropology. It is that: "Earth's climate has continued to change, and those changes have continued to shape human evolution."[67] In a remarkable modern reversal of that notion, the United Nations Internal Panel on Climate Change (IPCC) Working Group One (WK1) asserted in 2007 that:

> ... Since 1750, it is extremely likely that humans have exerted a substantial warming influence on climate. This RF [Radiative Forcing] estimate is likely to be at least five times greater than that due to solar irradiance changes. For the period 1950 to 2005, it is *exceptionally unlikely* that the combined natural RF (solar irradiance plus volcanic aerosol) has had a warming influence comparable to that of the combined anthropogenic [man made] RF.[68]

Given the discoveries shared in the prior chapters, it was suspected that there might be more to the story than contemporary diggers and weathermen were touting. First, some unseen periodic wave force is responsible for Maya solar system World Age Precession cycle Epochs. Second, Degrees of Maya solar system Precession perfectly predict Earth's Obliquity, Inclination, Eccentricity, and Human Change cycles. It seemed only logical to assume that the same wave force would be influencing Global Temperature Change and concentrations of atmospheric Long-Lived Greenhouse Gases (LLGGs). Over the last decade and a half the historical data needed to test those assertions has been collected, described, analyzed, and interpreted. So let's get to it.

800,000 Years of Global Climatic Data

Global Earth Temperature Changes are not at all equally distributed over the last 800,000 years. We know this from the analysis of deep Antarctic Ice core data openly posted on-line. For purposes of this investigation, the Antarctic Vostok Ice Core Deuterium Data for the last 420,000 Years[69] and the European Project for Ice Coring in Antarctica (EPICA) deep ice core data collected from East Antarctica Dome C were downloaded. The combined data set represented an accepted index of Global Temperature Change for the last 800,000 years.[70]

The Vostok and EPICA data sets are independently obtained indexes of Earth's very long term temperature and LLGG variations. Cross validation studies found the

106

two samples to be in complete agreement for the 420,000 years that they overlap.[71]

There is no question about the highly variable and decidedly inhomogeneous nature of historic temperature and LLGG data. No statistical test is needed to confirm this erratic variability. One glance at EPICA and Vostok data plots, like that depicted in Figure 20, reveals the dramatic longitudinal jitter of Global Temperature. Corresponding variations in atmospheric Long Lived Greenhouse Gases are significantly correlated with the plotted variations in Global Temperature.

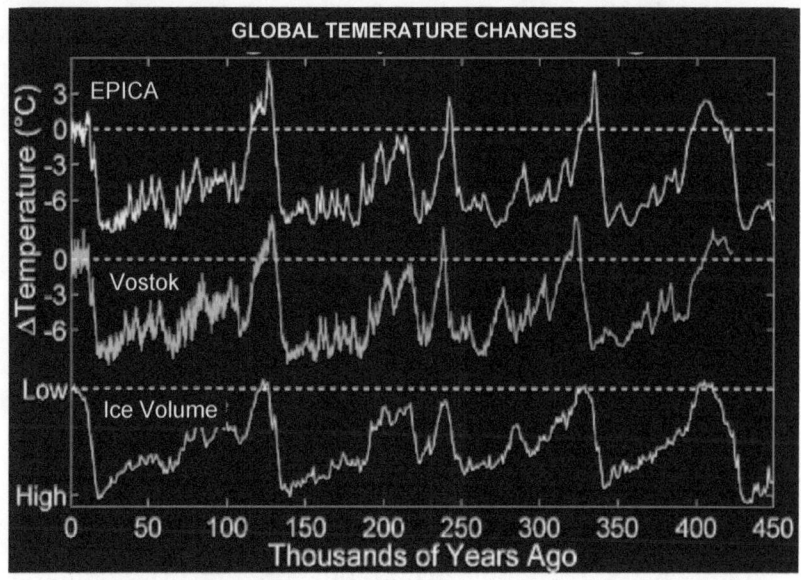

Figure 20 Vostok and EPICA Ice Core Temperature and Ice Volume Data

Source: Jouzel, J., Masson-Delmotte, V., Cattani, O., and Dreyfus, G. et al, 2007, Orbital and Millennial Antarctic Climate Variability over the Past 800,000 Years, Science 10 August 2007:Vol. 317. no. 5839, pp. 793 – 796

One recent NASA article summarizes the full scope of available Earth climate evidence as follows:

> From the oceans' depths to the polar ice caps, clues to the Earth's past climates are engraved on our planet. Sea sediments reveal how much ice existed in the world and hint at past temperatures and weather patterns. Ice cores also provide a glimpse of past temperatures and preserve tiny bubbles of ancient atmosphere [LLGGs]. Coral, tree rings, and cave rocks record cycles of drought and rainfall.
>
> ....
>
> Ocean cores show that the Earth passed through regular ice ages—not just the 3 or 4 recorded on land by misplaced boulders and glacial loess deposits—but 10 in the last million years, and around 100 in the last 2.5 million years.[72]

Ice core driven Global Temperature Change values were loaded into the database holding the dates of 282 millennia of Human Changes. Statistical correlation and regression analyses were first performed to discover the extent to which Global Temperature describes, explains, and predicts the dates of Human Change. After all, one cannot pick up a modern textbook on the origin of our species without reading that the changing climate has shaped man's evolution throughout human history. Finding a significant counting relationship between Global Temperature and the Dates of Human Change would tend to reinforce the prevailing contention. That view became the Alternate Hypothesis. That **Alternate Hypothesis states:** Global Temperature describes, explains, and predicts a significant portion of the variation in the dates of Human Change.

The EPICA Ice Core driven Global Temperature data was selected as the data source for Global Temperature Changes. The Vostok Ice Core Data could just have easily been substituted, as the two data sets have been shown equivalent for the 420,000 years contained in both samples.[73] The EPICA data analyzed extend from 281,895 years ago to 97 years ago. The ice core data sources are replaced by modern instrumented meteorological data starting in 1911.

Jouzel, Masson-Delmotte, Cattani, and Dreyfus, et al (2007) found that precession – *not Lunisolar precession* – and obliquity predict a significant portion of the variation in Global Temperature change over the last 800,000 years, as indexed by the EPICA and Vostok ice core data.[74] Given that, and prior reported results, the EPICA data were analyzed in three separate Date of Human Changes driven data sets. Set 1 consisted of all of the Human Change dates for the 282 millennia prior to 1911. Set 2 included only the Human and Temperature Changes from even numbered Epochs. Set 3 included only the data from odd numbered Epochs.

Weather Born Change Illusion

The all Epochs set produced a statistically significant negative correlation of -.55 (-0.544795877) between Global Temperature Change and the dates of

Human Change in Raw Years Ago. The sample of 280 cases of EPICA Ice Core data resulted in 278 degrees of freedom. A regression analysis of Data Set 1 found that Global Temperature Change describes, explains, and predicts only 29%, (an Adjusted R Square = 0.29427306) of the variation in the dates of Human Change measured in Raw Years Ago.

It was immediately suspected that the weak relation was due in part to the scarceness of fossils delineating the earliest dates of Human Change. Therefore, the Dates of Human Change were transformed into Period Years using the same procedure described earlier. The analysis of the Period Year transformed dates of Human Change boosted the negative correlation between Global Temperature Change and the dates of Human Change to -.84 (-0.839364574). That correlation is significant at the .001 level of confidence, and more in keeping with the prevailing view.

A regression analysis of the Period Year transformed data set yielded a significant Adjusted R Square of 0.703470058. That means that Global Temperature Change predicts 70 percent of the variation in the dates of Human Change measured in Period Years. Figure 21 depicts this significant predictive relation.

Given the results, the Alternate Hypothesis is retained. Global Temperature Change does predict a significant portion of the variation in Period Year transformed Dates of Human Change. That predictive relation is, however, severely reduced if the Dates of Human Change in raw years are substituted.

Obviously, part of the reduction in the prediction was due to the rarity of accepted fossils in the early periods of human history. It was suspected, however, that there is more to it, and there is.

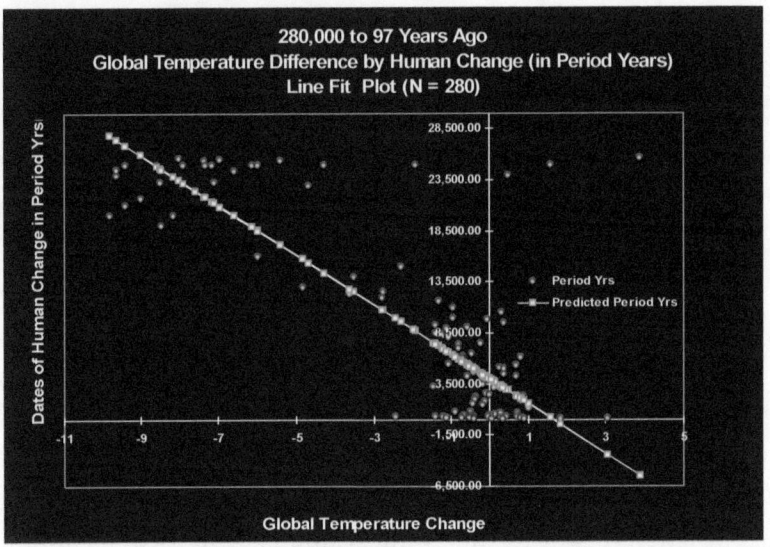

Figure 21 Dates of Human Change Predicted by Global Temperature Changes

Analysis of the 168 even numbered Epoch records produced a statistically significant correlation of -0.87 (-0.878764931) between Global Temperature Change and the Dates of Human Change in Raw Years. Clearly, dividing the data into even and odd Epochs improved the relations. Analysis of that even numbered Epoch set transformed into Period Years increased the correlation to -.91 (-0.910151617).

In anticipation of further analysis, Maya Degrees of Precession was included in the analysis. The resulting

111

correlation between Maya Degrees of Precession and Global Temperature Change exactly equaled the correlation between Global Temperature Change and the dates of Human Change. That equality of correlation remained, regardless of whether the Dates of Human Change were measured in raw Years Ago or Period Years. Clearly, the relation between Global Temperature Change and Human Change is equal to that of Maya Degrees of Precession and Global Temperature change in even numbered Epochs.

Analysis of the 112 records of the odd numbered Epoch set produced a statistically significant correlation of -0.67 (-0.671171908) between Global Temperature Change and Dates of Human Change in raw Years Ago. While superior to the all Epoch result, the odd Epoch correlation is 20 percent less than that found for even numbered Epochs.

Analysis of the odd Epoch Dates of Human Change in Period Years increased the correlation with Global Temperature Change to -.80 (-0.800094657). Although an improvement, that correlation is still eleven percent less than that found for the even numbered Epochs. Just as was the case for even numbered Epochs, the correlation between Maya Degrees of Precession and Global Temperature Change precisely equaled the correlation between Global Temperature Change and the Dates of Human Change for odd numbered Epochs.

In the 'Waves of Human change' chapter, Degrees of Maya Precession predicted 100 percent of the variation in the Dates of Human Change measured in Period Years. In the chapter, 'Waves of Celestial Motion,' Degrees of

Maya Precession predicted 100 percent of the variation in Earth's cycles of Obliquity, Inclination, and Eccentricity.

Given those earlier results, the relation between Maya Degrees of Precession and Global Temperature change is interpreted to be a *causal* relation. That interpretation makes the weaker overlapping relationship between Global Temperature change and the Dates of Human Change *associative, as opposed to causal.* That means that the true source of the correlation between Global Temperature Change and the Dates of Human Change is the result of a third source – the extra-solar system wave force precisely measured by Degrees of Maya Precession. That same wave force driving Precession, Obliquity, Inclination, and Eccentricity.

There is a simple analogy to the *associative* non-causal relation between Global Temperature and Human change. It is that while some birds may fly south in winter, the birds do not *cause* winter to occur. Nor do the birds returning in the spring produce the rising temperatures associated with the season. Taking Global Temperature as the birds in the analogy, it would be a mistake to gallop off to the ever popular conclusion that Global Temperature is the principal *cause* of Human Change. Conversely, it becomes equally dangerous to assume that Contemporary Human Change is significantly altering Global Temperatures. While Global Temperature Changes count with the dates of Human Change, these are not *causal* relations.

Historic Long Lived Greenhouse Gas Emissions

An examination of the variation in Long Lived Greenhouse Gases reinforces the case for an extra-solar system wave force; the more likely source of Global climate change.

As cited at the beginning of this chapter, The United Nations 2007 report by the IPCC (Intergovernmental Panel on Climate Change) Working Group ascribes the rise in Global Temperature since 1750 to Human Change. The specific Human Change the IPCC blames for this Global Warming is man-made increases in Long-Lived Greenhouse Gases. According to the United Nations IPCC Working Group, LLGGs include: carbon dioxide (CO_2), methane (CH_4), nitrous oxide (N_2O), halocarbons, and sulfur hexafluoride (SF_6). By far, the most prominent and predominant of the LLGGS pointed to by IPCC Scientists as the source of global warming is man's production of carbon dioxide (CO_2).

How much CO_2 could man and other forms of nature have historically contributed to the World? Having a database of internationally accepted CO_2 emission data from 250,000 years ago to 97 years ago, afforded the opportunity to find out.

From the onset of the analysis some of the IPCC assumptions were in serious doubt. Foremost among them is that 100 percent of the CO_2 emissions over the last 100 years are due solely to human land use and fossil fuel consumption.

Here's why. The ice core data from 250,000 years ago to 258 years ago (i.e., 1750 AD) show that CO_2

emissions have historically varied from 180 parts per million (ppm) to 287 ppm, with a global annual average of 258 ppm. Most of that history happened long before human beings could have altered LLGG levels. In more modern history, the maximum man could have contributed to the IPCC cited rise in CO_2 is 97 ppm, or about 34 percent – *never 100 percent.*

Another questionable IPCC assumption is that the rise in LLGGs since 1750 as represented by CO_2 is responsible for the observed increase in Global Temperature. According to the Jouzel, Masson-Delmotte, Cattani, and Dreyfus, et al (2007) analysis of 800,000 years of Ice Core data, CO_2 emissions are correlated with Global Temperature Change. But, the CO_2 levels tend to *follow* as oppose to lead Global Temperature Changes. *Causes* by necessity must *precede* effects. The documented lag in CO_2 makes it an *effect* as opposed to a *cause* of Global Ocean-Land Temperature Change.

The IPCC's crediting of manmade CO_2 as the source of Global Warming was looking more and more like a set of spurious correlation driven errors. Worse, that galloping diversion had to be masking the true source of Global Warming.

There is another oversight in the IPCC's thinking, as well, and it's a whopper. Parts per million (PPM) is a density measure. In physics and materials science, the density of a body (like Earth's atmosphere) is a measure of how tightly matter (like CO_2 gas molecules) is packed together within it. "The density of a body is the ratio of mass (m) divided by its volume (V)."[75] It is important to realize that both the matter of LLGGs and the *volume* of

Earth's atmosphere are free to vary. Earlier reported results strongly indicate that energy waves appear to have applied sufficient force on our solar system and planet to produce Precession, Obliquity, Inclination, and Eccentricity. It follows that the same compressive force should have produced significant variations in the volume of Earth's Atmosphere, as well as the matter within it. Hypothesis 7 puts that notion to the test.

Hypothesis H$_7$ asserts that from 250,000 to 100 years ago Degrees of Maya Precession describe, explain, and predict a significant portion of the variation in long-lived greenhouse gases (LLGGs), as represented by carbon dioxide (CO_2 in ppm) emissions extracted from Antarctic Ice Cores. **Hypothesis Null$_7$** argues that there is no statistically significant counting relationship between Degrees of Maya Precession and CO_2 emission levels from 250,000 to 100 years ago.

Analysis of 273 EPICA deep ice core data records from 250,000 to 100 years ago resulted in the rejection of Hypothesis Null$_7$. The analysis yielded a significant correlation (counting relationship) between Degrees of Maya Precession and Global CO_2 emissions of .91 (0.906286386). Hypothesis H$_7$ is retained.

A subsequent regression analysis revealed an Adjusted R-Square value of .82 (0.820695807), which is significant well beyond the .001 level. With near certainty, Degrees of Maya Precession predict 82 percent of the variation in CO_2 emissions for the years from 250,000 years ago to 100 years ago. The probability that this predictive relation is due to chance is 1×10^{-103} (that's

102 zeros after the decimal). Figure 22 depicts this predictive relation.

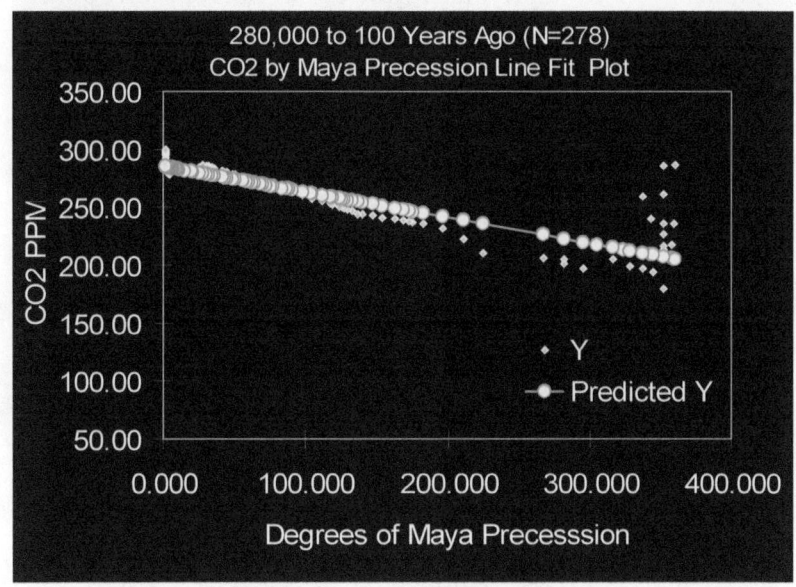

280,000 to 100 Years Ago (N=278)
CO2 by Maya Precession Line Fit Plot

Figure 22 Maya Degrees of Precession Prediction of CO_2 Levels

If CO_2 emission levels were even a candidate cause of Global Temperature change, then that CO_2 effect ought to be in evidence over the last 250,000 years of the history of the planet. Suspecting that to be the case, the notion was put to the test. **Hypothesis H_8** states that: from 250,000 to 100 years ago Global CO_2 Levels are significantly correlated with Global Temperature Changes. **Hypothesis Null$_8$** counters that there is no significant counting relation between CO_2 levels and Global Temperature Change.

Correlation and regression analyses of the 273 cases resulted in the rejection of Hypothesis Null$_8$.

Indeed, from 250,000 to 100 years ago Global CO_2 Levels are significantly correlated with Global Temperature Changes. The analysis returned a correlation of .79 (0.785722108).

The subsequent regression analysis yielded an Adjusted R-Square of .62 (0.615947273). This implies that 62 percent of the variation in Global Temperature is predicted by CO_2 levels.

It was noticed that for this same period Maya Degrees of Precession predicted 82 percent of the variation in CO_2. CO_2 is a gas contained within the volume of Earth's atmosphere. Consequently, the atmosphere and the CO_2 should respond more quickly and dramatically than Global Temperature to an externally imposed compressive wave force. Yet, numerous respected sources had reported that historically CO_2 lags Global Temperature. That conflict led to formulation of Hypothesis H_9.

Hypothesis H_9 asserts that from 250,000 to 100 years ago, Maya Degrees of Precession describes, explains, and predicts more of the variation in Global Temperature Change than that explained by Global CO_2 Levels. **Hypothesis $Null_9$** counters that Global CO_2 is a stronger predictor of Global Temperature Change than Maya Degrees of Precession.

The analysis resulted in the immediate rejection of Hypothesis $Null_9$. From 250,000 to 100 years ago, Maya Degrees of Precession is correlated .84 (0.838496795) with Global Temperature Change, while CO_2 is correlated .79 (0.785722108).

The regression analysis was equally telling. While Maya Degrees of Precession predicts 70 percent (0.702001067) of the variation in Global Temperature Change, CO_2 predicts only 62 percent (0.615947273) of that same variance. That is 8 percent less than is predicted by Maya Degrees of Precession. Hypothesis H_9 is retained; from 250,000 to 100 years ago, Maya Degrees of Precession predicts more of the variation in Global Temperature Change than does Global CO_2.

Given that Maya Degrees of Precession predicts 82 percent of the variance in CO_2, the counting association between CO_2 and Global Temperature Change is interpreted to be nothing more than an echo of the same variance explained by Maya Degrees of Precession. That makes the relation between CO_2 and Global Temperature associative, *not causal*. The extra-solar system wave force measured by Degrees of Maya Precession appears to be the true *cause* of the variation in Global CO_2 and Temperature Change.

The recent scientific literature provides partial independent confirmation. In 2007 in the refereed Journal *Science* J. Jouzel, V. Masson-Delmotte, O. Cattani, and G. Dreyfus et al, published the results of their analysis of the EPICA (European Project for Ice Coring in Antarctica) deep ice cores taken from the East Antarctica Dome C area.[76] The journal article was titled, "Orbital and Millennial Antarctic Climate Variability over the Past 800,000 Years." In that article, a group of more than thirty preeminent climatologists concluded that: "... the interplay between obliquity [the 41,000 year long cyclic variable-tilt and recovery of Earth's vertical axis of rotation] and precession accounts for the variable intensity of interglacial periods in ice core records."[77] In simple terms, *precession and obliquity* describe, explain, and predict a significant portion of the variation in Global Temperature change over the last 800,000 years.

These great leaders of climatology failed to divulge a key fact in the body of their milestone report. Contemporary climatologists use: "Babylonian, Hellenic, Chinese, and Islamic historical data to estimate the precession of the equinoxes."[78] Why do these science-minded weathermen use ancient historic astronomical data to estimate precession? The answer is that the calculated values of the Lunisolar model of precession are totally invalid and unusable for dates in the far distant past. Hence, the actual estimate of precession used by the cited climatologists was completely independent of Maya Degrees of Precession. Yet, those sources came to the same results and conclusions reported here. Importantly, the observations the climatologists used were based on time separated astronomical observations made by ancient peoples working an ocean away from the ancient Maya.

Current Epoch Motion, Temperature, and CO_2

The prior section dealt solely with the periods from 250,000 years ago to 100 years ago, and used the EPICA ice core data. The opportunity now presented was to replicate those findings using the independent instrumented measures of Global atmospheric CO_2 levels and Global Land-Ocean Temperature collected by climatologists from 1880 AD to 2007 AD. If the prior analysis results are valid and reliable, they should be approximately reproduced by the analysis of the data collected from 1880 to 2007.

Hypothesis H_{10} asserts that Degrees of Maya Precession describe, explain, and predict a significant portion of the variation in Global Temperature Change from 1880 AD to 2007 AD. **Hypothesis Null$_{10}$** argues that there is no significant counting relationship between Maya Degrees of Precession and Global Temperature change for that same time frame.

Hypothesis H_{11} asserts that Degrees of Maya Precession describe, explain, and predict a significant portion of the variation in long-lived greenhouse gases (LLGGs), as represented by atmospheric carbon dioxide (CO_2) levels from 1880 AD to 2007 AD. **Hypothesis Null$_{11}$** insists that there is no significant counting relationship between Degrees of Maya Precession and CO_2 levels over the time from 1880 to 2007.

Hypothesis H_{12} posits that Degrees of Maya Precession predict approximately the same amount of the variation in Global Temperature change as is predicted by CO_2 levels from 1880 AD to 2007 AD. **Hypothesis Null$_{12}$** counters that CO_2 levels predict substantially more of the variation in Global Temperature Change than is predicted by Degrees of Maya Precession over the time from 1880 to 2007.

Table 7 lists the correlation results relating to Hypothesis H_{10}, H_{11}, and H_{12}. The results include the correlations associated with Obliquity, Inclination, and Eccentricity, as well as those associated with the Hypotheses declared variables. The tabled results warrant the immediate rejection of Hypotheses $Null_{10}$, $Null_{11}$, and $Null_{12}$.

The correlations in the table for the 1880 to 2007 time frame confirm:

- H_{10}, Global Temperature centigrade (C) is significantly correlated with Precession;

- H_{11}, CO_2 is significantly correlated with Precession;

- H_{12}, the relation between Precession and CO_2 is approximately equal to the relationship between CO_2 and Global Temperature.

Table 7 Correlations for Hypotheses 10, 11, and 12

Period 1 Years 1880 to 2007 N = 128						
	Precess	Obliquity	Inclination	Eccentricity	Global Temp C	CO2 ppm
Precession	1.0000					
Obliquity	1.0000	1.0000				
Inclination	1.0000	1.0000	1.0000			
Eccentricity	1.0000	1.0000	1.0000	1.0000		
G. Temp C	0.8583	0.8583	0.8583	0.8583	1.0000	
CO2 ppm	0.9285	0.9285	0.9285	0.9285	0.8948	1.00

Note: CO2 ppm is substantially more highly correlated with Precession, Obliquity, Inclination, and Eccentricity than with Global annual Temperature.

The results clearly support the idea that a wave force has been exerting significant influence on the solar system and planet since at least 1880. Given earlier reported results, it is more likely that this progressive Poisson distributed wave pressure has been exerted on our planet for the last 508 years (i.e., since 1500 AD).

If this is as it appears, the relationship between CO_2 levels and Global Temperature change must be reinterpreted. The rise in concentrations of CO_2 and Global Temperature should be attributed to an extra-solar system wave force pressures. That is, the same extra-solar system wave force that has been driving Solar System Precession, and Earth's Obliquity, Inclination, and Eccentricity cycles for at least the last 250,000 years. In short, man-made increases in Long Lived Green House Gases are not the source of global climate change. Instead, the variation in Global Temperature change and LLGGs are but echo effects of a much greater cause.

Regression analyses of the data from 1880 to 2007 proved equally telling. In confirmation of Hypothesis H_{10}, an Adjusted R Square of 0.734544782 indicates that Maya Degrees of Precession predict 74 percent of the variation in modern Global Temperature. This result is consistent with the findings for the period from 250,000 to 100 years ago. Figure 23 depicts this predictive relation.

Consistent with Hypothesis H_{11}, an Adjusted R Square of 0.8207 indicates that Maya Degrees of Precession predicts 82 percent of the variation in CO_2 levels. This near term result is identical to the result obtained for the periods from 250,000 to 100 years ago. Figure 24 depicts the near term predictive relation.

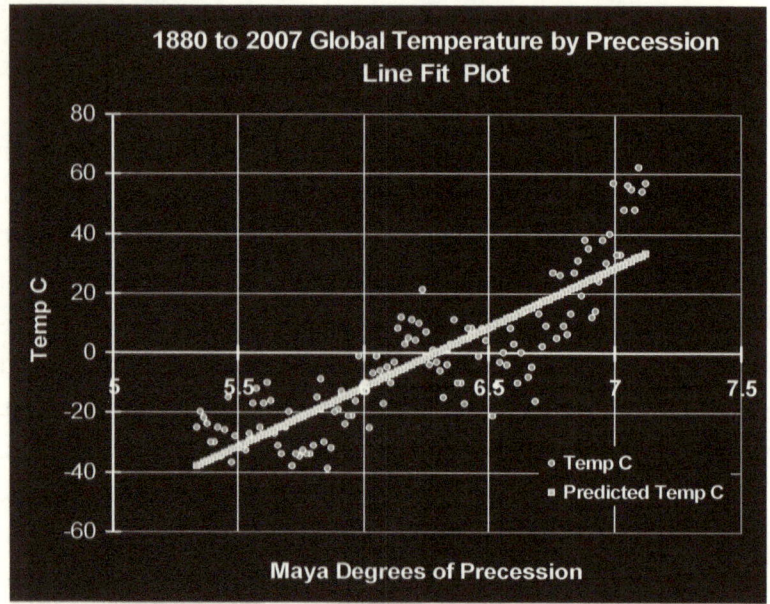

Figure 23 Maya Degrees of Precession Prediction of Global Temperature

Hypothesis H_{12} is also confirmed. Degrees of Maya Precession predict approximately the same portion of the variation in Global Temperature change as is predicted by CO_2 levels from 1880 AD to 2007 AD. There was less than 3.7 percent difference in these correlations and predictions. Hence, Hypothesis $Null_{12}$ is rejected. This result is interpreted to mean that CO_2 predicted not only the same amount of variation in Global Temperature Change as Maya Precession, but the *same* variation as predicted by Maya Precession. This result holds true not only for the time frame from 1880 to 2007, but for the entirety of the last 250,000 years of human history.

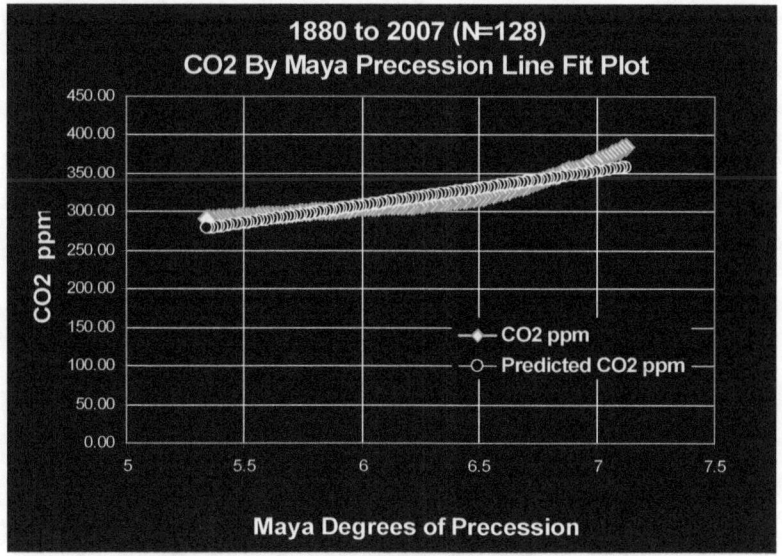

Figure 24 Maya Degrees of Precession Prediction of CO₂ Levels

Global Warming

From 1880 to 2007 CO₂ predicted just 3.7 percent more of the variation in Global Temperature than did Maya Degrees of Precession. Interestingly, for the time from 250,000 to 100 years ago, CO₂ predicted 8 percent less of the variation in Global Temperature than Maya Degrees of Precession.

It is not difficult to see why United Nations IPCC scientists might wrongly attribute all Global Warming since 1880 to the attendant human contribution of atmospheric CO_2. Still, such a galloping assumption is seriously misleading. In point of fact, the near term 3.7 percent of the variation in Global Land-Ocean Temperature attributable to atmospheric CO_2 may simply reflect man's contribution to the current Global Warming trend (Figure 25). In other words, all but 3.7 percent of the variation in historic global warming and cooling appears to be naturally occurring, historically in evidence, and other than man-made.

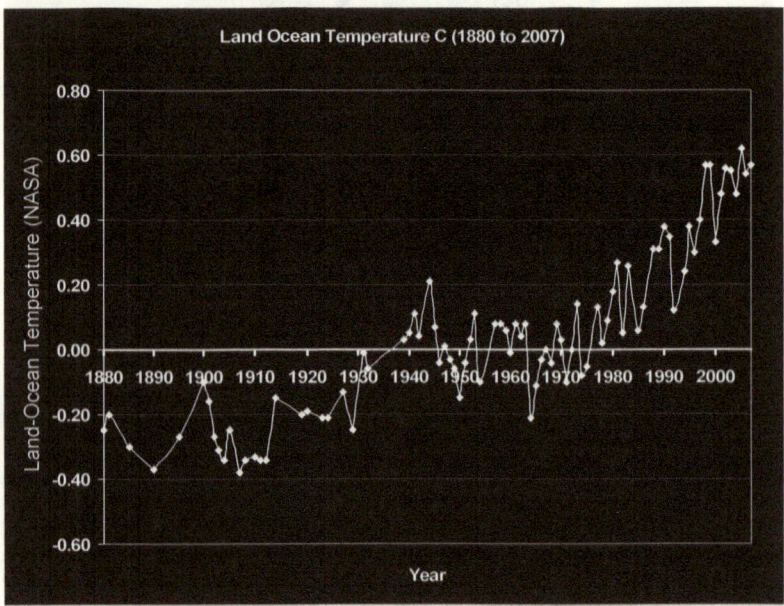

Figure 25 Global Land-Ocean Temperatures from 1880 to 2007

Source: Paulo Artaxo, Terje Berntsen, Richard Betts, David W. Fahey et al, 2007, Changes in Atmospheric Constituents and in Radiative Forcing, In: Climate Change 2007: The Physical Science Basis. Contribution of Working Group I to the Fourth Assessment Report of the IPCC; Cambridge University Press, Cambridge, United Kingdom and New York, NY, USA, p. 131

Global Warming remains a scary real fact of contemporary life. The very strong natural, as opposed to man-made, effects to Land-Ocean Temperature are startlingly apparent in Figure 25. In the main, however, the current Global Warming trend is neither man-made nor humanly-correctable.

The Unanswered Question

Dramatic World Age Precession cycles predict 282 millennia of Human Changes, solar system and Earth axial motions, atmospheric Greenhouse Gas variations, and Global Land-Ocean Temperature changes. Daunting questions result. They are: 1) what compressive wave force, 2) coming from where, drives those regular as clockwork World Age change cycles measured by Maya Degrees of Precession? The answers to those two questions are the subject of the next two chapters.

7

WAVES OF GRAVITATIONAL TORQUE

An Ancient Clue

Ancient cultures agreed; the force driving World Age cycles emerges from "the *Galactic Center*." A view echoed by the ancient peoples of Mesopotamia, Vedic East India, Assyria, Egypt, and China, and others. The early Mesoamericans were no exception. According to John Major Jenkins (2002):

> The Maya mythologized this location [the Galactic Center] as the womb of the Cosmic Mother ... Being adept astronomers and mathematicians, the *Izapans* ([occupants of] a city in the Mayan Empire) calculated forward to when the alignment between the [winter] solstice sun and the galactic heart would culminate, and they arrived at the year that we call A.D. 2012...[79]

Supposedly, having predicted the next so called *Galactic Center alignment*, the Maya made it the end-date of their Long Count Calendar. They are said to have called this date of Friday December 21st 2012 AD, *"the dawn of the Fifth Sun, "the great alignment,* [80 & 81] *"the end of space and time,"* and the time of *"the great transformation."*[82 & 83] Starting from that date, the Maya are assumed to have counted backwards some five 5,125.36 year long Epochs. They stopped when they'd delineated one World Age Precession cycle of 25,626.8 years.[84] Those five Epochs became the drivers of the Maya Long Count Calendar. The Maya are believed to have set the notional founding date of their great civilization to the origin of the fourth Epoch, August 11, 3114 BC; exactly 5,126 years from the all important Long Count calendar end-date.[85]

Initially the notions seemed laughable. Particularly doubtful was the idea that the Galactic Center could produce a wave force driving Maya World Age Precession cycles. The so called *Galactic alignment* was grossly defined and of dubious timing. Second, since the early 1990's, scientists at NASA's Very Long Baseline Interferometry (VBLI) Group had declared the Galactic Center much too far away and much too weak a gravitational source to in any way effect events in our solar system.

NASA's VLBI Group is the internationally recognized authority concerned with the:

- Motion of the Earth's tectonic plates
- Regional deformation and local uplift or subsidence.
- Definition of the celestial reference frame

- Variations in the Earth's orientation and length of day
- Maintenance of the terrestrial reference frame
- Measurement of gravitational forces of the Sun and Moon on the Earth and the deep structure of the Earth
- Improvement of atmospheric models[86]

NASA VLBI Group scientist Dan Mac Millan had answered an email from Walter Cruttenden, stating: "The answer to your question is that we do not account for the geometric effect of galactic rotation. It is a very small effect."[87]

Breakthroughs in astrophysics changed all that. One product of the Galactic Center was found to have the reach and influence needed to drive World Age Epochs. That discovery motivated the construction of the wave force criterion check-list depicted in Figure 26. The figure describes the specific requirements any candidate source of Maya World Age Precession cycles must satisfy to produce the results reported thus far.

Criteria

A viable candidate wave force must produce invisible waves. If the force driving Maya World Age cycles were directly visible within the electromagnetic spectrum, *astronomers* would have seen it, and they haven't. Given the effects of Maya World Ages and Epochs, any truly candidate source must produce Poisson or normally distributed (bell-shaped) wave peaks and valleys. Each such space time density altering wave must consist of one accelerating and one decelerating Epoch. Those Epochs must combine in wavelengths (measured peak to peak) of 10,250.7 light years, with peak-amplitudes (i.e. wave heights) of 5,125.36 light years.

- An invisible wave-like force
- Originating near or about the Galactic Center
- Self-propagating at the speed of light
- With a wavelength of ~ 10,250.7 light years
- With an amplitude of ~ 5,125.36 light years (1 Epoch)
- Produce Poisson Distributed Changes in Space-Time Density
  Constitute a potential source of:
  - Solar System Precession Cycles (5 Epochs)
  - Earth obliquity cycles (8 Epochs)
  - Earth inclination cycles (14 Epochs)
  - Earth eccentricity cycles (20 Epochs)
  - Solar system Z-Axis tilt (wobble)
  - Solar system Z-Axis vertical gain
  - Sun galactic orbital speed variation
  - Earth global temperature change
  - Earth Long Lived Greenhouse Gas variations
  - Changes in the rate of the human diffusion of Innovations (to include novel ideas, practices, objects, and physiological adaptations)

Figure 26 the Criterion Checklist for Candidate Waves

The sought wave force must also propagate at the speed of light (i.e., 299,792.46 kilometers per second). The light speed requirement constrains the time each wavelength takes to pass through our solar neighborhood to 10,250.7 calendar years. Each wave must originate at or about the Galactic Center (GC) and travel the full

26,627 light year distance to the orbital radius our Sun, with a new wave arriving once every 10,250.7 years. Thus, candidate force waves must form serial wave trains that began arriving at the Galactic radial distance of our Sun 2 to 3 billion years after the birth of the Galaxy. That is, long before the birth of our solar system 4.7 billion years ago.

Only one candidate wave force stood a chance of meeting these stringent criteria; a Galactic Gravitational Torque Wave (GTW). Two journal articles suggested it, and numerous related articles provided the evidence needed to confirm it.

The first article, written by the French team of F. Bournaud and F. Combes, appeared in 2002 in the journal *Astrophysics,* under the title, "*Gas Accretion on Spiral Galaxies: Bar Formation and Renewal.*" The second article, penned by Francoise Combs, was published in 2005 in Scientific American under the title, "*Ripples in a Galactic Pond.*" Those articles pointed to a plethora of research literature characterizing the bizarre nature and effects of *Galactic Gravitational Torque Waves.* Researchers refer to such waves as "density waves," "gravitational instabilities," and "compression waves."[88] Fortunately, those sources provided just the information needed to apply the wave force evaluation criteria to GTW.

An Invisible Wave-like Force; Gravity

Gravity – and any gravitational torques it asserts – is more than merely *invisible*. Gravity is the intangible relationship between two or more masses of energy or matter. Gravity and gravitational torques are distortions (the warping) of the very fabric of space-time geometry proper. Gravity is not a force in an inert and motionless space. Isaac Newton was as wrong about that as he was about the Lunisolar Model of precession. It took Albert Einstein to turn Newton's false notion of gravity on its head. In 2005, Physicist Michio Kaku put it this way:

> "In Einstein's universe, space and time is not the static arena Newton had assumed, but the dynamic, bending and curving in strange waves.
> Assume the stage of life is replaced by a trampoline net, such that the actors gently sink under their own weight. On such an arena, we see that the stage becomes just as important as the actors themselves.[89]

Thanks to Einstein we now know that *mass* is the measure of the amount of displacement (deformation or curvature) that one energy or matter based object imposes on the interdependent dimensions of space and time (space-time). Further, *gravity* is the effect of one mass's imposed space-time deformation on that of another proximate mass. By that definition, both energy and matter possess mass.

Sufficient Force; Torque

One on-line source simplifies the definition of torque this way:

Torque is a force that tends to rotate or turn things. You generate a torque any time you apply a force using a wrench. Tightening the lug nuts on your car's wheels is a good example. When you use a wrench, you apply a force to the handle. This force creates a torque on the lug nut, which turns the lug nut.

It is important to note that the torque units contain both a distance and a force (e.g., pound inches or pound feet). To calculate torque, you multiply that force by the distance from the center. In the lug nut example, if the wrench is a foot long and you put 200 pounds of force on it, you are generating 200 pound-feet of torque. If you use a 2-foot wrench, you only need to put 100 pounds of force on it to generate the same torque."[90]

Similarly, "In physics, a torque (τ) is a vector [a column of numbers describing both force magnitude and direction per unit time] that measures the tendency of a force to rotate an object about some axis."[91] "The magnitude of a torque is defined as force times [the length of] its lever arm."[92] Just as a force is a push or a pull, a torque can be thought of as a twist.

There is just such a lever bar rotating around the Galactic Center of the Milky Way. It is comprised of elder stars, and it imposes twisting torque forces on the very fabric of space-time. The overall length of that lever bar is approximately 25,626.8 light years, or about 12,813.4 light years per torque arm extension on either side of the Galactic Center. A torque wrench lever arm

121,224,123,837,366,832 kilometers long (about 810 million times the distance from the Earth to the Sun) should produce some pretty serious torque; and it does.

Waves

In general, "A wave is a disturbance that propagates through space and time, usually with the transference of energy."[93] Figure 27 describes a wave of constant amplitude (wave height).

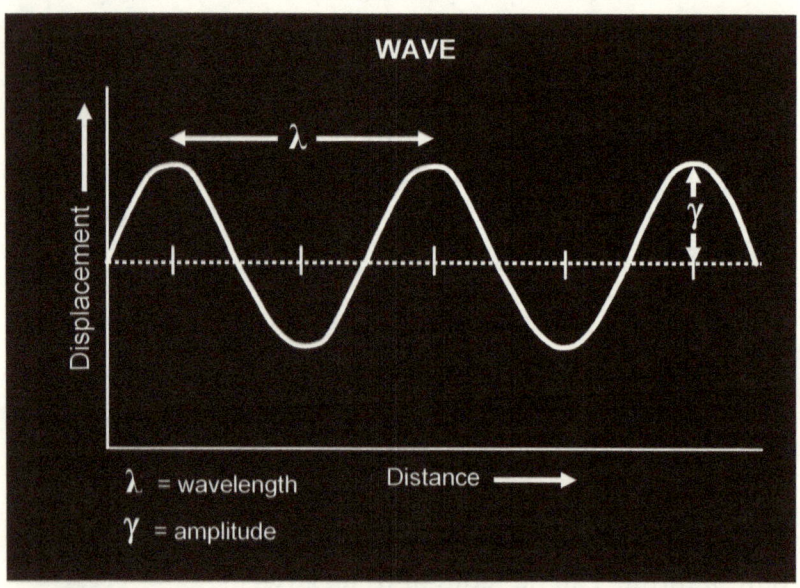

Figure 27 Example Description of a Wave
Source: Http://en.wikipedia.org/wiki/Wave, revisited January 7, 2008

In the figure, the amplitude of the wave denoted by γ (the Greek symbol *gamma*) is "the maximum vertical distance between the baseline and the wave. ... The wavelength denoted by [the Greek symbol *lambda*] λ is the distance between two sequential crests ([or zero crossings] or troughs)."[94]

The Galactic gravitational torque waves (GTW) of interest, have a wavelength (λ) of 648 million times the distance from the Earth to the Sun (which is 1 Astronomical unit or 1 AU), and amplitude (γ) of 324 million AU or about 5,125.36 light years. Not only that, but GTW do not merely propagate through space and time. Instead they deform (alter the density of) the literal fabric of space-time. There is a huge difference between waves that move in space and time (like light or sound waves), and a gravitational torque wave that pushes, stretches, and squeezes the very energy (mass) density of space-time itself. Conventional waves must obey the rules of physics, GTW determine what rules apply.

Gravity Waves

For some genuinely baffling reason scientists tend to distinguish between gravity waves and gravitational torque waves. The on-line Encyclopedia Wikipedia describes a *gravity wave* as follows:

> In physics, a gravitational wave is a fluctuation in the curvature of space-time which propagates as a wave traveling outward from a moving object or system of objects. Gravitational radiation is the energy transported by these waves. ... Although gravitational radiation [as opposed to gravitational toque waves] has not yet been directly detected, it has been indirectly shown to exist. This was the basis for the

1993 Nobel Prize in Physics, awarded for measurements of the Hulse-Taylor binary system."

...

In Einstein's theory of general relativity, the force of gravity is due to [the displacing deformation of the] curvature of space-time. This curvature is caused by the presence of massive objects. Roughly speaking, the more massive the object is, the greater the curvature it causes, and hence the more intense the gravity [displacement of space-time it imposes]. As massive objects move around in space-time, the curvature will change. If the objects move around in a certain way, ripples in space-time can spread outward like ripples on the surface of a pond. These ripples are gravitational waves."[95]

Often cited potential sources of gravity waves include: the big bang, a supernova, or two compact objects (white dwarfs, neutron stars, or black holes) closely orbiting each other. Gravity waves are characterized as a kind of radiation, somehow distinct from gravitational torque waves. The fundamental differences between the two are evasive. That confusion stems from the fact that both gravity waves and gravitational torque waves appear to satisfy a single operational definition.

Regardless, three characteristics of gravity waves are equally applicable to gravitational torque waves. First, "...gravitational waves can pass through any intervening matter without being scattered. Whereas light from distant stars may be blocked out by interstellar dust, for example, gravitational waves will pass through unimpeded."[96] Second, "The waves will spread out through the Universe at the speed of light, never stopping or slowing down."[97] Third, while gravitational waves are unaffected by (are not slowed, deflected, refracted, or reflected) by matter, the reverse is not true; matter is effected by gravitational waves. "However, the magnitude of this effect will decrease the farther the observer is from the source. In contrast to gravitational toque waves, any *gravity* wave effects expected to be seen on Earth will be quite small. Still, scientists are attempting to measure the effects of these gravity waves using extraordinarily precise experiments."[98]

Gravitational Torque Waves

Unlike the ever illusive gravity waves, Galactic Gravitational Torque Waves (GTW) have been indirectly observed and measured. GTW are in evidence in every observable spiral barred galaxy in the known universe. And, our Milky Way is a strongly barred spiral Galaxy.[99] [& 100]

Without GTW the spiral arms of galaxies like ours would wind up "like ropes [attached] around a fast turning winch,"[101] and end up wrapped around the Galactic Center within a couple of billion years.[102] Instead, thanks to the effects of GTW, the orbits and velocities of the dust, gas, stars, and dark matter in the Milky Way Galaxy are periodically re-synchronized with the Galactic Center. In short, as a direct result of GTW the spiral arm structure of our Galaxy was established and is periodically renewed, synchronized, and preserved.

Figure 28 depicts the latest NASA artist's conception of the top down "remapping" of the Milky Way. It was published in the June 21, 2008 issue of Science News. What is remarkable about this artist's conception is that instead of the four prominent spiral arms conventionally shown, the Galaxy is depicted as having just two dominant arms emanating from the ends of the very visible stellar bar. Those two gigantic arms clearly reflect the great stirrings and matter-distribution effects imposed by GTW.

GTW are products of the rapid rotation of the rectangular stellar bar extending outward from the center of the figure. GTW roles include Galactic disk generation and evolution, and new star formation and distribution. As a result, GTW have been extensively observed, measured, modeled, and simulated by modern scientists for nearly two decades.

In the cited definition of the term *torque,* it was noted that it is the force applied by turning a wrench on a lug nut. Well, in the Milky Way Galaxy the thing that generates the wrenching torque is the long rectangular bar of really old stars that fast rotate about the Galactic Center. The wrenching force produced by that fast turning stellar bar generates GTW that warp and compress the very fabric of space-time throughout the Galactic Plane. These huge fluctuations in the curvature of space-time propagate outward at the speed of light from the stellar bar to well beyond the edges of our Milky Way.

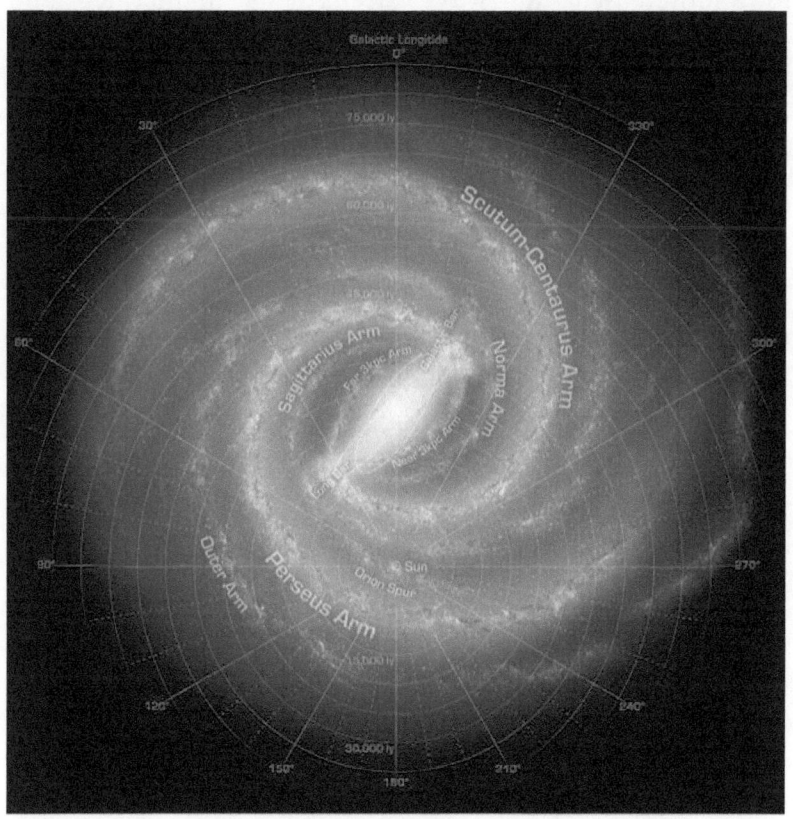

Figure 28 Artists Conception of the Milky Way Galaxy

Note: "Using infrared images from NASA's Spitzer Space Telescope, scientists have found that the Milky Way's elegant spiral structure, shown in the artist's rendition, is dominated by just two arms, instead of the four as previously believed." Source: *Ron Cowen, 2008, "Astronomers Remap the Milky Way," Science News, Vol. 173, No. 19, June 21, 2008, p.11*

GTW leverage torque on stars (like our Sun), gas, and dust and synchronize and distribute these objects throughout the Galaxy. The Galactic energy fields and matter so effectively communicate these waves from one to another that the wave strength is maintained and carried to well beyond the edge of our Galaxy. Some estimate that GTW extend to distances of 600 million light years or more from the Galactic Center.

Originating at the Galactic Center

At the center of our Milky Way Galaxy is a genuinely massive highly compressed star named "Sagittarius A*" (pronounced, Sagittarius A-Star).[103] According to those who ought to know, it is either a "super-massive black hole" (the prevailing view), or a "Dark Energy Star" (a hypothetical object first defined by quantum physicist George Chapline: 1999, 2004, 2005 and 2006 [104, 105, 106 & 107]). For reasons not at issue here, the Black Hole point singularity concept is fundamentally anomalous and inconsistent with a host of scientific observations. The Dark Energy Star concept is consistent with both theory and observation, but is about as popular as the notion of Black Holes was back in the 1970s. Funny isn't it ... how ridiculed *fringe theories* become sacred-doctrines in a few short decades, despite lingering contradictions.

According to Thomas Ott at the Max-Planck-Institute, "Sag A* rotates clockwise once every "16.8 ± 2.0 min".[108] Although a genuinely hasty rate, given Sag A*'s incredibly large mass, it is not at all unprecedented. The fastest rotation time ever observed for such a star is a blistering 400 times per second. That's once every 2.5 milliseconds (where a millisecond is 1,000th of a second).   Astronomers claim that rotation rate was recorded for the star named SAX J1808.4-3658.[109]

As it turns, Sag A* emits: positrons (antimatter electrons), x-rays (about 1 flare per day), gamma rays, radio waves (in highly varied amounts), and thermal (infrared) waves (1 flare up observed in 10 years).[110 & 111] Sag A* is also said to have a Hawking thermal radiation (as in the world famous cosmologist Stephen Hawking) which runs from 10-14 Kelvin's.[112]   All of these emissions are much too weak and far away to meaningfully alter the geometric motions of our solar system, let alone influence Earth.   In that sense the NASA VLBI scientists have it right.

Central to the present quest, however, is a different sort of Galactic Center emission. "Our Milky Way Galaxy, is a spiral barred galaxy," or so say: de Vaucouleurs 1964; Peters 1975; Cohen & Few 1976; Liszt & Burton 1980; Gerhard & Vietri 1986; Mulder & Liem 1986; Binney et al. 1991, Blitz & Spergel 1991; Weiland et al. 1994; Dwek et al. 1995; Binney, Gerhard, & Spergel 1997, Nakada et al. 1991; Whitelock & Catchpole 1992; Weinberg 1992; Nikolaev & Weinberg 1997; Stanek 1995; Sevenster 1996; Stanek et al. 1997.[113] At the center of our galaxy the long rectangular bar of elder stars rotates outrageously fast about Sag A*. The rotation of that bar generates invisible gravitational torque waves. Those propagating GTW twist, stretch, and squeeze the very fabric of space-time throughout the Galactic Plane. Figure 29 is a zoom in on that NASA artist's image of the Milky Way stellar Bar relative to the labeled spiral arms, and our Sun.

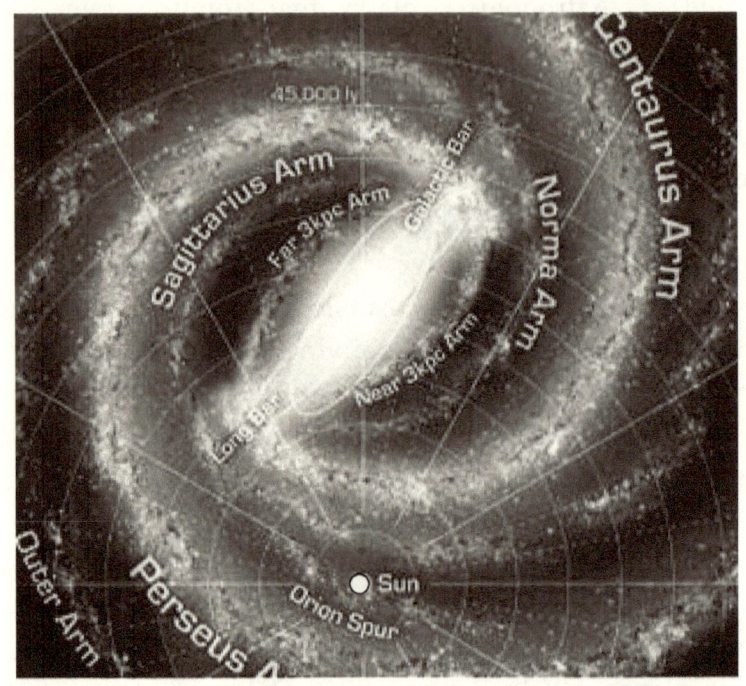

**Figure 29 Milky Way Galactic Stellar Bar
Relative to our Sun**
*Source:
http://en.wikipedia.org/wiki/Image:236084main_MilkyWay-
full-annotated.jpg, Revisited August 16, 2008*

So far, the top three items on the criterion checklist have been fulfilled. GTW constitute an invisible wave-like force, originating at or about the Galactic Center that self-propagate at the speed of light throughout the Galaxy. More importantly, no competing force is capable of challenging the GTW for the title of 'sole source of Maya Precession World Age cycles and Epochs.' The next step is to try GTW on for size.

Wavelength, Amplitude, and Pitch Angle

Do GTW have a wavelength of approximately 10,250.7 light years and amplitude of about 5,125.36 light years (i.e., 1 Maya Epoch)? Do those wave dimensions persist at and about the Galactic radius of our Sun? The answer to those questions is a resounding '*yes.*'

In 2007 astronomers measured the stellar bar that rotates about the Galactic Center. According to those estimates the stellar bar has:

> "a length of 25,440.197 light years (± 1,297.07 light years), giving it a radius from Sagittarius A* of 12,720.099 light years), a height of 3,913.877 light years, a width of 652.313 light years, and a leading edge pitch angle of approximately 43 degrees."[114]

The error of measurement in those dimensions is substantial. Other noteworthy star gazers, like the folks at the Jet Propulsion laboratory, set the bar length closer to "27,000 light years."[115] A Survey of the relevant literature found that the value 25,626.8 light years tends to fall near the center of the available estimates. That value also sits well within the reported error of measurement ascribed to available observations. Odd sort of coincidence don't you think – the bar length in light years is approximately equal to the duration of one Maya World Age cycle in solar years.

According to Françoise Combes (the first woman to win the French Science Prize), "the torques exerted by the stellar bar act as a gigantic stirrer,"[116] and "looks like a spinning lawn sprinkler."[117] The stellar bar generated GTW stir and compress (i.e., alter the mass density of) the galactic gas, dust, and stars of the galactic plane. These GTW maintain the galactic synchronization of stellar orbits of stars like our Sun and the Milky Way's spiral arms. That is how GTW manage to sustain the overall shape of our Galaxy.

Of particular interest is that GTW periodically resynchronize the orbits of stars like our Sun with the Galactic Center. Further, that the synchronizing GTW compress (twist, stretch, and squeeze) all the ordinary matter, dark matter, and dark energy that they pass through. Gravitation – the literal warping of space-time – is the only known force that acts equally on the masses of ordinary matter, dark matter, and dark energy. The other three forces (i.e., electromagnetism, the strong force, and the weak force) do not.

GTW alter the orbits of stellar objects proportionate to their mass and distance from the Galactic Center. As it turns out, our Sun orbits the Galaxy just *outside* a kind of resonant sweet-spot called "the Co-rotation Circle,"[118] which is also known as the "Galactic Habitable Zone" or the "Circumstellar Habitable Zone."[119] Objects that orbit closer to the Galactic Center and *inside* the Co-rotation Circle orbit Sag A* faster. GTW slow down the orbital speed of those stars.[120] Objects which orbit *outside* the Co-rotation Circle, like our Sun, rotate about the Galactic Center slower, and GTW speed up their orbital speed and synchronize them with the rotation of the stellar bar.[121]

Figure 30 illustrates the propagation of GTW from the stellar bar. This illustration symbolizes the gravitational wave effects that reach our Solar System.

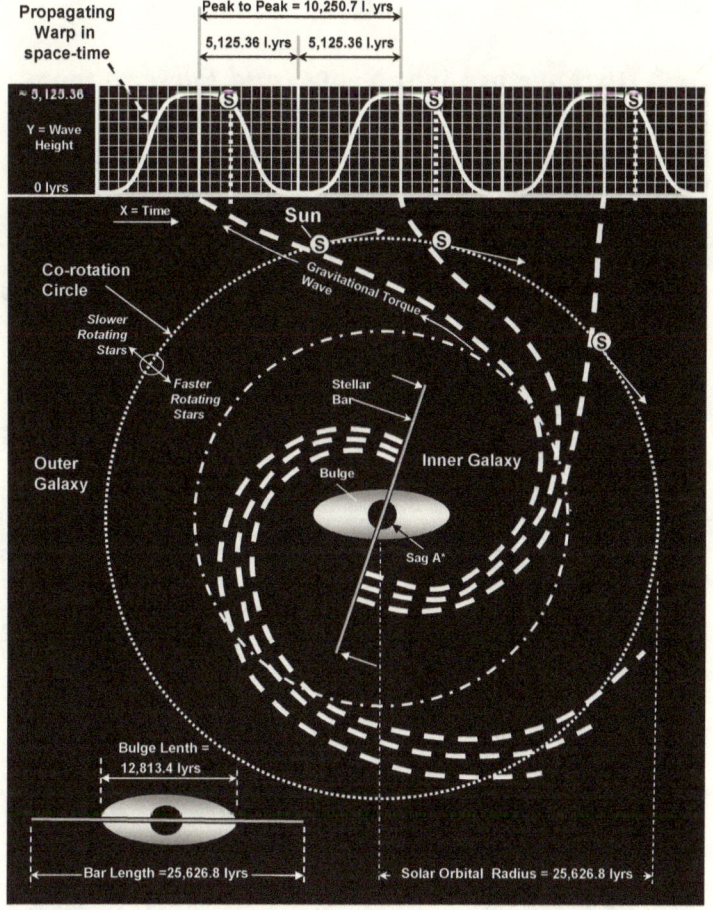

Figure 30 GTW Propagation and the Sun's Galactic Orbit about the Co-rotation Circle

Although the distance from our Sun to the Galactic Center is one of the most important measures in all astronomy, many different distances are reported for it in the scientific literature. Table 8 lists the various estimates of the distance from our Sun to the Galactic Center most frequently reported in the available literature.

Table 8 Distance in Light Years (*and Parsecs*) from Sun to the Galactic Center; *A comparison of the Estimates of the Earth's precession of the Equinoxes in Calendar Years*

Distance from Sun to Galactic Center Estimates			Precession of the Equinoxes Estimates	
Ref No	Estimated Distance in light yrs	Estimates in Parsecs	Ref No	Estimated Distance in years
1	23,157.1030	7,100.0000	1	25,765.0000
2	25,626.8000	7,857.2130	2	25,626.8000
3	27,723.2920	8,500.0000	3	25,800.0000
4	26,000.0000	7,971.6360	4	25,765.0000
5	26,000.0000	7,971.6360	5	26,000.0000
6	27,000.0000	8,278.2380	6	25,770.0000
7	26,092.5100	8,000.0000	7	25,800.0000
8	25,000.0000	7,665.0350	8	25,920.0000
9	26,000.0000	7,971.6360	9	26,000.0000
10	28,000.0000	8,584.8390	10	25,925.0000
11	24,853.1160	7,620.0000	11	26,000.0000
12	26,092.5100	8,000.0000	12	25,800.0000

Min =	23,157.1030	7,100.0000	Min =	25,626.8000
Max=	28,000.0000	8,584.8390	Max=	26,000.0000
Average =	25,962.1109	7,960.0190	Average =	25,847.6500
Var =	1,682,391.73	515,823.6500	Var =	14,193.3609
Std Dev =	1297.070441	397.6840	Std Dev =	119.1359

Note: Table References listed by number with the references cited at end of Endnotes Listing

148

Alongside the Sun distance numbers in Table 8 are the equally varied estimates of the current rate of precession reported in the literature. The Reference Numbers included in the table are the endnote listed sources that reported those estimates. The table specific endnotes are listed in a special section of endnotes located near the end of the book.

Differences in observation methods, instrumentation, analysis, and interpretation are likely to blame for all the differing estimates. These problems were well demonstrated in 2006 by Gilberto C. Gómez, not only for the distance to the center of the Galaxy, but for all the observation based distances used to characterize our Galaxy. Gómez summarized his findings this way:

> The error structure is such that, when kinematic [observed Galactic orbital] distances are used to develop a picture of the large-scale density distribution [of the Milky Way], the most significant features of the numerical model are significantly distorted or absent, while spurious structure appears.
>
> ...
>
> Nevertheless, it was soon realized that deviations from circular orbits, however small in absolute value, have a strong impact on how we see the Galaxy.[122]

The mean of the literature estimates for the distance from the Sun to the Galactic Center is approximately equal to the mean duration of precession. Both means are very close to the 25,626.8 year long constant of the Maya Precession Cycle. Odd coincidence don't you think?

Okay, some might object, after all the Sun's distance from the Galactic Center is measured in light years. In contrast, the duration of precession is counted in solar years. Sorry, the objection is made mute by the fact that gravitational torque waves travel at the speed of light. Hence, our waiting time for a GTW to travel to us from the Galactic Center reduces to plain old solar years. Getting straight to it, it appears that it takes approximately 25,626.8 solar years (one Maya World Age Cycle) for a GTW (or train of GTW) to reach our Sun from the stellar bar rotating around the Galactic Center.

This is only the beginning of freaky matches. As stated, the most recent measurement of the length of the Milky Way stellar bar is 25,440.197 light years. That bar length is spooky close to: the mean estimated distance from our Sun to the Galactic Center (estimated at 25,962.1109 light years ± 1297.07 light years), the mean estimate of the period of precession (which is 25,847.65 years ± 119.1359 light years), and the constant Maya World Age Precession Cycle duration of 25,626.8 years. Notice the trend here?

Hold on, there's more, and this time the literature actually suggests how gravitational torque waves produce the even Epoch synchronization and odd Epoch de-synchronization effects suggested by the Maya. In her 2005 article, "Ripples in a Galactic Pond," Françoise Combes states:

Amplified by proximity, the gravity [torque wave] of stars modifies the rotation speed of the ellipses [their galactic orbits]. Faster ones [orbits] slow down and slower ones speed up, so they bring themselves into synch. When a star enters the wave [a GTW] gravity locks it in, but only temporarily; after a while it becomes unlocked and exits.[123]

According to Françoise Combes, GTW accelerate and re-synchronize stellar orbital velocity about the Galactic Center, maintaining that synchronization for a period of "10,000 to 15,000 years."[124] The lower bound of the interval (10,000 years) is for objects like our Sun that rotate just outside the Co-rotation Circle. It is also the estimate of the GTW wave-length of 10,250.7 years, i.e., two Maya Epochs of 5,125.36 years each. Moreover, the acceleration of our Sun from a Galactic Orbital Speed of 210 km/s (kilometers per second) to 280 km/s by the even numbered Epoch Wave effects is practically guaranteed. As is the correspondent deceleration of our Sun's Galactic Orbital Speed during the odd numbered Epoch as the GTW departs the solar system.

Are you ready for another stunning coincidence? The radius of the Galactic stellar bar is approximately equal to the estimated diameter of the Galactic Bulge. The Galactic Bulge is this great football shaped clump of gas, dust, and stars that completely enshrouds the Galactic Center (i.e., Sag A*). The Galactic Bulge is estimated to have a major axis radius of 6,000 light years (± 250 light years), or a diameter of approximately

12,500 light years. That is very close to one half our Sun's radial distance from the Galactic Center.[125] It is also approximately one half the duration of one Maya World Age Precession Cycle (i.e., 12,813.4 solar years).

When summarized, these remarkable yet totally overlooked approximate equalities (symbolized as $\approx$) appear to be no coincidences at all. The equalities state that: the length of the Galactic Stellar Bar:

> $\approx$ Galactic Orbital Radius of our Sun
> $\approx$ Distance each gravitational torque wave
>   travels to reach our Sun
> $\approx$ Duration of one Maya Precession Cycle of
>   5 Epochs (2.5 GTW)
> $\approx$ Two times the diameter of the Galactic
>   Bulge
> $\approx$ Distance from the Galactic Center to the
>   edge of the Co-rotation Circle

What about GTW amplitude, does it meet the criterion? According to Xiaolei Zhang and Ronalad J. Buta (2007) among others, the amplitude of a GTW is:

> ... An exponential vertical density distribution with a scale height h_z of one fourth the radial scale length h_r (deGrijs 1998; Laurikainen et al 2004).[126]

If we define h_r as a radial scale length of 25,626.8 light years (our Sun's approximate Galactic radius) and divide by 4 as prescribed, we get a calculated GTW height (h_z) of 6,406.7 light years. That is just 1,281.34 light years greater than the criterion wave height of

5,125.36 light years; well within the magnitude of the error typically associated with such measurements. Still, there are other factors to consider.

At the Galactic Co-rotation radius, the GTW pull of matter and stellar angular momentum inward toward the Galactic Center reverses direction. At that radius a resonance is achieved which causes the GTW to stop pulling and start pushing matter and angular momentum out toward the edges of the Galactic plane. Yu N. Mishurov and I. A. Zenina describe the offset of our Sun from the Co-rotation Circle as:

> ... The Sun is situated very close to [just outside] the corotation [pronounced co-rotation] resonance where the rotation velocities of the disk and of the spiral pattern coincide. The displacement Delta R of the Sun from the corotation circle is: Delta R ~ 0.1 kpc [0.1 kilo parsec equals 326.156 light years].[127]

We can safely assume that the resonant reversal at the Co-rotation radius (at around 326.156 light years less than our Sun's Galactic orbital radius) smoothes off and reduces the height of a GTW in our solar neighborhood. According to Fresneau, A. Vaughan, and R. W. Argyle (2005), in the region of our Sun just outside Co-rotation:

> The stellar wave remains smooth, with a small density perturbation [compression] of about 25% in amplitude (wave height) and an average surface density of 10 million solar masses per square parsec."[128] Quillen

153

(1994), and Fresneau and Argyle (2005) and others put this wave amplitude (height) reduction and wave density compression closer to "20%".[129]

The criterion wave height of 5,125.36 light years is equal to the radial distance 25,626.8 times 20 percent. That is precisely the reduction in wave height (amplitude) prescribed by Quillen (1994) and Fresneau and Argyle (2005). While others suggest an amplitude reduction of 25 percent (resulting in a GTW wave height 1,281.34 light years greater), the criterion estimate of 5,125.36 is still well within acceptable error limits). That makes the Maya inspired wave height of 5,125.36 light years very reasonable indeed. What's more, it could well be smack on perfect.

In 2001, Xiaolei Zhang stated:

> ... The energy and angular momentum transfer between the basic state matter [pre GTW condition] and the spiral density wave [e.g., the GTW] is achieved through a temporary local gravitational instability at the spiral arms (Zhang 1996). The [current] length scale of this instability at the solar neighborhood is about 1 Kilo parsec [i.e., 3,261.564 light years], which coincides with the [observed] length scale of the giant molecular and HI complexes near the Galactic spiral arm regions."[130]

The current "gravitational instability" that Zhang is referring to represents the central region of the GTW, which he describes in the same paper (consistent with the criterion) as being *"bell-shaped."*[131]

Authors: M. Lopez-Corredoira, A. Cabrera-Lavers, T. J. Mahoney, P. L. Hammersley, F. Garzon, C. Gonzalez-Fernandez, in their paper (*"The Long Bar in the Milky Way; Corroboration of an old hypothesis"*), offer up:

> ... two new analyses that corroborate recent and earlier claims concerning the existence in our Galaxy of a long flat bar with approximate dimensions 7.8 kpc [kilo parsecs] x 1.2 kpc x 0.2 kpc and a *position angle [pitch angle of the leading edge]* of approximately 43 degrees.[132]

It is here suggested that the Galactic stellar bar "position angle" is approximately equal to the leading edge pitch angle of each GTW the stellar bar generates. Zhang's 3,261.564 light year length of the gravitational instability also appears approximately correct. In an article in Scientific American, Kate Wong (2005) put the angle of the Galactic stellar bar at a "45-degree angle to the galaxy's main plane."[133] According to Wikipedia's page on the Milky Way, "The Galaxy's bar is thought to be about 27,000 light-years long, running through its center at a 44 ± 10 degree angle to the line between the Sun and the center of the Galaxy."[134]

The asserted notion that the position angle of the leading edge of the stellar bar may be approximately equal to the leading edge pitch angle of a GTW is not at all unfounded. According to Zhang and Buta (2007), the

Co-rotation phase shift (change of numeric sign discussed earlier):

> ... is largest for pitch angles close to 45 degrees and is zero for pitch angles of 0 degrees and 90 degrees. ... For intermediate Hubble type galaxies [like the Milky Way] ... the phase shift distribution changes sign at corotation ... which leads to ... the rapid steepening of the sinusoidal wave profile [135] ...

Figure 31 depicts a Hypothetical GTW with Zhang's central length estimate placed within it.

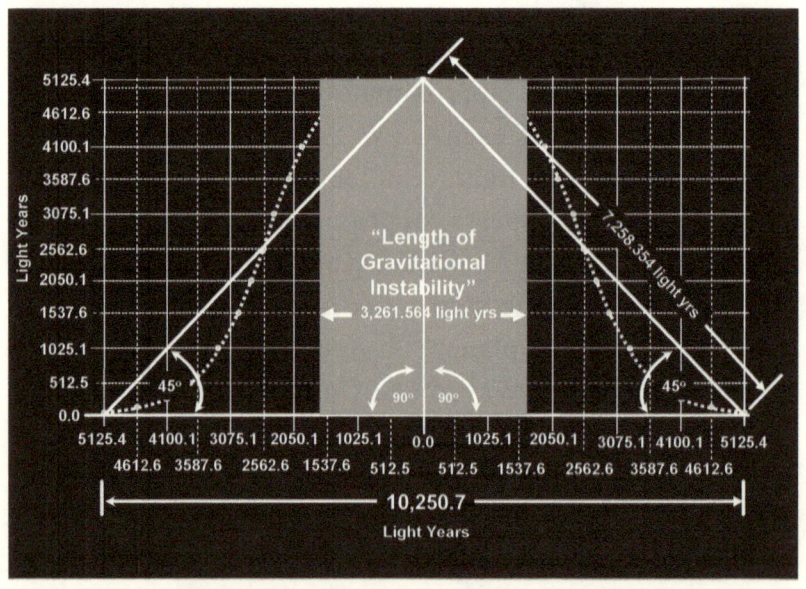

Figure 31 Geometry of a Hypothetical GTW

The figure depicted wavelength, height, and pitch angle tend to confirm that GTW meet the dimensional wave force criteria.

There is another quick reasonableness check for the GTW height dimension. The thickness of the Galactic Disk at or about the position of our Sun is estimated at "2,300 to 2,600 light-years."[136] Assuming that the gravitational wave height extends above and below the Galactic disk, the criterion meeting wave would have 2,562.68 light years of its vertical wave-height within the Disk, and extend 1,281.34 light years above and below the disk. These values fit within the observation limits of so-called "velocity streaming effects" observed in our solar neighborhood. The Radial Velocity Experiment (RAVE) website[137] reports observations of star streaming effects at these very altitudes above the Galactic Plane.[138]

Poisson distributed

According to Dehnen (1999 & 2003)[139] Ergeny and Gedalin (2000), Korchagin, Girard, Borkova, Dinescu, and Altena (2003) [140], and Zhang (2006)[141] GTW can be modeled using the Poisson distribution. Girard, Borkova, Dinescu, and Altena (2003) state that:

> The total surface density of all gravitating matter can be determined from the Poisson equation once the strength of the gravitational field F is estimated (Blinnery & Maerrifield, 1998)."[142]
>
> ...

The local dynamic density [of the solar neighborhood] can be determined then from the Poisson equation ..."[143]

Figure 32 depicts the quantitative profile of a single GTW built using Poisson and normal distribution mass function generators.

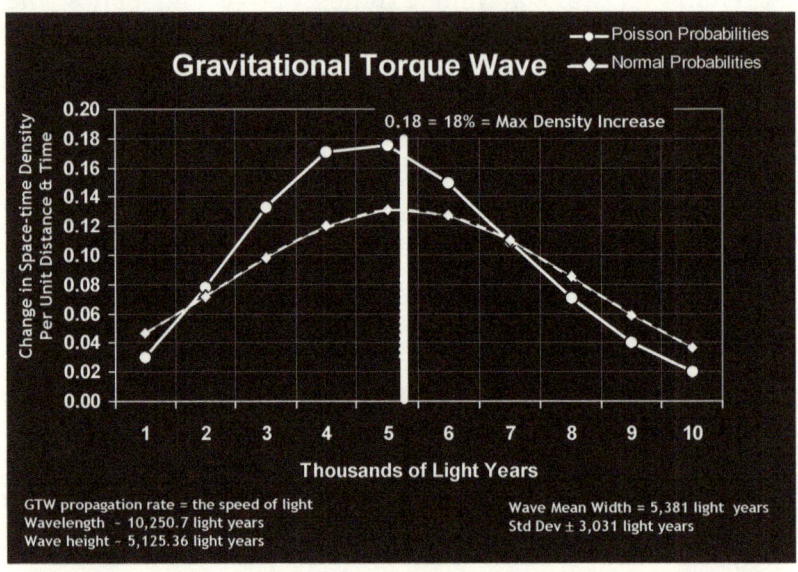

Figure 32 Profile of a Gravitational Torque Wave

As shown in the upper middle of the figure, the maximum space-time density change (compression) imposed by a Poisson distributed GTW is 18 percent. That is just 2 percent less than the Quillen (1994) and Fresneau and Argyle (2005) estimates. As you may recall, those sources suggested that a GTW imposes a 20 percent wave amplitude reduction and density change at or about the Co-rotation radii. Given the results of recent solar neighborhood surveys, as well as the available theoretical sources, the Quillen et al estimate of maximum GTW density change appears to be very well supported, and a very close match to the Figure 32 GTW Profile.

The Poisson distributed wavelength in Figure 32 is 10,250.7 light years, which is in direct compliance with the wavelength criterion. The wave height (amplitude), however, is slightly less than the 5,125.36 light year criterion. The wave height discrepancy is exceedingly small. At just ± 102.51 light years, the wave height discrepancy is approximately equal to the error of measurement of related astronomical observations.

Given the evidence cited, GTW are declared to be excellent matches to the pre-defined wave force evaluation criteria. However, one GTW does not make a complete Maya World Age Precession Cycle. The multi-GTW wave train requirements are the subject of the next chapter.

8

WAVES OF WORLD AGES

Sourcing and Forcing Trains

Waves, even the surface waves in Earth's oceans, come in *trains*. Here, the term 'trains' refers to a line of relatively uniform waves coupled together to form a serial progression. Each wave in a train multiplies the squeezing and stretching density changing effects imposed by prior waves.[144] Surface ocean-waves only act on the surface particles situated at relatively shallow depths. Other kinds of waves – like sound and pressure waves – compress and expand all the particles in their path proportionate to particle mass. Gravity Waves and Gravitational Torque Waves alter the density of space-time around and within every particle they pass through. Hence, *trains* of individual GTW multiply the density of every level of the universe they contact from invisible vacuum dark energy to the largest aggregations of ordinary matter.

Figure 33 depicts a single Poisson distributed GTW. The figure lists both the single and multi-wave-train phenomena GTW must relate to and predict. This chapter tests Gravitational Torque Waves in terms of those descriptive, explanatory, and predictive wave force evaluation criteria.

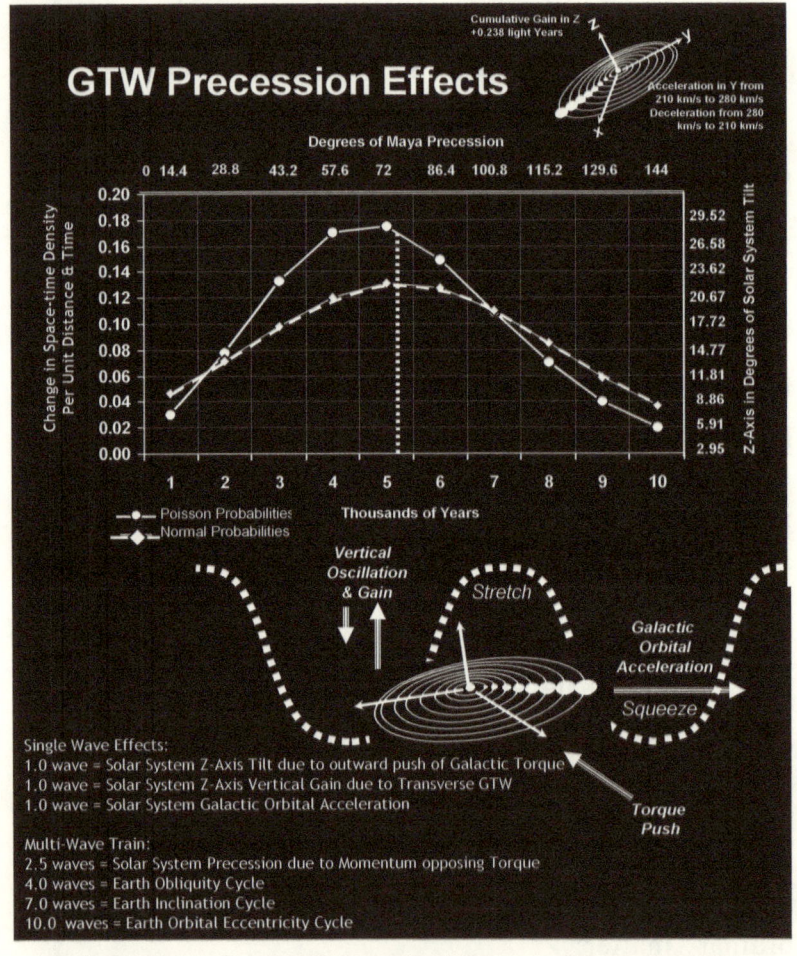

Figure 33 GTW Sourcing and Forcing

Testing Gravitational Torque Waves

To meet the evaluation criteria, GTW emanating from the Galactic Center must impose Poisson distributed space-time density changes (Compression strengths) that describe, explain, and predict a significant portion of the variation in:

- Solar System Precession and the Dates of Human Change
- Earth Obliquity
- Earth Inclination
- Earth Eccentricity
- Global Temperature Change
- Global Long Lived Greenhouse Gas Levels as represented by CO_2

If GTW Density Compression equals – is near perfectly correlated with – Maya World Age Precession, then it is assumed that same GTW Compression will also predict Solar System Z-Axis Tilt, Z-Axis Gain, and Galactic Orbital Speed. Therefore, those measures need not be directly tested.

From 285,012 BC to 1500 AD

Hypothesis H_{13} asserts that for any given Maya World Age Epoch of 5,125.36 years occurring between 285,012 BC and 1500 AD, GTW Compression is significantly correlated with:

$H_{13.1}$: Solar System Precession and hence the
Dates of Human Change

$H_{13.2}$: Earth Obliquity

$H_{13.3}$: Earth Inclination

$H_{13.4}$: Earth Eccentricity

Alternate and Hypothesis Null$_{13}$ states that the Center of the Galaxy is much too far away and much too weak a gravitational source to have any influence whatsoever on the geometric motions of the solar system, planet, and ourselves.

The database for this test contained Degrees of Maya Solar System Precession, Earth Obliquity, Inclination, and Eccentricity cycle values. That database covered the time from 285,012 BC to 1500 AD divided into 512 year records. The database needed was that described for the analyses related in the chapter, 'Waves of Climate Change.'

To that database was added the Poisson distributed values of GTW Density Compression. Then, the database was subdivided into two sub-databases. The first sub-database contained only even numbered Epochs. The second contained only the odd numbered Epochs. Correlation analyses were run and the results are presented in Table 9 for even numbered Epochs and in Table 10 for odd numbered Epochs.

Given the correlation results, **Alternate Hypothesis Null$_{13}$** is rejected. Consistent with investigative Hypothesis H_{13}, it is immediately obvious that the Center of the Galaxy *is not* too far away and too weak a gravitational source to influence the geometric motions of the solar system and ourselves. Moreover,

GTW Compression is significantly correlated with Maya Degrees of solar system Precession, and Earth's Obliquity, Inclination, and orbital Eccentricity.

Table 9 Correlation Matrix for Even Epochs

Correlation Matrix: 285,012 BC to 1500 AD Even Numbered Epochs				
	Compression	Precession	Obliquity	Inclination	Eccentricity
Compression	1.00				
Precession	0.98	1.00			
Obliquity	-0.98	-1.00	1.00		
Inclination	-0.98	-1.00	1.00	1.00	
Eccentricity	0.98	1.00	-1.00	-1.00	1.00

Table 10 Correlation Matrix for Odd Epochs

Correlation Matrix: 285,012 BC to 1500 AD Odd Numbered Epochs				
	GTW Compression	Precession	Obliquity	Inclination	Eccentricity
Compression	1.00				
Precession	-0.99	1.00			
Obliquity	-0.99	1.00	1.00		
Inclination	0.99	-1.00	-1.00	1.00	
Eccentricity	0.99	-1.00	-1.00	1.00	1.00

Overall, the Dates of Human change are near perfectly correlated with Degrees of Solar System Precession. Further tests confirmed that the correlation of GTW Compression with the Dates of Human Change equal the correlation between GTW Compression and Precession.

In regression analyses run for each *Even-numbered* Epoch, GTW Compression predicts 95 percent of the variation in Degrees of Maya Precession. That predictive relation – which leaves only 5 percent of the variation unexplained – is significant well below the .001 level of confidence, and extends to 6 decimal places. Regression analyses run for *Odd numbered* Epochs found that GTW Compression predicts 97 percent of the variance in Degrees of Maya Precession. That predictive relation of 0.97 is significant to well below the .001 level of confidence (i.e. the calculated significance value is 0.00000021).

The minor difference in the predictive relations between the even and odd numbered Epochs appears to reflect the sensitivity of Precession, Obliquity, Inclination, and Eccentricity to the odd Epoch extension of the Poisson distribution. In odd Epochs, the Poisson distributed wave form has an extended period of relaxed GTW compression.

The current Epoch (i.e. the period from 1500 AD to 2007) is an even numbered Epoch. It afforded the opportunity to validate the analysis results. Before that was done the hypotheses were expanded to include Global Temperature Change and the variation in Long Lived Greenhouse Gases as represented by CO_2.

From 1880 to 2007

Hypothesis H_{14} states that for the elapsed portion of the current Even Epoch from 1880 to 2007, Gravitational Torque Waves (GTW) emanating from the

Galactic Center will impose Poisson distributed space-time density changes (Compression strengths) that describe explain and predict a significant portion of the variation in current period:

$H_{14.1}$: Solar System Precession and the Dates of Human Change
$H_{14.2}$: Earth Obliquity
$H_{14.3}$: Earth Inclination
$H_{14.4}$: Earth Eccentricity
$H_{14.5}$: Global Temperature Change
$H_{14.6}$: Long Lived Greenhouse Gas Levels as represented by CO_2

Alternate Hypothesis Null$_{14}$ counters that for the elapsed portion of the current Even Epoch (from 1880 to 2007) Gravitational Torque Wave Compression is not significantly related to the cited measures of celestial motion, variation in Global Temperature, or LLGG levels.

Table 11 relates the correlation analysis results. Given those tabled results, Hypothesis Null$_{14}$ is rejected. GTW compression is perfectly correlated (coefficient = 1.00) with Maya Degrees of Precession, Obliquity, Inclination, and Eccentricity. GTW Compression is also significantly correlated with Global Temperature Change and atmospheric CO_2.

Regression analysis confirms that for the current Epoch from 1880 to 2007, GTW imposed Poisson distributed space-time Compression describes, explains, and predicts 100 percent of the current period variance in: $H_{14.1}$: Solar System Precession and the Dates of Human Change, $H_{14.2}$: Earth Obliquity, $H_{14.3}$: Earth Inclination, and

$H_{14.4}$: Earth Eccentricity. Moreover, those same GTW Compression strengths are correlated 0.86 with and predict 74 percent of the variation in ($H_{14.5}$) Global Temperature Change. GTW compression Strength is also correlated .93 with and predicts 87 percent of the variation in ($H_{14.6}$) Long Lived Greenhouse Gases as represented by CO_2. The $H_{14.6}$ result indicates that GTW Compression effects are the likely source of the variation in CO_2, which cannot possibly be the source of GTW Compression.

Table 11 Correlation Matrix for 1880 to 2007

From 1880 to 2007 N = 128	GTW Compr	Precess	Obliq	Inclin	Eccen	Glob Temp	CO2 ppm
GTW Compression	1.00						
Precession	1.00	1.00					
Obliquity	1.00	1.00	1.00				
Inclination	1.00	1.00	1.00	1.00			
Eccentricity	1.00	1.00	1.00	1.00	1.00		
Global Temp.	0.86	0.86	0.86	0.86	0.86	1.00	
CO2 ppm	0.93	0.93	0.93	0.93	0.93	0.89	1.00

CO_2 is correlated 0.89 and predicts 79 percent of the variation in Global Temperature. As stated, GTW Compression is correlated .86 and predicts 74 percent of the variation in Global Temperature. The four percent higher correlation and five percent more variance explained by CO_2 in Global Temperature is *insignificant* in any sense of the term – statistical or otherwise. Still, these small differences replicate a prior result and are instructive. The 4 percent higher correlation and 5 percent higher prediction of Global Temperature by CO_2, likely reflects the recent human contribution to LLGGs and Global Climatic change. Beyond that, the widely overemphasized relation between CO_2 and Global Temperature change is little more than a redundant representation of the effects of Gravitational Torque Wave Compression of Earth's atmospheric volume and global climate. Once again the results indicate that the overwhelming majority of Global Warming is neither man-made nor humanly-correctable.

These findings of course stand in direct opposition to those touted in the 2007 Working Group report published by the United Nations (UN) Intergovernmental Panel on Climate Change (IPCC). That report states that:

> "... Anthropomorphic [meaning *man-made*] LLGHGs [*long-lived greenhouse gases*, and most particularly "carbon dioxide (CO_2), methane (CH_4), and nitrous oxide (N_2O)"] remain the largest and most important driver of climate change." [151]

As previously stated, between 250,000 and 258 years ago, atmospheric CO_2 has varied ± 29.743 ppm, a total range of 59.486 ppm. That variation in atmospheric CO_2 is exceedingly close to the maximum density change imposed by a Poisson distributed gravitational torque wave. The fact that GTW Compression correlates with and predicts the majority of variation in atmospheric CO_2

in that time, simply confirms the recurrent presence of trains of GTW in our solar system.

The recent rise in atmospheric CO_2 from 1880 to 2007 represents a total increase (over the historic maximum CO_2 density) of about 25 percent. Assuming that a Gravitational Torque Wave has been present in our solar system since 1500 AD, at least 1 to 2 percent of that near term 25 percent increase in atmospheric CO_2 should be attributable to the density compression effects of that GTW. The current GTW is, however, but one of a train of GTW.

Trains of GTW multiply the density compression effects over Maya World Age Cycles. Hence, the amount of variation in the density of atmospheric CO_2 attributable to cumulative GTW effects should be at least 18 percent (i.e. one full GTW) greater than the 1 to 2 percent attributable to the current GTW.

Given that at present a 25 percent increase in atmospheric CO_2 is in evidence, we must deduct the current and prior wave residual density effects (a total of 20 percent). That leaves us with only a 5 percent unique non-GTW induced increase in atmospheric CO_2. That 5 percent increase is most likely modern man's contribution to total current global atmospheric CO_2 *density*. That is not man's total output of CO_2. Rather, it is just a reasonable estimate of man's recent contribution to global atmospheric CO_2 *density*.

Most importantly here, the modern increase in atmospheric CO_2 and Global Ocean-Land Temperature provides further evidence of the presence of a GTW in our solar system since 1500. A GTW that is now compounding the effects of prior GTW to produce human, temperature, atmospheric LLGG density, and Precession indexed solar system motion changes.

Confirmation from the Edge

On August 31, 2007 another possible sign of the current and prior presence of a GTW in our solar system occurred. Voyager 2 radioed back to Earth data describing the crafts fifth turbulent shock encounter. According to Ron Cowen, this happened at the place:

> "Where the solar wind – the Sun's hot supersonic wind of protons and other charged particles, which carves a bubble in space extending well beyond Pluto – slams into cold interstellar space and abruptly slows."[153]

There was a big surprise in that Voyager 2 report. The message came at a distance of around "83.7 astronomical units (1 AU is the average Earth-Sun separation)."[154] A second surprise was that altogether Voyager 2 reported a total of five shocks that mark "multiple crossings" of the edge of our solar system, which indicates that the southern extent of the solar system is not the "steady structure that is predicted by the simplest theory."[155]

Launched in 1977, Voyager 2 is the sister craft of Voyager 1. Voyager 1 reached the northern fringes of our solar system in 2004. Curiously, at the northern bound of the solar system Voyager 1 reported only one shock encounter. That single encounter occurred at a distance of 1.6 billion kilometers (10.695 AU) farther from the Sun than the five shock encounters reported by the south bound Voyager 2. The multiple Voyager 2 measurements strongly suggest "that the solar system is lopsided;" and seriously compressed in the southern extent, as depicted in Figure 34.

NASA scientists speculate that the distortion of the southern solar system may be the result of "a series of supernova explosions about 10 million to 20 million years ago ..."[156] We propose a different explanation.

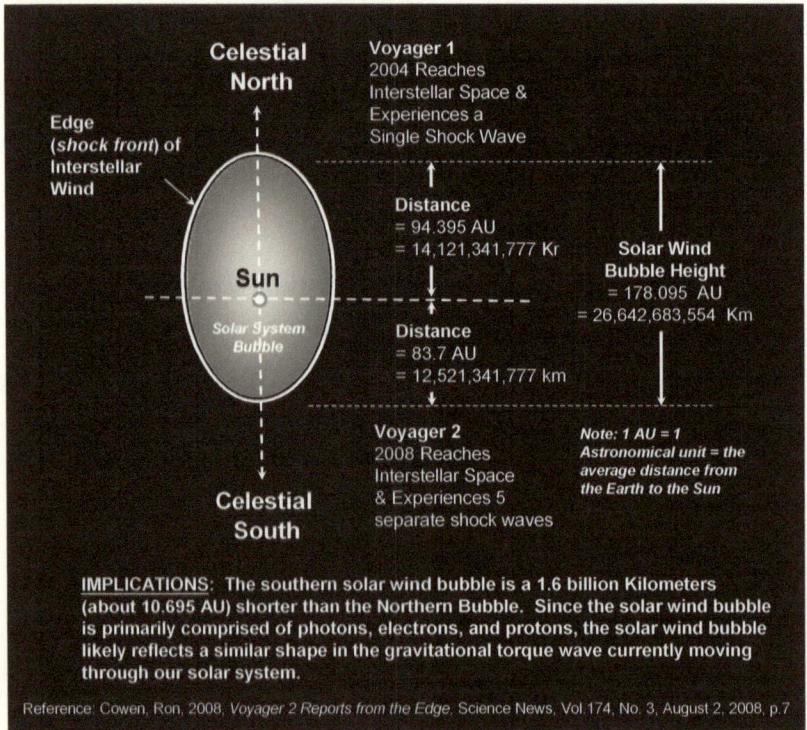

Figure 34 Voyager 1 and 2 Encounters with the Edges of the Solar System

The solar system and its wind bubble are situated in what is called the "celestial sphere." A left over from antiquity, the celestial sphere has Earth as its center and "is still considered a convenient frame of reference."[157] The solar system and celestial sphere north are tilted 60 degrees from Galactic north,[158] at a distance of about 67 astronomical units above the Galactic Plane.[159] Celestial-north and its solar system solar wind bubble lean sharply into their clockwise orbit about the Galactic Center. This is illustrated in Figure 35.

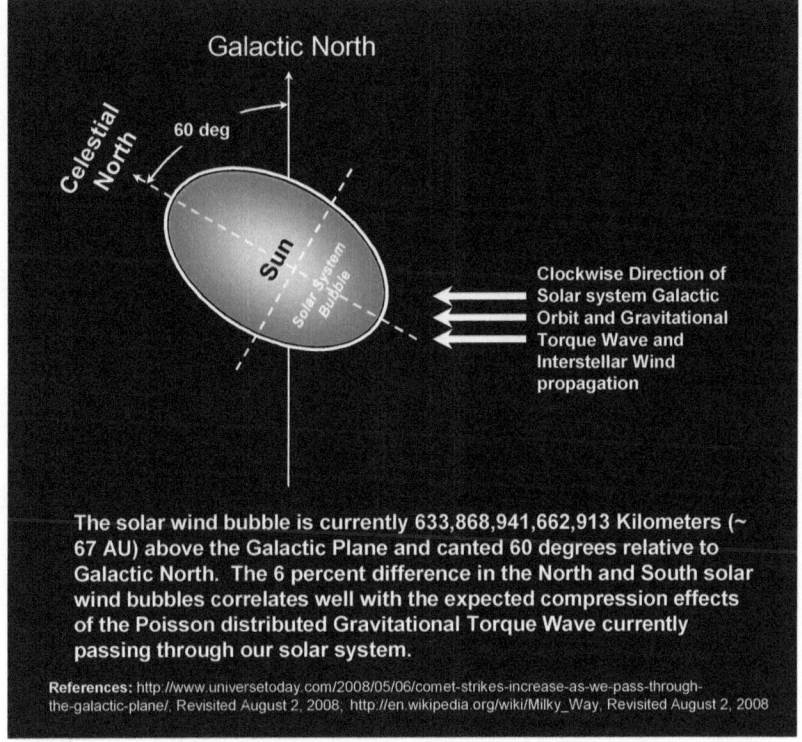

Galactic North

Celestial North

60 deg

Sun

Solar System Bubble

Clockwise Direction of Solar system Galactic Orbit and Gravitational Torque Wave and Interstellar Wind propagation

The solar wind bubble is currently 633,868,941,662,913 Kilometers (~ 67 AU) above the Galactic Plane and canted 60 degrees relative to Galactic North. The 6 percent difference in the North and South solar wind bubbles correlates well with the expected compression effects of the Poisson distributed Gravitational Torque Wave currently passing through our solar system.

References: http://www.universetoday.com/2008/05/06/comet-strikes-increase-as-we-pass-through-the-galactic-plane/, Revisited August 2, 2008; http://en.wikipedia.org/wiki/Milky_Way, Revisited August 2, 2008

Figure 35 the Evidence of GTW Compression Effects

There is a 6 percent difference between the Sun-to-maximum extent of the northern solar wind bubble and the Sun-to-maximum southern wind bubble extent. That difference is only 1.6 percent greater than the space-time compression value predicted for a Poisson distributed Gravitational Torque Wave entering our solar system around 1500 AD. What is more, that 6 percent difference is literally identical to the Poisson distribution predicted current period increase in the rate of Human Change. There is always the possibility that it is just an amazing coincidence. Given the results reported in this book, it is suggested that it is not.

What are the implications for the future of our solar system? Could the GTW in evidence be the source of the Voyager detected 6 percent reduction in the southern extent of the solar system? The answer is not as simple as it might seem. The complexity of the problem results from the unknown effects of trains of GTW.

If the southern solar system compression is solely the result of the current GTW, then over the remainder of the 5,125.36 years now unfolding that deformation could shorten the southern extent of our solar system by an additional 44 percent, from 83.7 AU to 46.9 AU. If the deformation in the south is the result of multiple GTW, then the increase in the deformation will be much smaller.

It is suggested that the last four to five Epochs produced the overall compression in evidence in the southern end of the solar system. Why; the south bound Voyager 2 reported five separate shock fronts. In contrast, the north bound Voyager 1 reported only one shock front.

The bottom line is that there is every indication that a Poisson distributed GTW is now present in our solar neighborhood, and it has been here since 1500 AD. Further, it and predecessor Epoch GTW have dramatically impacted our solar system, the motions of Earth, our global climate, and our selves. The current gravitational wave has already produced mammoth changes in the diffusion of Human Innovations and our global climate. Still, those impacts represent less than 6 percent of the total magnitude of change that GTW has in store for us.

In 2008 this work was rudely interrupted by something that provided yet more dramatic evidence of the presence of a GTW, and the prescient validity and reliability of the Maya Long Count System. The next chapter shares it.

9

WAVES OF ECONOMIC CHANGE

Crashing Clues

In 2008 the World's attention was abruptly diverted to a set of genuinely scary economic events. Around the globe housing prices fell like an avalanche of loose shale. Mortgage foreclosures skyrocketed. Credit tightened and financial lending threatened to halt altogether. Mortgage backed securities were suddenly declared "toxic" and written down to zero. The big international Investment banks went commercial, up for sale, or bankrupt. Government leaders raced to "bailout" failing financial institutions, as global stock markets crashed and shattered historic records. In short, global market capitalism abruptly went from overheated exuberance to stone cold paralysis.

Stunned and baffled, we had to keep reminding ourselves that over the last 500 years the amount of market trading (the number of times shares or their equivalents changed hands) had continued to climb, "roughly doubling every eleven years."[160] Despite the occasionally burst bubble, recession, and a difficult to remember depression, "this [accelerating] pace of change shows no signs of letting up."[161]

Economic History Lessons

A bit of on-line research quickly revealed that historically all attempts to smooth the temporal oscillations (volatility) in global free markets has failed – miserably. The literature is clear. New economic upsurges are born on the double-edged sword of *"innovation."* According to Steve Forbes, "... most financial innovations, properly used, enhance our standard of living by helping create and deploy capital more efficiently."[162] Further, financial innovations are the same sorts of novel objects, ideas, and practices addressed in the 'Waves of Human Change' chapter of this book. And, like all man-made innovations, economic novelties are adopted and diffused through social network transactions.

Proactive actions to smooth market volatility simply fail or exacerbate market jitters. That is because all such interventions are born of prior innovations and existing practices. Such measures constitute little more than out of date band-aids slapped on gaping new wounds. The same spears may pierce production and consumption anew, but they inevitably strike places that existing shields do not cover. As a result, booms, busts, and back-fit repairs blemish our economic complexion like pernicious outbreaks of adolescent acne. What is more, they do our socio-economic life just about as much good.

New bear markets, recessions, or depressions result from failures to constrain the misuses and abuses of emerging market *innovations*.[163] The consequences of such oversights reach far beyond the disruption and damage of market transactions. As was shockingly obvious in the fall of 2008, the misuses of financial innovation destabilize economically bound regional, national, and global social systems. Cutting right to it, the abuse of economic innovations threatens the very foundations of modern civilization. It is no secret that the poisonous roots of World War II were firmly planted in the toxic soil of the Great Depression.

There is another key lesson repeatedly gleaned from our sordid economic history. Volatile booms and busts are innate in human free market trading. As Alan Greenspan noted in his 2008 edition of his book ("The Age of Turbulence)," "History is replete with waves of self-reinforcing enthusiasm and despair; innate human characteristics not subject to a learning curve."[164] Similarly, Steve Forbes has insisted that, "Excesses are normal in an innovative economy. When something new comes along, hordes of investors want in. But the inevitable bust will come."[165] Economists warn that such financial rocks and shocks will become increasingly frequent as we continue to automate market processes and accelerate the diffusion of ever multiplying innovations.[166]

Building Perspective

When the bottom fell out of the known economic world in 2008, the discoveries discussed in earlier chapters had already been made. Maya World Age Epochs had been shown to accurately count out the progress of Galactic Gravitational Torque Waves. Waves that propagate from the domain of elemental space-time all the way up to the realm of aggregate ordinary matter. Those Poisson distributed GTW compression effects had already been modeled. GTW had described, explained, and predicted: 1) the great history of human change and innovation, and 2) the variations in Earth's climate and the celestial motions of our planet and solar system.

Those descriptive and predictive relations, however, occurred on time scales ranging from thousands to hundreds of thousands of years. So far, but a single exception had been found; Degrees of Maya Precession, and thus GTW compression, predict nearly all the variation in human change from 1500 AD to the present; a period of only 508 years.

The staggering on-paper financial losses of 2008 begged a new question. Might GTW compression influence human events and outcomes in shorter timeframes – events like contemporary economic trends? The ancient Maya certainly thought so. They designed their Long Count Calendar to predict sub-cycles of change ranging from a single day to World Ages spanning 9,360,000 days. The subtleties and dramatic predictive utility of the five Maya Sub-cycles is the subject of the next chapter, 'Wavelets of Synchronization.' This chapter addresses the relationship of Maya World Age Precession and GTW Compression to the record of annualized economic changes over the last 508 years.

Stochastic Mysticism

Today's scientists relegate all ideas, objects, and practices like the Maya Long Count calendar to the dungeon of primitive religion and astrology. They denounce all such practices as the literal antithesis of *objective* science. Of course those same scientists view modern economics as a "pseudo-science" as well. They assert that the behaviors of financial markets are hopelessly unpredictable "*stochastic* processes."

> A stochastic process is one whose behavior is non-deterministic in that a ... next state is determined both by the process's predictable actions and by a random element. Stochastic crafts are complex systems whose practitioners, even if complete experts, acknowledge that outcomes result from both known and unknown causes.[167]

There are serious contradictions to the supposed random nature of human economic markets. Those contradictions tend to indicate that market trading is not as irrational as some scientists like to think it is. For example, take the following niggling fact.

> ... Stock market operators who regularly appear in the Forbes 400 [Richest People]

list made their fortunes working as full time businesspeople, most of who received college educations and adhered to a strict stock picking philosophy they developed at a relatively early age. If 'throwing darts at the financial pages' were as effective an approach to investment as deliberate financial analysis, one would expect to see casual, part time investors appearing in [the Forbes] rich lists as frequently as professionals like George Soros and Warren Buffett."[168]

It is hard to deny that, "a small number of investors ... have outperformed the market over long periods of time in a way which is difficult to attribute to luck, including Peter Lynch, Warren Buffett, George Soros, and Bill Miller."[169]

Trained-trader success is not the only contradiction to the presumed unpredictably stochastic nature of free market behavior. There's a mountain of evidence that magnificently denies it. Specifically, the overwhelming upward trend in most major economic indicators over the last 508 years denies it. Take for example the per-person (per capita) consumption in the United States from 1889 to 2004 (Figure 36). That trend characterizes the very idea of market economics and consumption, as quoted in the following.

Economics is the social science that studies the production, distribution, and *consumption* of goods and services ... In supply-and-demand analysis the price of a good coordinates production and *consumption* quantities.[170] In economics, *consumption* is the primary motivating force in the wealth or utility maximizing paradigm. ... All activities are directed towards consumption, either of traditional goods and services, or of personal and perhaps unique activities.[171]

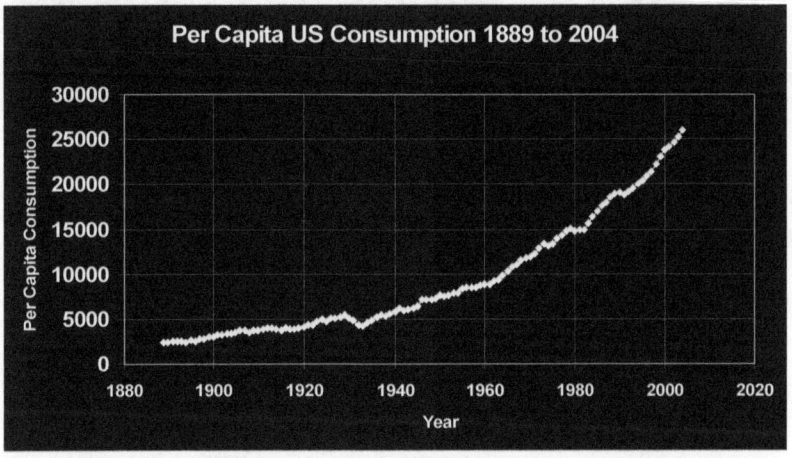

Figure 36 Per Capita US Consumption
Source: Robert J. Shiller, Stock Market Data Used in "Irrational Exuberance" Princeton University Press, 2000, 2005 [Website]

US per capita consumption is the average consumption per person in the United States. The long term upward progress of US Per Capita Consumption is not only blatantly obvious, but plainly other than irrationally stochastic.

The Real Gross National Product (GNP) of the United States from 1890 to 1974 shows a similar determined upsweep, as depicted in Figure 37.

183

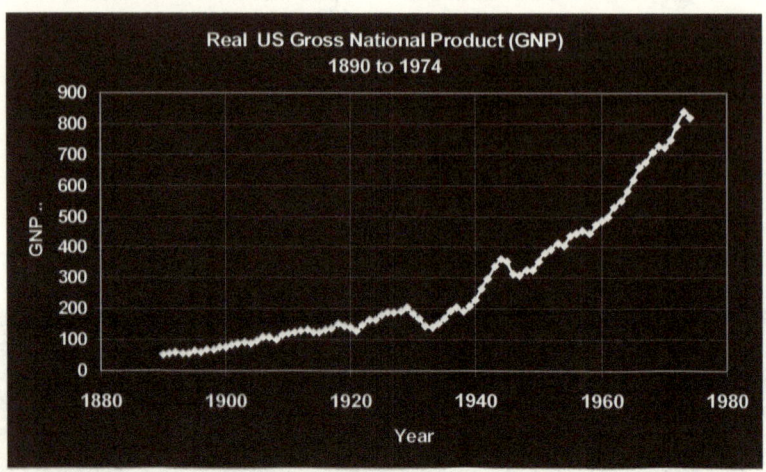

Figure 37 Real US Gross National Product
Source: Robert J. Shiller, Stock Market Data Used in "Irrational Exuberance" Princeton University Press, 2000, 2005 [Website]

The Gross National Product (GNP) is the value of all the goods and services produced in the economy, plus the value of the goods and services imported, less the goods and services exported. The United States used GNP in its national accounts until 1992. In that year GNP was replaced by Gross Domestic Product (GDP) as the measure of choice.[172]

Real US Consumption (Figure 38) is the measured Gross US Consumption corrected for the genuinely stochastic effects of inflation. As a result, the Real US Consumption is a slightly smoother plot, and not surprisingly, it exhibits the same decidedly upward trend as Gross US Consumption.

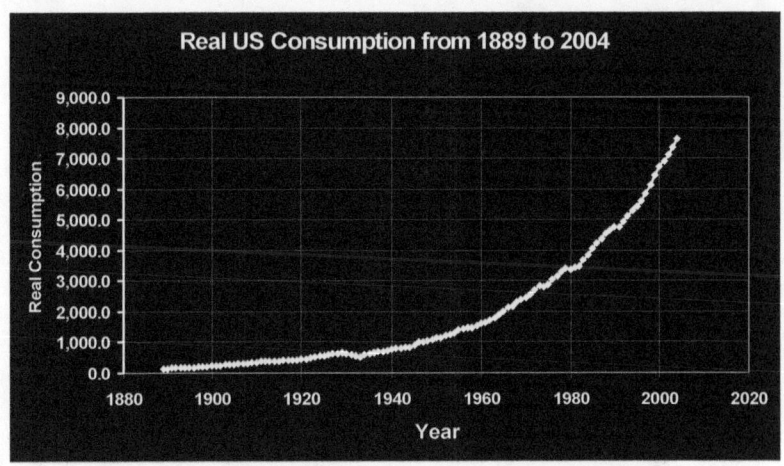

Figure 38 Real US Consumption

Source: Robert J. Shiller, Stock Market Data Used in "Irrational Exuberance" Princeton University Press, 2000, 2005 [Website]

Gross Domestic Product or "GDP is defined as the total market value of all final goods and services produced within the country in a given period of time (usually a calendar year). It is also considered the sum of a value added at every stage of production (the intermediate stages) of all final goods and services produced within a country in a given period of time. It is given a money value."[173]   Figure 39 depicts the US GDP from 1920 to 2007, which repeats the less than purely *stochastic* upward curve seen in the prior figures.

US Money Supply is the total amount of money in billions of US dollars within the US economy over the course of a given year. Again, as is clear in Figure 40, that Money Supply trends determinedly upward.

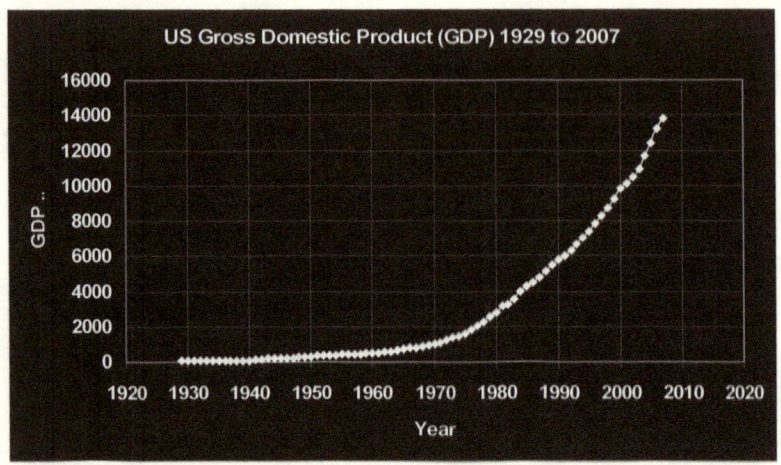

Figure 39 US Gross Domestic Product (GDP)

Source: US Bureau of Economic Analysis National Income and Product Accounts

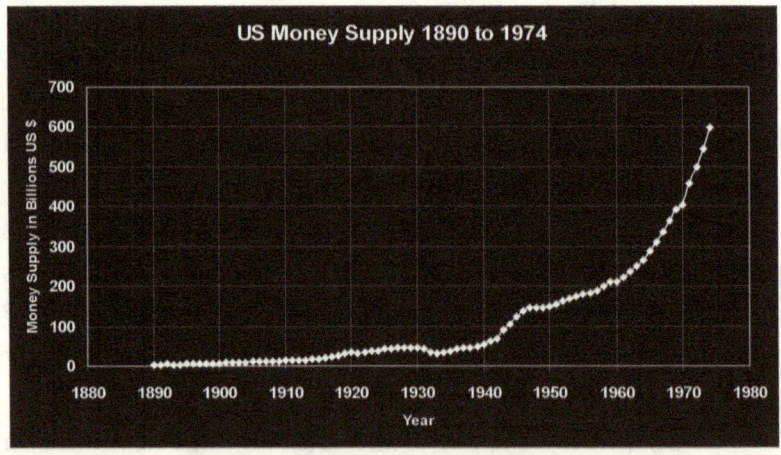

Figure 40 US Money Supply

Data *Source: Hipel and Mcleod (1994); Hyndman, R.J. (n.d.) Time Series Data Library, http://www.robhyndman.info/TSDL, Revisited October 26, 2008*

Unlike the economic indicators charted thus far, the *Velocity of Money* exhibits a distinctive downward trend. As shown in Figure 41, the Velocity of Money is inverted relative to the prior measures.

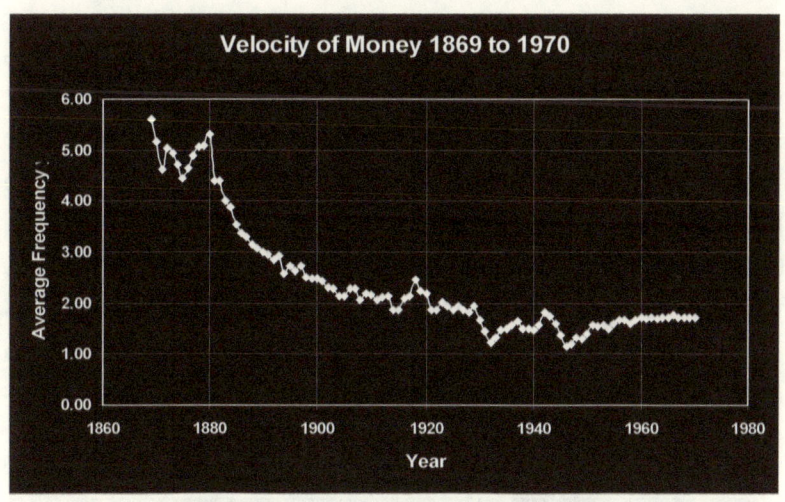

Figure 41 Velocity of Money
Sources: Hyndman, R.J. (n.d.) Time Series Data Library, http://www.robhyndman.info/TSDL, Revisited October 27, 2008 and Hipel and Mcleod (1994).

According to a recent article by Steve Forbes, the Velocity of Money represents:

> How quickly money is used. If everyone is clutching cash, velocity goes to near zero, a situation we are experiencing today [in February 2009]. In a Germany-style hyperinflation, volatility goes berserk as everyone gets rid of rapidly depreciating money. ... There is an enormous lag between what a central bank does and the impact of its actions on the market place. ... if too

much money is printed, it will be several months before that's reflected in the prices you pay at the supermarket or gas station, [and hence in a change in the velocity of money].[174]

The US history of the Velocity of Money appears strongly systematic, as opposed to unpredictably stochastic. The significant lag in the changes of the Velocity of Money and Money Supply effects, which Forbes cites, make the prediction of time-series trends in these measures much more difficult to estimate, weakening associated correlations and predictions.

The US Consumer Price Index (CPI) plotted in Figure 42 "is a measure of the average price of consumer-goods and services purchased by households.

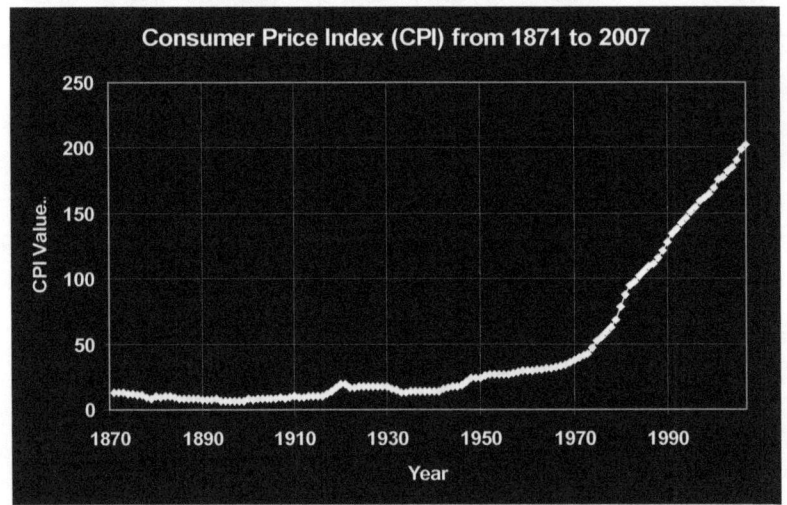

Figure 42 US Consumer Price Index

Source: Source: Robert J. Shiller, Stock Market Data Used in "Irrational Exuberance" Princeton University Press, 2000, 2005 [and related data Website]

CPI is one of several price indices calculated by US national statistical agencies."[175]   It is copiously clear that CPI is, on the whole, not the least unpredictable.

The longest in use economic measure found is Annual Global Hourly Wages. As depicted in Figure 43, beginning as far back as 1260, Annual Global Hourly Wages (measured in English pounds) has been tracked and reported.

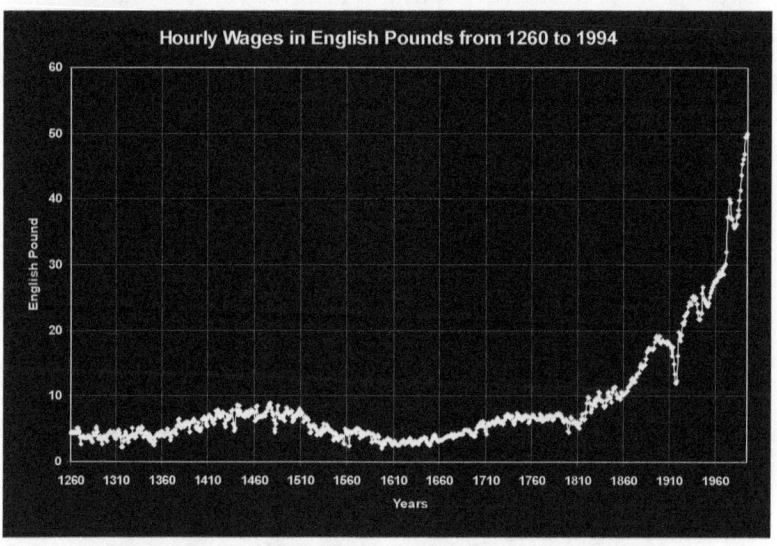

Figure 43 Global Hourly Wages

Data Source: Wheelwright and Hyndman, R.J. (n.d.) Time Series Data Library, http://www.robhyndman.info/TSDL, Revisited October 26, 2008 Makridakis

Global Hourly Wages exhibit a very long term upward trend. The long rise feebly begins in the early 1600s, surges after 1660, and rises quickly after 1810. Interestingly, prior to 1600, global wage changes do look rather unpredictable. Something clearly started around

1500 and took hold between 1615 and 1660. Whatever it was, has dominated annual wage trends ever since.

As shown in Figure 44, the Dow Jones Industrial Average (DJIA) Adjusted Annual Closing Price exhibits a distinctly upward slope. Similar charts are readily available from the cited source for the NASDAQ Index, Standard and Poor's (S&P) 500 Index, 30 Year US Treasury Bonds, and so on. All of those charts show the same upward trend in trading volume and annual closing share Price as exhibited by the Dow. Correspondent chart trends exist in international trading exchange data as well.

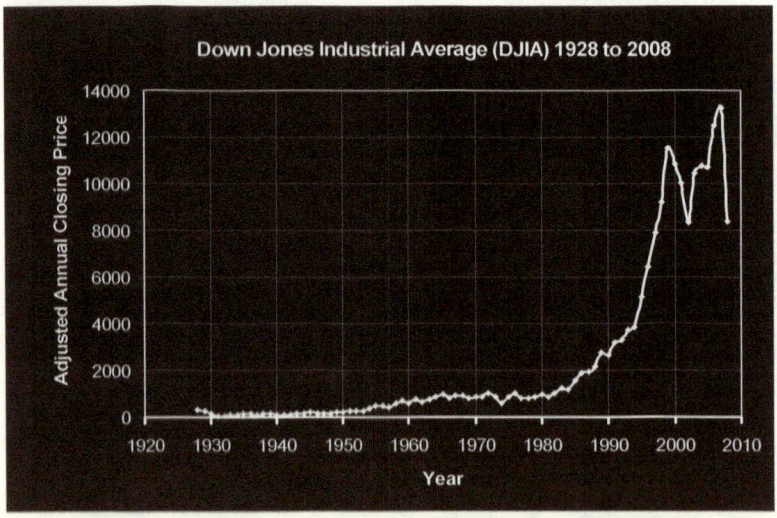

Figure 44 Dow Jones Industrial Average (DJIA) Adjusted Annual Closing Price
Source: Stockcharts.com, Revisited October 27, 2008

Figure 45 depicts the US Composite Stock Index. Common shares of stock convey corporate voting rights, where equity shares do not. The Composite stock index is comprised of all the shares listed on US Exchanges.

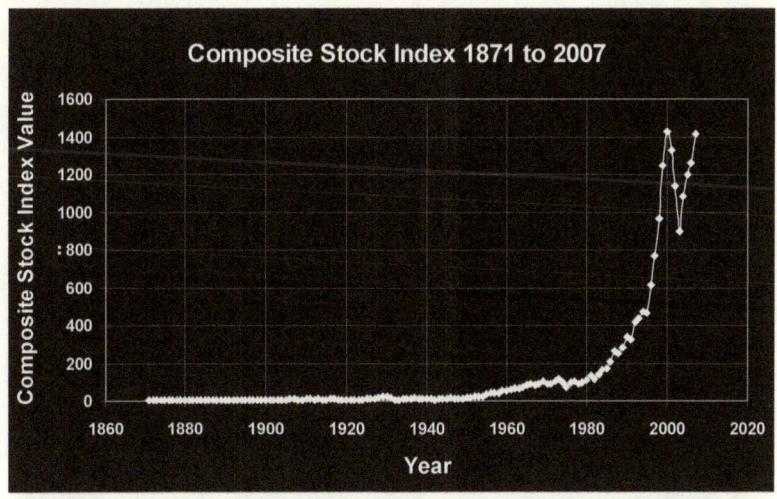

Figure 45 Composite Stock Price Index from 1871 to 2007

Source: Hyndman, R.J. (n.d.) Time Series Data Library,
http://www.robhyndman.info/TSDL, Revisited October 27, 2008 and
Hipel and Mcleod (1994).

From 1871 to 2007 the overwhelming trend of the Composite Stock Index is progressively upward. The dramatic down turn in 2003 represents a now historic recession likely still fresh in memory. The housing boom followed with a protracted period of overtly easy lending. That boom renewed the strong upward trend in trading through 2006. Then, the housing market bubble began to collapse and the resulting deep dive of the 2007 credit crises becomes clearly visible. The catastrophic collapse of 2008 had not yet reached history making depth when the data plotted here were collected and reported. Still, if trading history serves, the DJIA and other indexes will again recover to trace yet another round of innovation, abuse, and boom-born buying. Then, within the usual period of 6 to 10 years, yet another bubble will burst and cut another notch in this great graph of greed and excess.

Figure 46 depicts DJIA Annual Trading Volume. This is the number of shares that traded hands by the end of each year. The plot strongly echoes the upward slope of the charting of adjusted annual closing price of DJIA shares.

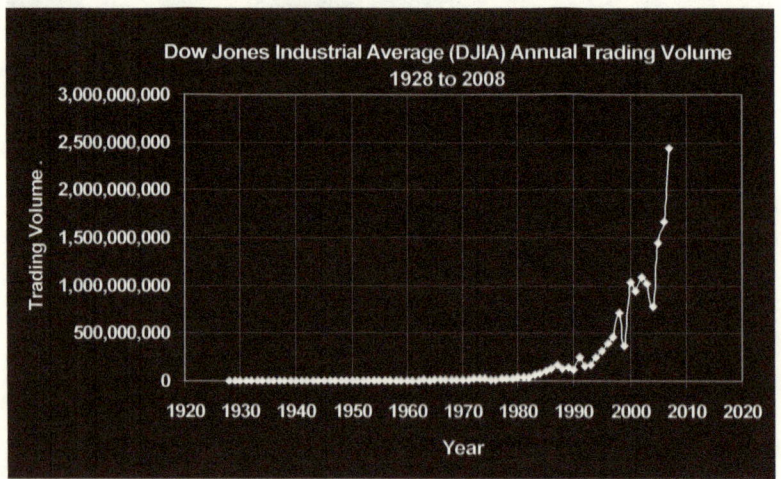

Figure 46 Dow Jones Industrial Average (DJIA) Annual Trading Volume
Source: Stockcharts.com, Revisited October 27, 2008

Figure 47 plots a very different sort of process; the US National Debt Outstanding. US debt turned sharply upward after World War II, tended seriously vertical in the late 1950s, and became outrageously steep around 2001. National debt exhibits the work of deliberate

192

human interventions. In fact, there is not one unexplained tick on the entire plot. Every dot represents a specific act of known US Federal Reserve borrowing. As I write this, the US is borrowing more than $8 billion a week from the likes of China, Japan, and various developing nations in order to cover the National Account.

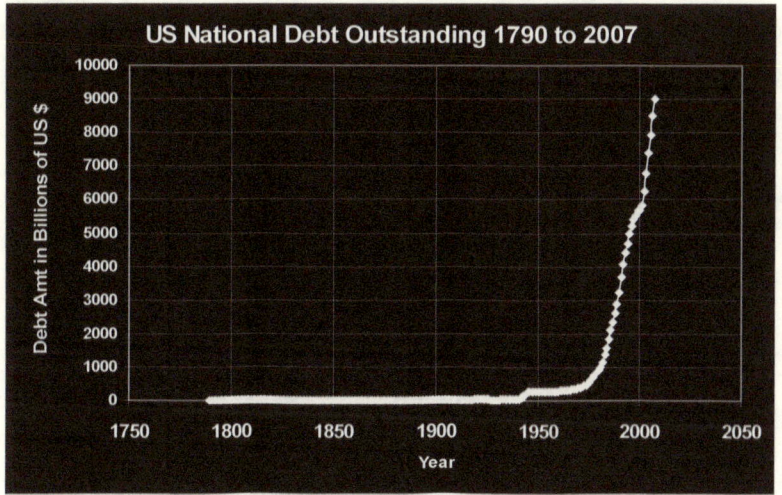

Figure 47 US National Debt Outstanding
Source: US Treasury Department;
http://www.treasurydirect.gov/govt/reports/pd/histdebt/histdebt.htm,
Revisited October 28, 2008

The evidence is abundant. Market innovations in supply, prices, wages, and consumption all exhibit dramatic long term upward trends. Any genuinely unpredictable stochastic processes will produce and plot equally likely declines and increases. A truly unpredictable stochastic process should trace a twitchy horizontal line that sustains no systematic upward or downward trends.   No such random plotting is in evidence in the measures discussed thus far, which include the primary leading economic indicators.

To exemplify the point, Figure 48 displays the Annual Percent of inflation in the United States from 1914 to 2007. This chart shows precisely what one would expect from an unpredictable stochastic process. That is, an erratic jagged line with no persistent upward or downward trend.   Still, even that is something of a deception.

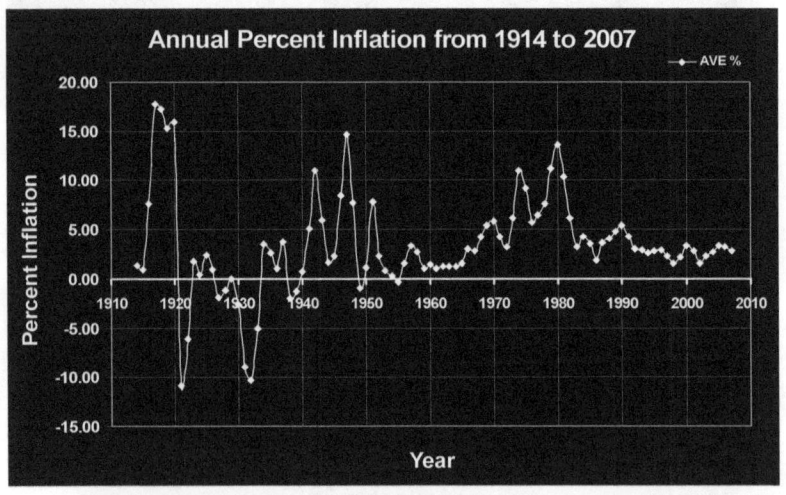

Figure 48 Annual US Percent Inflation
Source:
http://inflationdata.com/inflation/Inflation_Rate/HistoricalInflation.
aspx?dsInflation_currentPage=7, Revisited October 28, 2008

Leading economist and Nobel Prize winner Milton Friedman established that inflation is "strictly a monetary process. If a central bank prints too much money, you're in trouble, if it doesn't, money retains its value."[176]   In other words, increasing the supply of money decreases the purchasing power of currency and *inflates* the amount of money needed to make and buy things. In this regard, inflation is not unpredictable at all. If you know when the central bank printed more money, how much it printed, and you factor in the months of lag time it takes for inflation to hit the market, you can predict the outrageously disproportionate damage and duration of inflation. What you are really doing, however, is predicting the peculiar behavior of the US Federal Reserve Chairman and his collaborators, as well as the calibrated lag to impact. Inflation is simply the product of a particularly nasty form of human behavior.

Testing Gravitational Torque Wave Effects

Given the obvious contradictions to the prevailing view of econometric measures other than inflation, Hypotheses H_{15} and H_{16} were formulated and tested.

Hypothesis H_{15} states that: Maya Degrees of Precession (and hence Gravitational Torque Wave (GTW) Compression) are significantly correlated (count with) the variation in the:

$H_{15.1}$: US Per Capita Consumption (in Yr. 2000 dollars) from 1889 to 2004
$H_{15.2}$: Real US Gross National Product (GNP) from 1890 to 1974
$H_{15.3}$: Global Hourly Wages (in English Pounds) from 1871 to 1994
$H_{15.4}$: US Real Consumption (in Yr. 2000 Dollars) from 1889 to 2004
$H_{15.5}$: US Gross Domestic Product (GDP) from 1929 to 2007

$H_{15.6}$: US Money Supply from 1890 to 1974

$H_{15.7}$: US Velocity of Money from 1869 to 1970

$H_{15.8}$: US Consumer Price Index (CPI) from 1871 to 2007

$H_{15.9}$: Global Hourly Wages (in English Pounds) from 1500 to 1994

$H_{15.10}$: Dow Jones Industrial Average (DJIA) Annual Closing Price from 1928 to 2007

$H_{15.11}$: US Common Stock Price from 1871 to 1970

$H_{15.12}$: US Real Stock Price Index from 1871 to 2007

$H_{15.13}$: US Stock Price Index from 1871 to 1994

$H_{15.14}$: Dow Jones Industrial Average (DJIA) Trading Volume from 1928 to 2007

Hypothesis Null$_{15}$ counters that: the variation in the financial and trading behavior of human beings is random and unpredictable, and quantitatively unrelated to Degrees of Maya Precession and GTW imposed density changes.

Hypothesis H_{16}: asserts that Gravitational Torque Wave (GTW) imposed space-time density changes (Compression strengths), describe, explain, and predict a significant portion of the variation in:

$H_{16.1}$: US Per Capita Consumption (in Yr. 2000 dollars) from 1889 to 2004

$H_{16.2}$: Real US Gross National Product (GNP) from 1890 to 1974

$H_{16.3}$: Global Hourly Wages (in English Pounds) from 1871 to 1994

$H_{16.4}$: US Real Consumption (in Yr. 2000 Dollars) from 1889 to 2004

$H_{16.5}$: US Gross Domestic Product (GDP) from 1929 to 2007
$H_{16.6}$: US Money Supply from 1890 to 1974
$H_{16.7}$: US Velocity of Money from 1869 to 1970
$H_{16.8}$: US Consumer Price Index (CPI) from 1871 to 2007
$H_{16.9}$: Global Hourly Wages (in English Pounds) from 1500 to 1994
$H_{16.10}$: Dow Jones Industrial Average (DJIA) Annual Closing Price from 1928 to 2007
$H_{16.11}$: US Common Stock Price from 1871 to 1970
$H_{16.12}$: US Real Stock Price Index from 1871 to 2007
$H_{16.13}$: US Stock Price Index from 1871 to 1994
$H_{16.14}$: Dow Jones Industrial Average (DJIA) Trading Volume from 1928 to 2007

Hypothesis Null$_{16}$ argues that: the variation in the financial and trading behaviors of human beings is fundamentally random and quantitatively unrelated to any and all GTW Compression imposed density changes.

The fundamental premise of the two multi-indexed hypotheses may come as a bit of a shock to traditional economists. That premise is that the collective market trading behavior of human beings over recent centuries, as well as their economic processes and products, are largely determined by the synchronizing influences of a solar system present Gravitational Torque Wave. Specifically, that a GTW entered our solar system around 1500 AD and continues to impose Poisson distributed levels of density compression. Levels of compression correlated with Degrees of Maya Precession and the diffusion of human innovation.

The data used to test these hypotheses was taken from the sources named under the previously presented figures 36 through 47. Degrees of Maya Precession and magnitudes of Poisson distributed GTW compression values were derived as described in earlier chapters.

Table 12 lists the correlation analysis results returned by the tests of Hypotheses $H_{15.1}$ through $H_{15.14}$. All of the correlation coefficients returned for Hypotheses $H_{15.1}$ through $H_{15.14}$ are significant at or beyond the .001 level of confidence. **Hypothesis Null$_{15}$** is rejected, and **Hypotheses $H_{15.1}$** through **$H_{15.14}$** are retained.

The correlation coefficients for Degrees of Maya Precession and GTW Compression are nearly identical for every economic variable tested. As a result only one of these two candidate predictors, (i.e., GTW Compression) was needed to conduct the subsequent regression analyses required to test Hypotheses $H_{16.1}$ to $H_{16.14}$. This is the case, as statistically identical results are achieved regardless of which of the two predictors is used.

Table 13 lists the results of the regression analyses performed to test Hypotheses $H_{16.1}$ to $H_{16.14}$. The tabled correlation and regression results are presented in descending order, based on the percent of the variance described, explained, and predicted by GTW Compression Effects. Sample sizes and sample periods are those given for the correlations in Table 12.

Table 12 Hypothesis H_{15} Correlation Analysis Results

Rank	Variable	Compression	Precession	N	Years	
1	Per Capita Consumption	0.93575	0.935572	116	From	To
2	Real GNP	0.920841	0.920841	85	1889	2004
3	Global Wages	0.912539	0.912539	124	1890	1974
4	Real Consumption	0.896497	0.896249	116	1871	1994
5	US Gross Domestic Product (GDP)	0.873858	0.872583	79	1889	2004
6	Money Supply	0.84681	0.84681	85	1929	2007
7	Velocity of Money	-0.842835	-0.842835	102	1890	1974
8	Consumer Price Index (CPI)	0.804817	0.804255	137	1869	1970
9	Global Wages	0.785206	0.785206	495	1871	2007
10	Dow Jones Closing Price	0.761759	0.760151	80	1500	1994
11	Com Stk Price	0.738579	0.738579	99	1928	2007
12	Real Stock Price Index	0.730474	0.729834	137	1871	1970
13	Stock Price Index	0.700953	0.700953	124	1871	2007
14	Dow Jones Trading Vol.	0.632418	0.629845	80	1871	1994
15	US National Debt	0.5703	0.569028	218	1790	2008

Table 13 Hypotheses $H_{16.1}$ to $H_{16.4}$ Regression Analysis Results

Rank Order	Variable	Correlation w/GTW Compression	Percent Variance Explained
1	Per Capita Consumption (in Yr 2000 dollars)	0.9357	87.45%
2	Real GNP (Gross National Product US)	0.9208	84.61%
3	Global Wages (*Subsample*)	0.9125	83.14%
4	Real Consumption (in Yr 2000 Dollars)	0.8965	80.20%
5	US Gross Domestic Product (GDP)	0.8739	76.06%
6	Money Supply	0.8468	71.37%
7	Velocity of Money	0.8428	70.75%
8	Consumer Price Index (CPI)	0.8048	64.51%
9	Global Wages	0.7852	61.58%
10	Dow Jones Industrial Avg Adj Closing Price	0.7618	57.49%
11	Com Stk Price	0.7386	54.08%
12	Real Stock Price Index	0.7305	53.01%
13	Stock Price Index	0.7009	49.13%
14	Dow Jones Industrial Average Trading Vol.	0.6324	39.99%
15	US National Debt	0.5703	32.21%

GTW Compression describes, explains, and predicts a significant portion of the variation in all of the Hypothesis 16 cited economic variables at the .001 level of confidence

Prediction of Per Capita Consumption

Gravitational Torque Wave Compression predicts 87 percent of the variation in US Per Capita Consumption (Figure 49) for the period from 1889 to 2004, a sample size of 116 years.

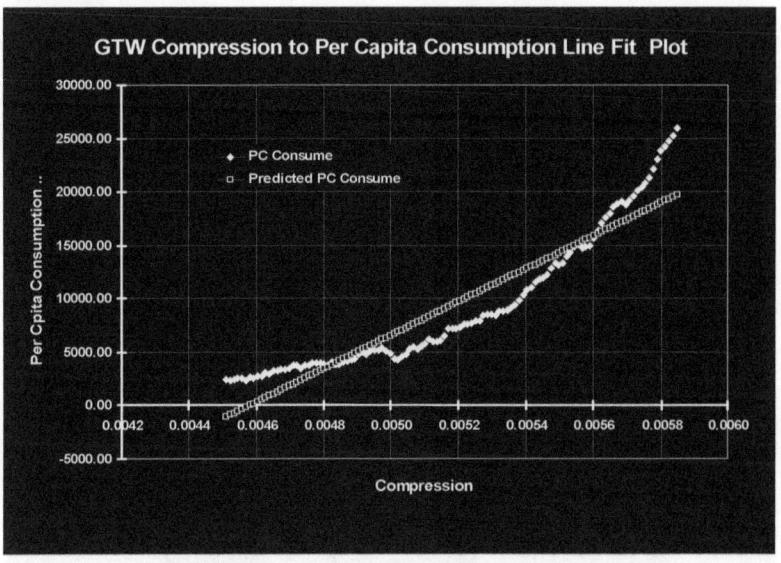

Figure 49 GTW Prediction of US per Capita Consumption

The proportion of variation explained in Per Capita Consumption is significant to over 52 decimal places and cannot be attributed to sampling errors.

Only 13 percent of the variation in US Per Capita Consumption is not predicted by Gravitational Torque Wave compression effects. The remarkably high level of GTW Compression effects associated with per capita consumption suggests that prevailing tendencies in human consumption behaviors are significantly amplified by GTW Compression. In short, as GTW Compression increases, so does the trend in per capita consumption.

Prediction of Real US Gross National Product

Gravitational Torque Wave Compression predicts 85 percent of the variation in Real US Gross National Product (GNP), for the period from 1890 to 1974, a sample size of 85 years. Figure 49 depicts this predictive relation.

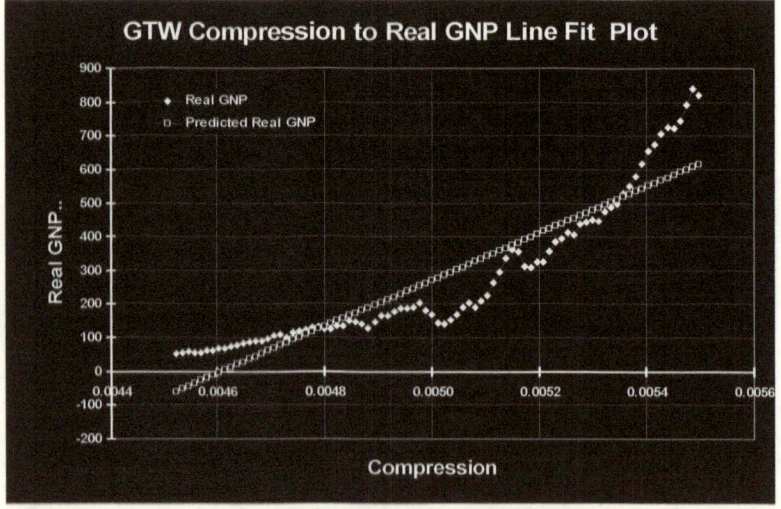

Figure 50 GTW Prediction of Real US Gross National Product

The prediction is significant to over 35 decimal places and cannot be attributed to sampling errors. Only 15 percent of the variation in US Per Capita Consumption is not predicted by Gravitational Torque Wave

Compression effects. The remarkably high level of apparent GTW Compression influence on production suggests that prevailing production tendencies are significantly amplified by GTW Compression effects. In short, as GTW Compression increases, so does production.

Prediction of Global Wages

Gravitational Torque Wave Compression predicts 83 percent of the variation in Global Wages for the period from 1871 to 1994, a sample size of 124 years. The prediction is significant to over 48 decimal places and cannot be attributed to sampling errors. Figure 51 depicts this predictive relation.

Only 17 percent of the variation in Global Annual Wages is not predicted by Gravitational Torque Wave Compression effects. The remarkably high level of GTW Compression effect on wages suggests that the prevailing tendencies in wage setting behavior are significantly amplified by the effects of GTW Compression. In short, as GTW Compression increases in value, so does the tendency to increase amount paid to labor.

As repeatedly stated in this book, a GTW entered our solar system around 1500 AD. As such, the wave build up of Gravitational Torque Compression was minimal prior to the late 1600s. GTW effects did not become strongly pronounced until the late 1800s. Accordingly, the regression analysis was rerun to relate GTW Compression to Global wages using all the data from 1500 to 1994, a sample of some 495 years.

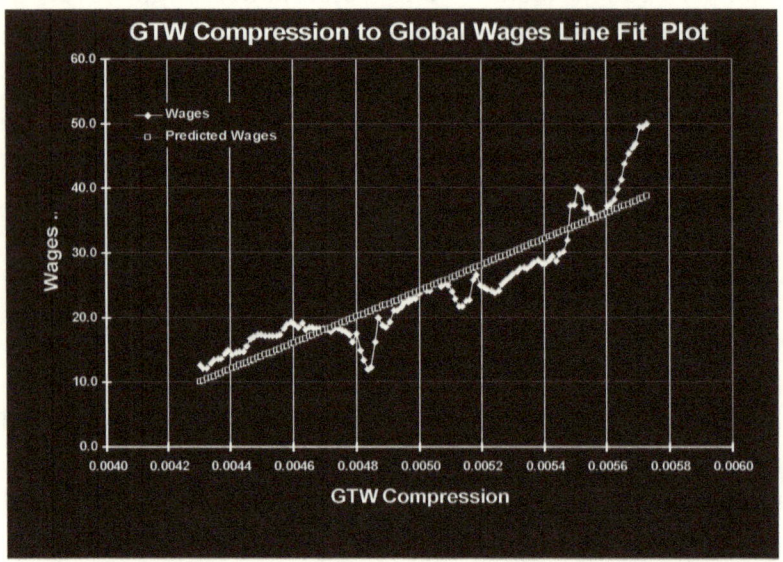

Figure 51 GTW Prediction of Global Wages

The correlation for the greater sample was reduced to .79. Further, only 62 percent of the variation in Global Wages was explained by GTW Compression for the larger sample period. Although GTW Compression still predicted a portion of the variation in Global wages significant to 104 decimal places, the amount of variation explained was substantially reduced.

Figure 52 depicts the line fit of GTW Compression from 1500 to 1994. The plot clearly shows why the prediction was so severely diminished for the longer term sample. The very small effect of the GTW Compression for the period from 1500 to 1600 seriously weakened the prediction. This reduction in the predictive relation suggests that the GTW Compression effect is decidedly real. It also suggests that our assumption as to when the

GTW began to exert an influence on our planet appears to be accurate.

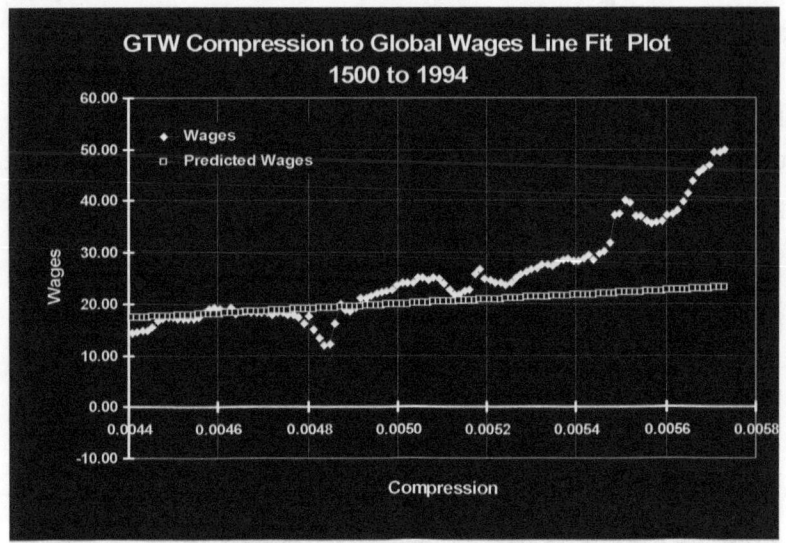

Figure 52 GTW Longitudinal Prediction of Wages

The specific Maya Long Count Calendar year 12 *Baktun*, 0 Katun, 0 *Tun*, 0 *Uinal*, and 0 *kin* (by convention written 12.0.0.0.0) forms what the Maya considered to be a condition of "resonant sub-cycle synchronization."[177] The specific Maya date of 12.0.0.0.0 corresponds to our Gregorian calendar date of Wednesday, October 14, 1615.

Taking the Maya perspective, 115 years was cut off the global wage sample (i.e., the period from 1500 to 1615). The analysis was then rerun to see if the predictive relation between GTW Compression and Global wages improved. Indeed, it did. The correlation between GTW Compression and Global Wages for the period from 1615 to 1994 increased to 0.87, and the proportion of variance explained increased to 75 percent, an improvement of 13 percent.

Figure 53 plots this improved predictive relation. It is apparent that the more tightly the sample matches the effective period of the current GTW, the stronger the predictive relation becomes.

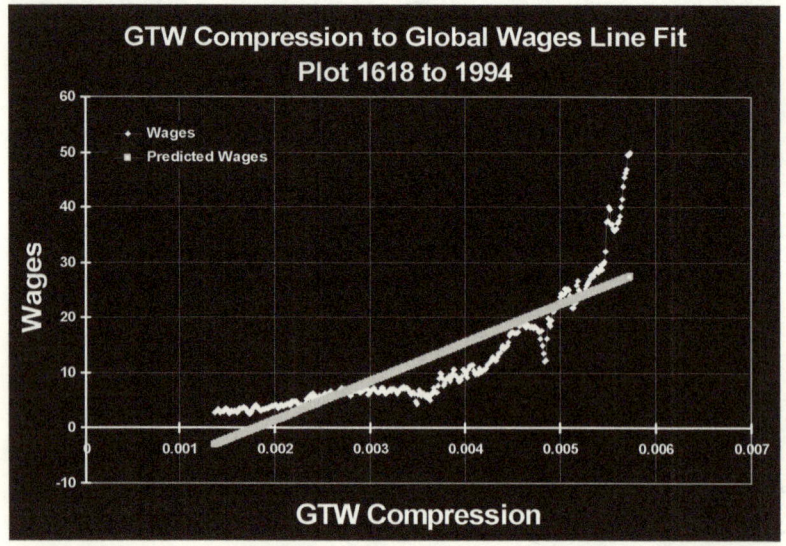

Figure 53 GTW Compression Prediction of Wages From 1618 to 1994

What's more, the Maya resonant synchronization date of 1615 coincides exactly with the earliest emergence of the predictive relation. Could this Maya prescience be mere coincidence? Given the Maya assertions previously validated, that was very much in doubt. The next chapter, 'Wavelets of Synchrony' vindicates that stance and exposes the amazing predictive potential of the Maya Sub-cycles of the Long Count Calendar system. The predictive relations from this chapter will be improved twice over in the course of that next chapter. Thus, the GTW predictions of economic variables in this chapter constitute a kind of GTW predictive baseline for that work.

Prediction of Real US Consumption

As depicted in Figure 54, Gravitational Torque Wave Compression predicts 80 percent of the variation in Real US Consumption, a sample of 116 years extending from 1884 to 2004. The prediction is significant to over 30 decimal places and cannot be attributed to sampling errors. Only 20 percent of the variation in Real US Consumption is not predicted by Gravitational Torque Wave Compression effects. The remarkably high level of apparent GTW Compression effect on real Consumption suggests that prevailing tendencies in consumption are significantly amplified by GTW Compression. In short, as GTW Compression increases in value, so does the trend in real consumption.

GTW Compression explains 11 percent more of the variation in Per Capita US Consumption than in Real US Consumption. This difference for a nearly identical period is likely due to raw nature of Real US Consumption data, and the inherent smoothing calculation imposed by economic analysts on the Per Capita Consumption data.

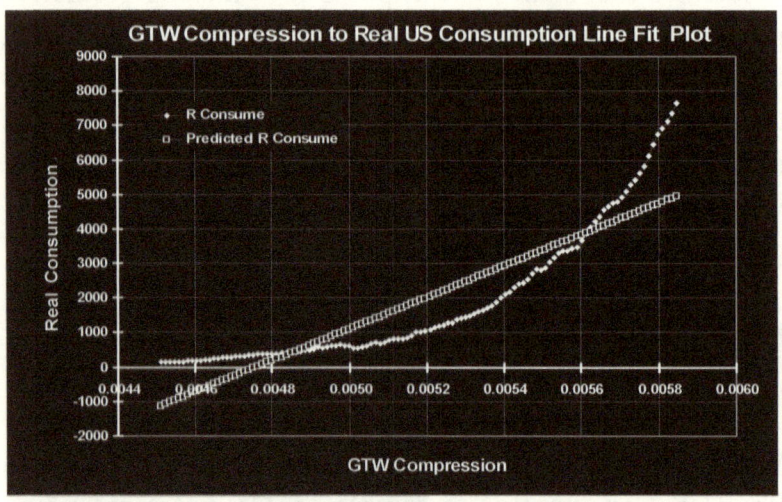

Figure 54 GTW Prediction of Real US Consumption

Prediction of US Gross Domestic Product

GTW Compression predicts 76 percent of the variation in US Domestic Product (GDP) for the period from 1929 to 2007, a sample of 79 years. Figure 55 depicts this predictive relation. The prediction is significant to over 25 decimal places and cannot be attributed to sampling errors. Only 24 percent of the variation in GDP is not predicted by GTW Compression. The remarkably high level of apparent GTW Compression influence on domestic production suggests that prevailing tendencies in production behavior are significantly amplified by GTW Compression. In simple terms, as GTW Compression increases in value, so does the trend in domestic production.

GTW Compression explains 9 percent more of the variation in Gross National Product (GNP) than in Gross Domestic Product (GDP). This difference is likely do to the difference in the period over which GNP and GDP were sampled, and government imposed changes in the definition of the two measures. While the two measures are similar, the classic GNP was sampled from 1890 to 1974. The newer GDP measure was sampled from 1929 to 2007.

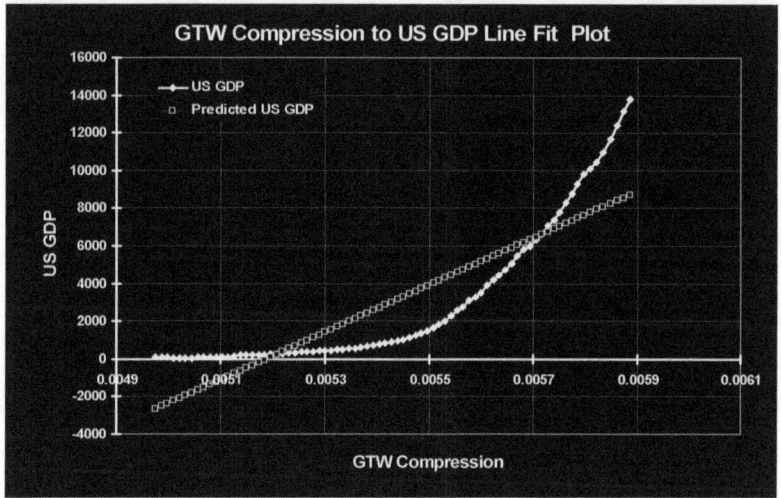

Figure 55 GTW Compression prediction of US GDP

Prediction of US Money Supply

GTW Compression predicts 71 percent of the variation in the US Money Supply for the period from 1890 to 1974, a sample size of 85 years. The prediction depicted in Figure 56 is significant to over 23 decimal places and cannot be attributed to sampling errors.

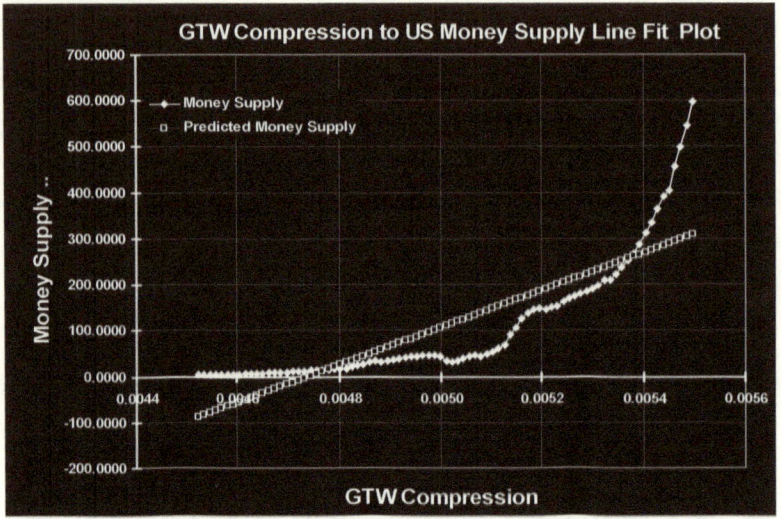

Figure 56 GTW Prediction of US Money Supply

Only 29 percent of the variation in the US Money Supply is not predicted by GTW Compression. The results suggest that prevailing tendencies to increase the supply of money are significantly amplified by GTW Compression effects. In short, as GTW Compression increases in value, so does the desire of the Federal Reserve to print more money. Part of this is attributed to the decision in the 1960s to decouple the US Dollar from the price of Gold. That move to the age of "Fiat Money" (money not tied to any tangible commodity) seriously contributed to the false sense of liquidity, interest rate meddling, and unconscionable lending practices that produced the crash of 2008. Still the predictive relation with GTW Compression is undeniable.

Prediction of the Velocity of Money

Gravitational Torque Wave Compression is negatively correlated (-.84) with, and predicts 71 percent of the variation in the US Velocity of Money for the period from 1869 to 1970, a sample size of 102 years. The prediction depicted in figure 57 is significant to over 27 decimal places and cannot be attributed to sampling errors.

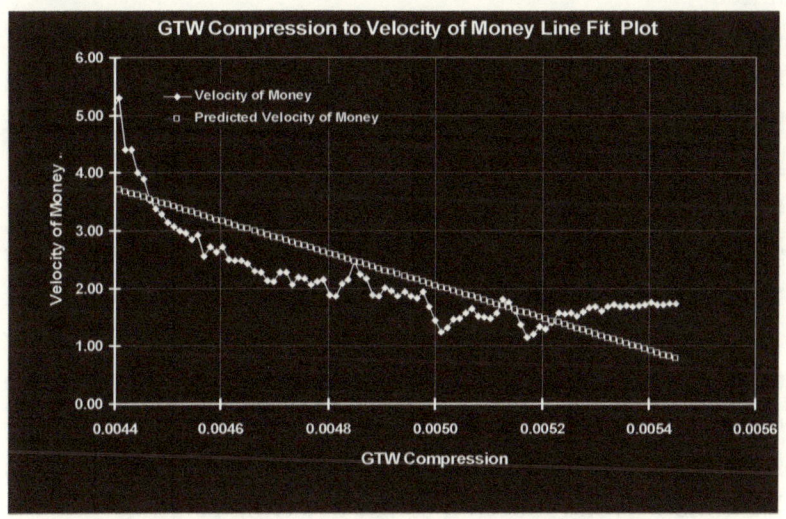

Figure 57 GTW Prediction of the Velocity of Money

Only 29 percent of the variation in the Velocity of Money is not predicted by Gravitational Torque Wave Compression effects. The remarkably high level of apparent GTW Compression effects on the velocity of money suggests that prevailing tendencies in money retention and spending behavior are significantly amplified by GTW Compression effects. In short, as GTW Compression increases in value the Velocity of Money (the rate at which cash is used) decreases in value.

This indicates that cash is increasingly clutched (saved) as GTW Compression increases. Clearly the last two US Federal Reserve Chairman agreed, as they have done everything in their power to leverage money out of savings and back into the active economy. The result of their efforts has been just the reverse. In short they failed.

The crash of 2008 and its carry over to 2009 put US Consumers on a course to unprecedented saving and historically dramatic levels of reduced consumption. It is assumed that the predictive relation between GTW Compression and the Velocity of Money is actually greater than that found. It is known that changes in the Supply of Money and the resultant Velocity of Money lag sourcing events by many months. Hence, the counting relation between GTW compression and these two measures is reduced by the offset in the times of the sourcing cause and the measured economic effects (i.e. a lag-time driven reduction of the counting relations).

Prediction of the US Consumer Price Index

GTW Compression predicts 65 percent of the variation in the US Consumer Price Index (CPI) for the period from 1871 to 2007, a sample of 137 years. The prediction depicted in Figure 58 is significant to over 30 decimal places and cannot be attributed to sampling errors.

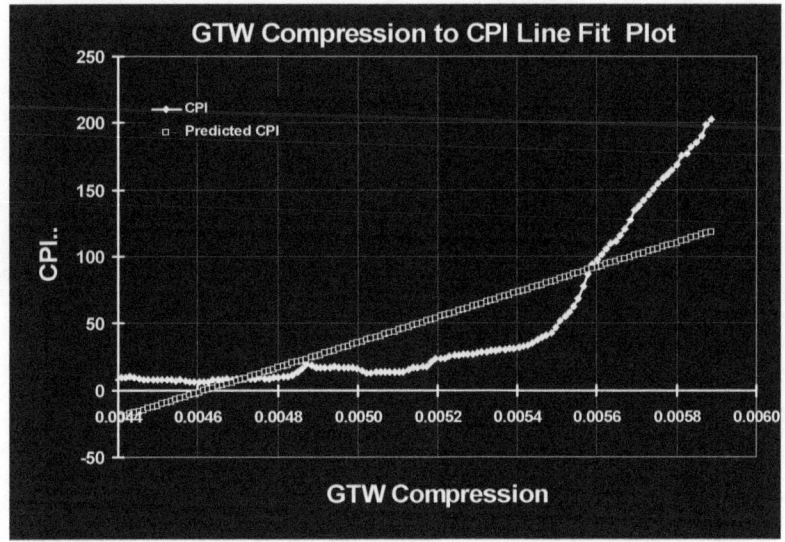

Figure 58 GTW Prediction of the US CPI

A full 35 percent of the variation in US CPI is not predicted by GTW Compression. This unexplained variation is assumed to represent the extent to which local (non GTW) influences determine the average price of consumer goods and services purchased by households. The significant level of apparent GTW Compression effects on such prices suggests that prevailing tendencies in pricing behavior are significantly amplified by GTW Compression. In short, as GTW Compression increases, so do the prices of consumer goods and services that are purchased by US households.

Purging Perplexity

The idea that human innovations in production, consumption, pricing and wages are unpredictable for the last half millennium is plainly false. Since 1615 (the last Maya Long Count resonant synchronization), GTW Compression (and hence Degrees of Maya World Age Precession) predict a significant portion of the variation in human production, consumption, pricing, wages and the movement of money. GTW predicts from 71 to 87 percent of the variation in the most tightly monitored indexes of human market behavior. This trend is clearly visible in the previously presented plots of those measures.

The arrival of a Galactic GTW within our solar system around 1500, and the effects of that progressing wave, significantly amplified human economic transactions. The variation in economic indexes resulting from GTW compression in no way denies the existence or the exercise of free will. Rather, it simply suggests that prevailing tendencies in human behavior are significantly amplified by the effects of GTW Compression.

GTW compression does not determine what specific innovations we make in our economy. Instead, it influences the *frequency and magnitude* of leading economic indicator values that trace the effects of those innovations we adopt. Given the breadth and depth of the GTW present in our solar system, the effects of that wave are expected to continue. We've little recourse but to envision the exponential amplification of human economic behavior continuing for at least the next two millennia.   Thus, for the foreseeable future GTW Compression will continue to be the principal causal force driving the rates of innovation, supply, and demand.

GTW Compression driven Maya World Age Epochs have proven strong predictors of Human behavior. This is as true for the distant past, as it is from the Age of Enlightenment to the crash of 2008. The Maya Long Count, however, contains still more tools. Tools that can give the correlations and predictions in this chapter a serious booster shot.   The next chapter introduces those tools and exposes their significant value adding utility.

10

WAVELETS OF
SYNCHRONIZATION

Maya Sub-cycles

Figure 59 depicts the now familiar plot of hourly wages in English Pounds from 1260 to 1994.

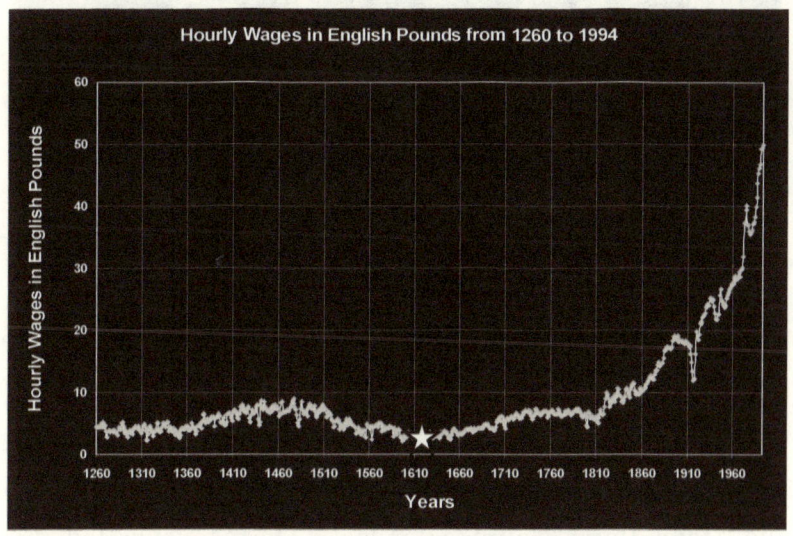

Figure 59 Long Term Trend in Global Hourly Wages Beginning in 1615

A pronounced upward trend in wages begins around 1615 (the white star in Figure 59) and continues through 1994. Gravitational Torque Wave Compression – and hence Maya Degrees of Precession – is .87 correlated with that upward rise in Wages. Figure 60 shows how the GTW rate of Compression predicts 75 percent of the variation in that 376 year long gain in wages.

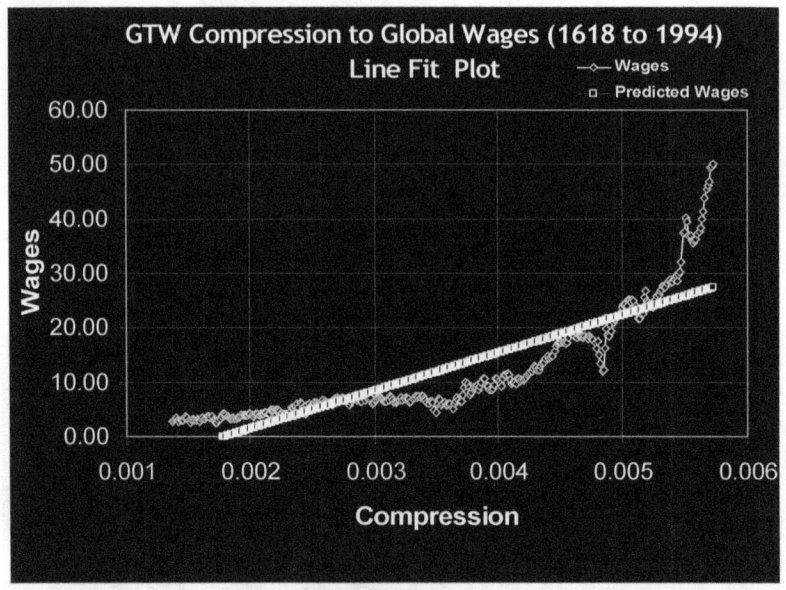

Figure 60 GTW Compression and Degrees of Maya Precession Prediction of Global Hourly Wages from 1618 to 1994.

As stated earlier, the date of Wednesday, October 14, 1615 is one of a special group of dates predicted by the Maya Long Count calendar. The Maya called this unique class of dates "resonant synchronizations."[178] On resonant synchronization dates, four of the five gear-like sub-cycles of the Maya "Long Count Calendar" reset and align to zero.

The most important Long Count dates to the Maya were the great Epoch cycles of Precession that punctuate each World Age. But, the Maya also held sacred the sub-cycle resonant synchronization dates within those Epochs. A Long Count *(Baktun)* resonant synchronization date occurs once every 144,000 days (once every 394.26 of our Gregorian calendar years). The last such Long Count Synchronization occurred on Wednesday, October 14, 1615.

What if the simultaneous start of the rise in hourly wages and the Maya sub-cycle resonant synchronization of 1615 was no chance coincidence? What if the subordinate Maya Long Count sub-cycles describe, explain, and predict a significant amount of the unexplained variance in other measures of human behavior. Most specifically, behaviors like the rate of human innovation and Global hourly wages?

It suddenly looked as if the five nested sub-cycles of the Maya Long Count calendar had been intentionally designed to scale down the subtle effects of Maya Precession. Effects of things that tend to occur over shorter time periods than Maya World Ages, Waves, and Epochs. That particular insight motivated a closer look at the five Sub-cycles of the Maya Long Count Calendar.

Unique Maya Day

The Maya Long Count calendar is not just another Earth, Moon, and Sun referenced day counting system.

Unlike our Gregorian calendar, the Maya's Long Count *day* counting sub-cycles have no celestial reference. The Maya created many calendars for basic time keeping purposes (e.g. the *tzolkin*, the *haab*, and the Venus Round, and others). They didn't need the Long Count for that. Archeologists, anthropologists, and Maya scholars agree. Ancient Maya shaman and priests used the Long Count Calendar almost exclusively for "*divination*"[179] (i.e., explanation and prediction).

The Long Count sub-cycles count days like any other calendar, but these are no ordinary days. What's more, each Maya day unit of each cycle is qualitatively, and most likely quantitatively, weighted.

Synthia and Colin Andrews pointed out in 2008 that the Maya Sub-cycles do something other calendars do not. The Maya sub-cycles are linked to a mysterious definition of the ebb and flow of the "quality" of time. The Maya called this mysterious wavelike rhythm of nature, "*k'ul*."[180] Each numeric value of each sub-cycle day is associated with a glyphic symbol representing its *k'ul*. Scholars have long insisted that Maya shaman and priests *subjectively* interpreted the sub-cycle interactions with these glyphic symbols. Those same scholars assert that through such subjective interpretations the Maya shaman made short and long term predictions of things to come. Just how the shaman went about this interpretive process has remained a secret since at least 100 years before the arrival of the Spanish in the Americas.

It is highly unlikely that *k'ul* is a pictographic method of subjective wild-guessing. The Maya worked outrageously hard for centuries to make the objective relation between Galactic waves of change and Maya Precession payoff. What's more gravity and gravitational torques *scale*. Hence, there is every reason to suspect that the Maya Sub-cycles of the Long Count represent the quantitative sub-cycle-scaling of Solar System Precession and GTW Compression.

As repeatedly demonstrated, Degrees of Maya Precession are nearly perfectly correlated with GTW Compression. The Maya, however, didn't work in degrees, and they didn't know the specifics of GTW Density Compression. Instead, the Maya worked in days, and had learned from experience that Solar System Precession describes the past, explains the present, and predicts the future.

What if the Maya's enigmatic *k'ul* factor is a quantitative – as opposed to subjective – scaling of GTW Compression? As it turned out, that assumption is almost certainly right in terms of the sub-cycle scaling of GTW Compression. The assumption is, however, flat out wrong relative to the most likely meaning of *k'ul*. What was about to be discovered is that the Maya likely derived their quantitative measure of *k'ul* in order to extend and supplant the predictive accuracy of their Long Count Sub-cycles.

The ancient Maya would have formulated an objective and quantitative predictive system completely consistent with their Long Count World Age Epochs, the "Great Cycle," and the "Grand Cycle" of World Age Precession.[181] It is highly likely that the Maya Long Count Sub-cycles count the GTW Compression *cause* in days, while the *K'ul* metric measures *the response* of things to it. The product of the Precession Effect of GTW Compression times the associated K'ul response weighting factor would thus account for the resonant interaction effects that result.

The Maya Long Count Day

This study began with a detailed examination of the basic counting unit of the Maya Long Count Calendar – the Maya day. Unlike contemporary measures of a day, the Maya Long Count day makes little reference to the Earth, Moon, or Sun. To understand the true value of this genius masterstroke, one has but to contrast the Maya *day* with our own.

Our scientists set one day equal to one rotation of the Earth about its axis. They declare each Earth rotation equal to 86,400 seconds, which equates to 1,440 minutes or 24 hours. The Maya, in stark contrast, held tight to the view that everything in the universe varies. They summarily dismissed celestially bound approaches to the measurement of their Long Count day constant. After all, the purpose of the Long Count system was prediction. To make it work they needed an invariant chronographic baseline against which all observable and measurable variations could be described, explained, and predicted.

An ancient Maya shaman would view our modern definition of a day as fundamentally flawed– which it is – and here's why. The rotation rate of the Earth about its axis does *not* consistently equal 86,400 seconds. One modern day is *not* uniformly 24 hours long, and our scientists know it well.

Because of the way the second is defined, the *mean* length of a day [one Earth rotation] is currently [estimated to be] about 86,400.002 seconds, and increases by about 1.7 milliseconds (thousands of a second) per century (an average [taken] over the last 2,700 years).[182]

Shoddier still, every major seismic event like the recent Sumatra and Chilean earthquakes and Tsunami, shift the mass of the planet. That alters the axis and rotation rate of the planet, which changes the length of our day. The problem with our varying day, however, is much worse than that.

The International Bureau of Weights and Measures (BIPM) currently define a second as '... the duration of 9,192,631,770 periods of the radiation corresponding to the transition between two hyperfine levels of the ground state of the caesium 133 atom.'[183]

A problem with that was reported by D. Castelvecchi (2008), J.H. Jenkins, E. Fischbach, J.B. Buncher, J.T. Gruenwald, D.E.Krause, and J.J. Matte (2008), among others. The supposed constant decay rate (half life) of the protons and neutrons (nuclides) comprising the atoms of the Periodic Table of Elements is now in question. "Radioactivity may in fact not proceed at its own *constant* pace."[184]   Bottom-line, Earnest Rutherford's long ago discovered *half lives* of atoms may not be the constants he took them for. Accordingly, a second, as defined by the BIPM, may not be a reliable time reference for measurements of extreme precision and extreme duration. What this means – best case – is that our day is less than a reliable measure of time. Worst case, our measures of a second and a day are hopelessly confounded with the very things they are supposed to consistently index. In short, our measure of a day could be fundamentally invalid. For example:

> "The original length of one day, when the Earth was new about 4.5 billion years ago, was about six hours as determined by computer simulation. It was 21.9 hours 620 million years ago."[185]

The clever Maya calculated outrageously large numbers and event durations. It was essential they make every effort to minimize all such noise in the most fundamental unit of measurement in their Long Count day keeping system. To do that, they simply defined one day as a perfect circle. One cycle around that circle (a day) they called a *"kin."* The only external celestial reference the Maya used to establish their Long Count day was the winter solstice.

> The winter solstice occurs some time between December 20 and December 23 each year in the northern hemisphere, and between June 20 and June 23 in the southern hemisphere, during either the shortest day or the longest night of the year.[186]

The Maya used the winter solstice (the longest day of the year in their neck of the woods) to start and calibrate their fundamental circular gear-like *day* cycle. By this simple method, they achieved a rigorously calculated constant day counting system almost totally independent of celestial sources of variation. One that is extraordinarily reliable, and fundamentally valid.

We today automatically divide the Maya day cycle (a circle) into 360 degrees, or into the normalized unit values of the unit circle ranging from zero to 1. The ancient Maya divided their unit day circle into 20 units. Those units represent the individual Maya numeric symbols ranging from 0 to 20, the values that comprise the Maya's basic numbering system.

Scholars invariably begin any discussion of the Maya Long Count Calendar cycles with the next larger cycle, the 20 day cycle that the Maya called a *Uinal*. Hence, scholars totally avoid having to address any and all sub-divisions of a single day. The Maya, however, were consummate mathematicians and clearly described their day as a divided *cycle* they called a *Kin*.

Maya Long Count Sub-cycles

Including the *Kin* (day) cycle, there are *five* Maya Sub-cycles in the Long Count Calendar system. The duration of each of the other four sub-cycles is measured in days (*Kin* cycles). Figure 61 indicates the way these Long Count Sub-cycle gear calculations intermesh to produce the Maya Long Count. At the center of the figure the *Kin* (day) cycle is represented simply by the number

1. 20 *Kin* (days) form the next larger outer cycle called a "*Uinal*."[187]   360 days form the still larger outer cycle called a "*Tun*."[188] 7,200 days forms the next larger cycle the Maya called a "*Katun*."   144,000 days form the largest outer cycle, which the Maya called a "*Baktun*."[189]

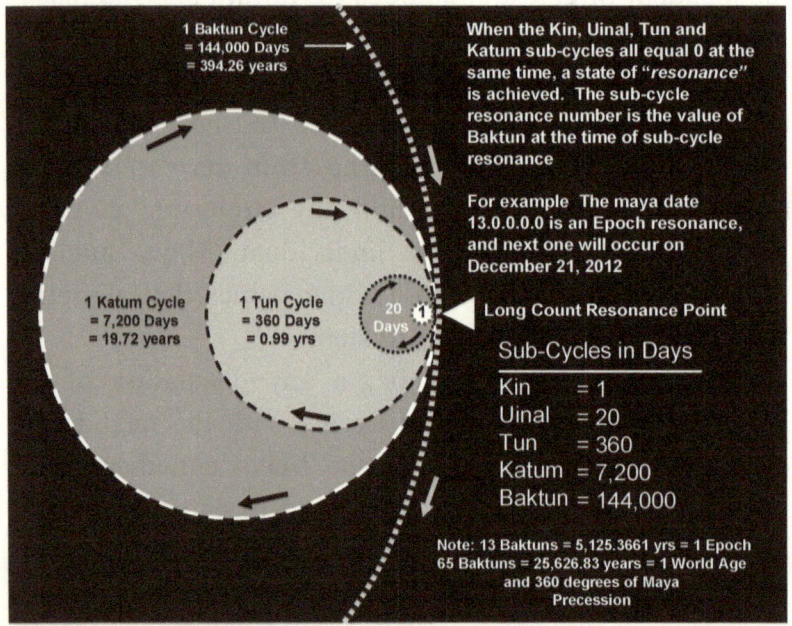

1 Baktun Cycle
= 144,000 Days
= 394.26 years

When the Kin, Uinal, Tun and Katum sub-cycles all equal 0 at the same time, a state of "*resonance*" is achieved. The sub-cycle resonance number is the value of Baktun at the time of sub-cycle resonance

For example The maya date 13.0.0.0.0 is an Epoch resonance, and next one will occur on December 21, 2012

1 Katum Cycle
= 7,200 Days
= 19.72 years

1 Tun Cycle
= 360 Days
= 0.99 yrs

20 Days

Long Count Resonance Point

Sub-Cycles in Days

Kin	= 1
Uinal	= 20
Tun	= 360
Katum	= 7,200
Baktun	= 144,000

Note: 13 Baktuns = 5,125.3661 yrs = 1 Epoch
65 Baktuns = 25,626.83 years = 1 World Age and 360 degrees of Maya Precession

Figure 61 Maya Long Count Calendar Sub-cycles

The Maya Epoch and World Age cycles are computed Baktun cycle *results*, as opposed to being directly clocked sub-cycles. A total of 13 Baktun cycles equals one full Maya Epoch of 1,872,000 Maya days. 26 Baktun cycles equates to a single GTW with a wavelength of 3,744,000 days. 65 Baktun cycles equates to a single Maya World Age of 9,360,000 days.

The start and end date of a complete Baktun cycle is called a *resonant synchronization* point. The completion of 13 Baktuns equals 1 Epoch, which is a *major resonant synchronization point*.

Table 14 lists the individual names that the Maya gave the five nested gear-like sub-cycles. Beside each name is the number of days (*Kin*) that count out each sub-cycle. Also given in the table is the number of days in each sub-cycle divided by 20, which is the major base of the Maya numbering system. All of these sub-cycle durations are multiples of twenty, save one. The Tun cycle is equal to 18 times 20, which equals 360 days, as opposed to 20 times 20 which equals 400. This clever exception insures that when 1 Baktun cycle is complete, exactly 20 Katun cycles, 400 Tun cycles, 7,200 Uinal cycles, and 144,000 Kin cycles have been clocked.

Table 14 Breakout of Each Long Count Sub-cycle

Maya Sub-Cycle *(Gear)* Name	Duration in Maya Days (Gear Teeth)	Days Divided by Base 20 # System
Kin	1	.05
Uinal	20	1
Tun	360	18
Katun	7,200	360
Baktun	144,000	7,200

Whenever the Baktun cycle reaches a new whole integer value, the other four sub-cycles (the kin, Uinal, Tun, and Katun) align to exactly zero. Just as each *Epoch* resonant alignment was sacred to the Maya, so too was each end of a Baktun cycle.[190] By convention, the Maya date signifying the end of the current Epoch, what the Maya called a "Great Cycle," is written "13.0.0.0.0."[191] That number represents: "13 Baktun, 0 Katum, 0 Tun, 0 Unial, and 0 kin."[192] Five Epochs equals one complete Maya Precession Cycle, and 25,626.8 of our Gregorian calendar years. The Maya shaman also called their "World Age" a "Grand Cycle."[193]

Sub-cycles in Perspective

Not surprisingly, the cumulative counts of each of the five sub-cycles for one complete *Baktun* cycle correlate perfectly with one another (i.e., they produce pair-wise product moment correlation coefficients of 1.00). Further, the cumulative sub-cycle counts for one complete Baktun cycle correlates near perfectly with Degrees of World Age Precession, and the magnitudes of GTW Compression. Hence, for purposes of description, explanation and prediction (i.e., correlation, analysis of variance, regression, etc) within a completed *Baktun* cycle, any one of the sub-cycle measures can be substituted for any other.

Once properly synchronized, the basic utility of the sub-cycles is that they partition progressively smaller amounts of time. Within their respective time frames, each sub-cycle retains the descriptive, explanatory, and predictive validity, reliability, practicality and economy of the other cycles. The Maya sub-cycles thus form a nearly perfect clock and calendar system, indexing Maya

Precession, and hence GTW Compression. Thus, each of the Maya Sub-cycles is a robust universal measure of the *cause* of change.

Indexing the *K'ul* Response and Resonance

The last chapter demonstrated that GTW (and hence Degrees of Maya Precession) predict a significant amount of the variation in most leading modern economic indicators. Impressive to be sure, but there is serious doubt that those results would knock the proverbial socks off an ancient Maya shaman. The shaman would want more than just a good predictive measure of the *cause* of change. He would also want a measure that predicts the *response effect* elicited by the source of change. After all, a Maya shaman was in the business of predicting the future, as well as describing and explaining the past.

The ancient Maya and their predecessors painstakingly observed, measured, and recorded the heavens and their surrounding natural environment for hundreds, if not thousands, of years. Then, they diligently mined and analyzed those observations for hundreds more. Their purpose was to create the most precise and accurate predictive Long Count calendar system in the ancient world. These were not the sort of people to abandon their rigorous quantitative methods and suddenly revert to making subjective guesses. By that line of reasoning, it was theorized that the Maya must have measured the *response* of things to the cyclic force indexed by Precession. Therefore, what the Maya called *k'ul* would have been an observation driven quantitative objective measure. We've no way to know how the Maya actually derived their measure of *K'ul*. From what they left us, and our newly acquired knowledge of GTW Compression, here is how it was done for this study.

The process started with an examination of the Poisson and Normal distributions of a single Gravitational Torque Wave. Figure 62 plots those distributions over 26 Baktun cycles (one GTW).

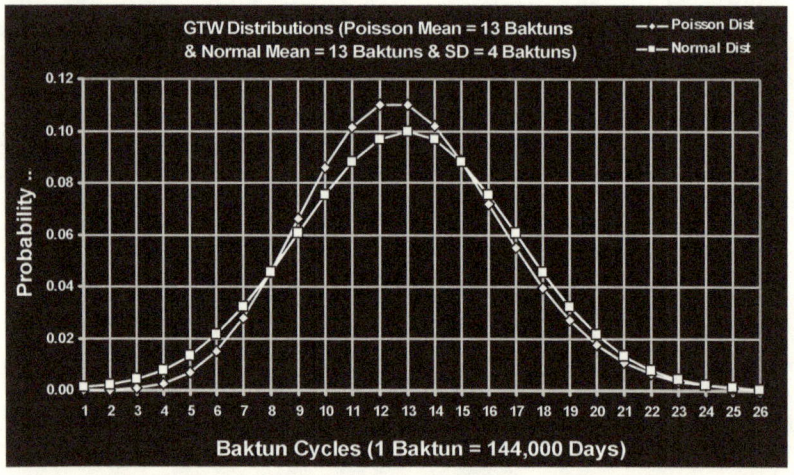

Figure 62 Poisson and Normal Mass Distribution of a Gravitational Torque Wave

That GTW distribution has a mean of 13 *Baktuns*, a standard deviation of 4 *Baktuns*, and a total wavelength of 26 *Baktuns*. The ratio of the mean to the standard deviation is 13 divided by 4, which equals 3.25. Resetting the full wavelength of the GTW to a value of 1.0 (i.e., normalizing the time scale), the distribution mean became 0.5 and the standard deviation 0.30 (i.e. 1.00 divided by 3.25 equals .307). It was decided to round down to 0.30, as opposed to rounding up to 0.31. This was done because the Maya reputedly hated fractional divisions.

Next, the product moment correlation between GTW Compression and the smoothed and extended rates of Human innovation were calculated. A smoothed Human Innovation *rate* (meaning the 5th order polynomial trend in the innovation rate) extending from Wednesday, October 14, 1615 – the last Maya resonance date – to Friday December 21, 2012 (the next resonance date) was derived. Figure 63 depicts the resulting smoothed trend line for the predicted Human Innovation Rate. The 5th order polynomial equation is included in the header.

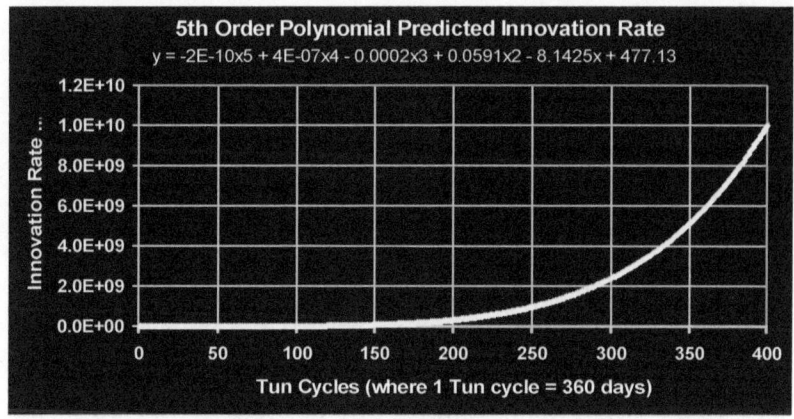

Figure 63 Fitting the Innovation Rate

As a result of the smoothing and extrapolation to 2012, the correlation between GTW Compression and the smoothed rate of human innovation dropped. The correlation went from a perfect relation of 1.00 to 0.82. That is 18 percent less than the correlation reported in

Chapter 2 between the current-period observed innovation *dates* (in raw Years Ago) and Degrees of Maya Precession. As a result, Long Count Sub-Cycle Days predict only 67 percent of the variation in the *smoothed* innovation rate trend for the period from 1615 to 2012. The smoothed expected innovation rate trend thus became an excellent test criterion for this study of *K'ul*. To satisfy a Maya shaman, any truly viable measure of *K'ul* would have to substantially improve the percent of variation explained in the smoothed trend of Human Innovation.

The test database defined one normalized *Baktun* cycle, valued from 0.0 to 1.0 *Baktun* (the period from Wednesday October 14, 1615 to Friday December 21, 2012). Each of the 400 records in the database equaled one Maya *Tun* cycle of 360 days. The *K'ul* Index was generated using a cumulative normal distribution generator. For each *Baktun* cycle value from 0 to 1 the distribution generator produced a corresponding cumulative normal probability value. Each probability value generated was produced using a distribution mean of 1 and a standard deviation of .30. These cumulative normal probability values became the estimate of the *K'ul* Response, (the *effect*) imposed by GTW Compression (the *cause*).

Figure 64 depicts the plot of the *K'ul* Response Effect Index relative to GTW Compression. The nearly flat curve plotted along the bottom axis of the figure represents the excruciatingly slow clime of GTW Compression.

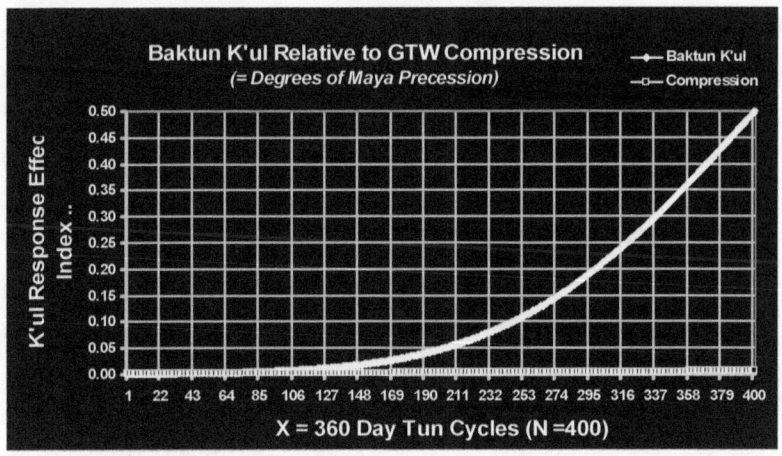

Figure 64 K'ul Response Effect Index

Testing the *K'ul* Resonance

The first *K'ul* test hypotheses is **Hypothesis H$_{17}$,** **which** asserts that: *Response Effect K'ul* is significantly correlated with the smoothed Rate of Human Innovation for the current Maya Baktun Cycle. **Hypothesis Null$_{17}$** counters that there is no significant relationship between Response Effect *K'ul* and the smoothed rate of human innovation for the current Maya *Baktun* Cycle.

Hypothesis H$_{18}$ states that: *Response Effect K'ul* describes, explains, and predicts, more of the variation in the expected Rate of Human Innovation than GTW Compression as measured in the Maya *Baktun* Cycle units. **Hypothesis Null$_{18}$** simply counters that the Maya *Baktun* Cycle unit measure of GTW Compression is a better predictor of the rates of expected Human Innovation than Response Effect *K'ul*.

The product moment correlation of the *K'ul* Response Effect Index with the expected smoothed Rate trend in Human Innovation was calculated. *Response Effect K'ul* correlated 0.98 with the smoothed Rate of Human Innovation. This correlation is significant to 290 decimal places. The correlation represents a 16% improvement over the .82 correlation between the Rate of Human Innovation and GTW Compression measured in Maya *Baktun* Cycle units. Hypothesis H_{17} is thus retained and Hypothesis $Null_{17}$ rejected.

A regression analysis was run to determine the portion of variation in the rate of Human Innovation described, explained, and predicted by the *K'ul* Response Effect Index. The prediction was significant to over 290 decimal places. *Response Effect K'ul* describes, explains, and predicts 96 percent of the variation in the expected Rate of Human Innovation (see Figure 65). Hypothesis H_{18} is retained and $Null_{18}$ rejected. Response Effect *K'ul* explains 29 percent more of the variation in the smoothed Rate trend in Human Innovation than GTW Compression measured in Baktun sub-cycle units. Quite simply, quantitative *K'ul* Effect beat the predictive pants off of sub-cycle day indexed GTW Compression.

The test demonstrates that the Maya could have conceivably – and most likely did – develop a quantitative measure of the *K'ul* Response Effect. It seemed likely that the meticulous Maya shaman would have tried to push their predictive toolkit even further. After all, they had demonstrated a consistently obsessive-compulsive preoccupation with the concept of *resonant synchronization*. What's more, the ancient Maya numbering system works in powers of 5, 10, 15 and 20. Clearly, the Maya were no strangers to multiplication.

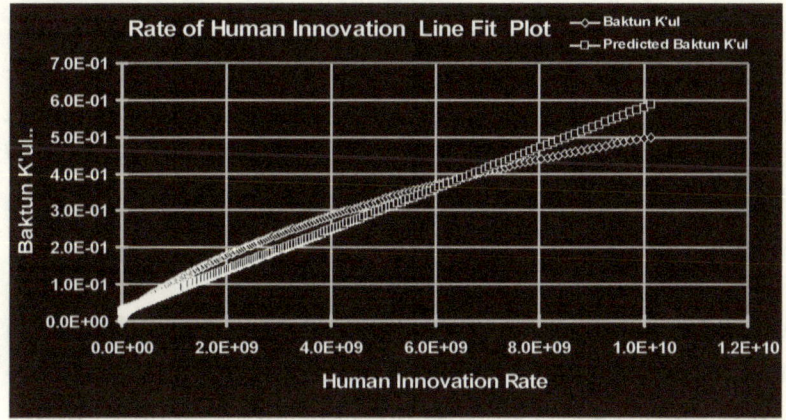

Figure 65 K'ul Response Prediction of Human Innovation

If Sub-cycle indexed GTW Compression and *K'ul* Response Effect each predict a significant portion of the variation in the expected Rate of Human Innovation, wouldn't the two multiplied together predict even more? True, the two measures do overlap and explain much of the same variation.   But, a single composite *K'ul Resonance* predictor formed by combining those two measures should form an even better universal predictor. That less than subtle point could not have escaped generations of math-mad Maya.

A new K'ul Resonance predictor variable was added to the test database. That was accomplished by simply row-wise multiplying Sub-cycle indexed GTW Compression times the *K'ul* Response Effect values.

Then, Hypotheses H_{19} and H_{20} were formulated to test the predictive accuracy of the new *K'ul* Resonance measure.

Hypothesis H_{19} states that: *K'ul* x Sub-cycle Compression (*K'ul* Resonance) is significantly correlated with the expected Rates of Human Innovation for the current Maya *Baktun* Cycle. **Hypothesis Null$_{19}$** counters that: there is no significant relationship between *K'ul* Resonance and the expected rates of human innovation during the current Maya *Baktun* Cycle.

Hypothesis H_{20} states that: *K'ul* Resonance describes, explains, and predicts, more of the variation in the expected Rates of Human Innovation than either sub-cycle indexed GTW Compression or Response Effect *K'ul* for the current Maya *Baktun* Cycle. **Hypothesis Null$_{20}$** argues that Response Effect *K'ul* is a better predictor of the expected rates of Human Innovation than *K'ul* Resonance.

Table 15 relates the Hypothesis H_{19} product moment correlation results. Those results mandate the rejection of Null$_{19}$ and retention of Hypothesis H_{19}. The *K'ul* Resonance measure is near perfectly correlated with the expected Rates of Human Innovation.

The true gain is the improvement in the amount of variation explained (and hence predicted) in the expected Rates of Human Innovation by *K'ul* Resonance. Figure 66 depicts the near perfect fit of *K'ul* Resonance to the smoothed Rate of Human Innovation.

K'ul Resonance predicted 98.89 percent of the variation in the smoothed Human Innovation trend for the current *Baktun* cycle. That predictive relation is significant to a number of decimal places approaching infinity.

Table 15 Product Moment Correlations

Correlations			
	Expected Innovation Rates	*Baktun K'ul* Resp. Effect	*K'ul Resonance*
Expected Innovation Rates	1.000		
Baktun K'ul Resp. Effect	0.9818	1.0000	
K'ul Resonance	0.9945	0.9962	1.0000

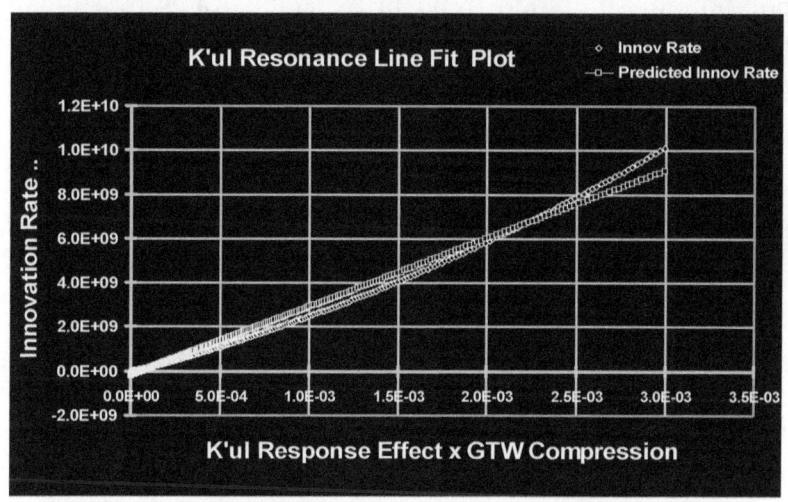

Figure 66 K'ul Resonance Prediction of the Rate of Human Innovation

K'ul Resonance improved the portion of variation explained in the expected Rate of Human Innovation by 32 percentage points, over that explained by Sub-cycle indexed GTW Compression. The measure also improved

237

the variation explained by 2.5 percent over that predicted by *K'ul* Response Effect. Hence, Hypothesis H_{20} is retained, and Hypothesis $Null_{20}$ rejected. The proportion of variation in the smoothed trend *not* explained by the composite *K'ul* Resonance measure is a meager 1.11 percent. Now that is a level of prediction that would satisfy an ancient Maya shaman.

Validating the *K'ul* Resonance

It was worried that the prior test might not have been representative. Specifically, that the 5th order polynomial smoothed trend in expected Rates of Human Innovation might not be fully representative. So to verify the test results, the Global Annual Wages from 1615 to 1994 were added to the test mix.

The resulting product moment correlation analysis spanned the 379 Gregorian calendar years from 1615 to 1994 and revealed a very interesting hierarchy of relationships. Those relationships are listed in Table 16.

The squared correlation coefficients reveal that *K'ul* Resonance predicts 98.9 percent of the variation in the Rate of Human Innovation and 96.24 percent of the variation in Global Annual Wages. Equally surprising the smoothed expected Rate of Human Innovation predicts 96.04 percent of the variation in Global Annual Wages. These results could only have happened if the smoothed Rates of Human Innovation were highly representative, the *K'ul* Resonance is a strong predictor, and the relations no mere localized success.

Table 16 Correlation Test with Global Annual Wages

Correlations	Innovation Rate	Global Annual Wages	K'ul Response Index	K'ul Res
Innovation Rate	1.0000			
Global Annual Wages	0.9801	1.0000		
K'ul Response Index	0.9818	0.9769	1.0000	
K'ul Resonance	0.9945	0.9810	0.9965	1.0000

The predictive relation between the expected Rates of Human Innovation and Global Annual Wages solidly reinforces one of the basic tenants of modern economics. Specifically, that *innovation* is the principal driver of economic supply, demand, and compensation. The addition to that theory resulting from this study is that *K'ul* Resonance captures the underlying source of the variation in future rates of human innovation.

Compression and Resonance Distributions

Distribution generators were used to produce Figure 67. The figure projects one normally distributed GTW 10,250.7 years, and depicts the Poisson distributed *K'ul* Response Effect on the left side of the figure. The Ku'l Response Effect distribution was calculated using a mean of 5 *Baktun* cycles. The GTW Compression normal distribution was calculated using a mean of 13 *Baktuns and* a standard deviation of 4 Baktuns, within a wavelength of 26 *Baktuns*.

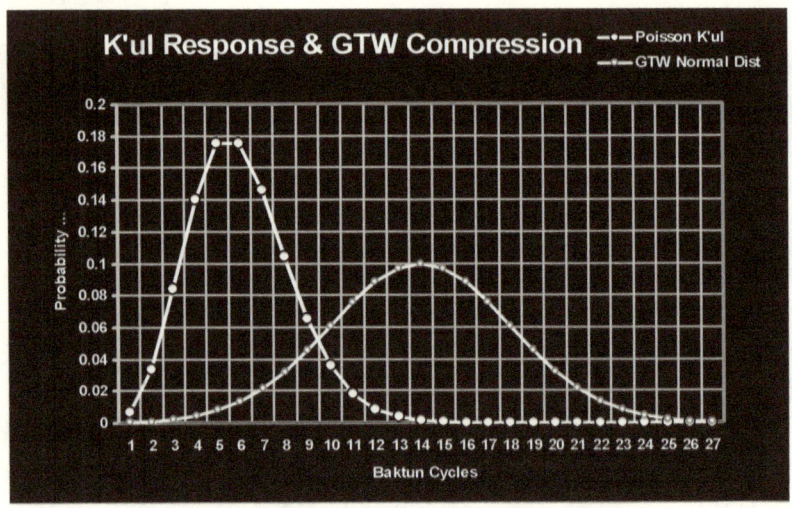

Figure 67 Pyramids of GTW Compression and Response

Upon generating Figure 67, the eerie similarity of the plotted distributions to the shapes of pre-Maya and Maya era pyramids became apparent. The distributions echo the pyramids built by the Maya Olmec predecessors (i.e. the broad bell-shape of the early Mesoamerican pyramids) and the later vertically spiking shape of the Maya pyramids. Photo 1 portrays these pyramid images and exemplifies these striking similarities. Matches suspected to be more than mere chance coincidences.

K'ul Improved Predictions

Might the *K'ul* Response Effect and *K'ul* Resonance measures improve the prediction of the economic indexes addressed in the prior chapter?

Hypothesis H_{21} and H_{22} were formulated and tested to find out.

Photo 1 (left) Tikal's Pyramid of the Great Jaguar; (right) Olmec, Pyramid of the Sun

Hypothesis H_{21} asserts that: *Response Effect K'ul* is a better predictor of the variation in the economic indicators listed in Table 17, than Sub-cycle indexed GTW Compression alone. **Hypothesis Null$_{21}$** counters that Sub-cycle indexed GTW Compression is the better predictor of the economic indicators in Table 17.

Hypothesis H_{22} states that: *K'ul Resonance* is a better predictor of the variation in the economic variables in Table 17, than *Response Effect K'ul*. **Hypothesis Null$_{21}$:** argues that *Response Effect K'ul* is the superior predictor of the economic variables in Table 17.

Each economic indicator constitutes a unique sub-hypothesis under the stems of Hypotheses H_{21} and H_{22}. The Velocity of Money variable addressed in the last chapter was excluded from these tests. The reason for that exclusion is that economists have shown that

241

Velocity of Money lags changes in the Supply of Money. Thus, Money Supply (SH$_6$) provided a more reliable test of the hypotheses.

Table 17 Sub-Hypotheses

Num	Varialble	Cases	Sampled Years From	To
SH $_1$	Global Annual Wages	380	1615	1994
SH $_2$	Per Capita Consumption (in Yr 2000 dollars)	116	1889	2004
SH $_3$	Real GNP (Gross National Product US)	85	1890	1974
SH $_4$	Real Consumption (in Yr 2000 Dollars)	116	1889	2004
SH $_5$	US Gross Domestic Product (GDP)	79	1929	2007
SH $_6$	Money Supply	85	1890	1974
SH $_7$	Consumer Price Index (CPI)	137	1871	2007
SH $_8$	Dow Jones Industrial Avg. Adj. Closing Price	80	1928	2007
SH $_9$	Common Stock Price	99	1871	1970
SH $_{10}$	Real Stock Price Index	137	1871	2007
SH $_{11}$	Stock Price Index	124	1871	1994
SH $_{12}$	US National Debt	219	1790	2008

Table 18 provides the results of the regression analyses performed for each sub-hypothesis of Hypothesis H$_{21}$ and H$_{22}$.

For each of the economic indexes, the percentage of variation described, explained and predicted by *K'ul* Response Effect was substantially greater than that of Sub-cycle day indexed GTW Compression. Hypothesis

H_{21} is therefore retained at the .001 level of confidence for SH_1 through SH_{13}. Hypothesis $Null_{21}$ is rejected.

Table 18 Regression Results H_{21} and H_{22}

Num	Economic Index	Percent Variation Explained		
		GTW	*K'ul* Response	Resonant *K'ul*
SH 1	Global Annual Wages (In English Pounds)	74.65%	86.59%	96.24%
SH 2	Per Capita Consumption (in Yr 2000 dollars)	87.45%	89.79%	93.45%
SH 3	Real GNP (Gross National Product US)	84.61%	86.49%	90.07%
SH 4	Real Consumption (in Yr 2000 Dollars)	80.20%	83.02%	87.65%
SH 5	US Gross Domestic Product (GDP)	76.06%	77.96%	80.70%
SH 6	Money Supply	71.37%	73.68%	78.29%
SH 7	Consumer Price Index (CPI)	64.51%	68.41%	75.39%
SH 8	Dow Jones Industrial Avg. Adj. Closing Price	57.49%	60.76%	63.53%
SH 9	Common Stock Price	54.08%	56.81%	62.86%
SH 10	Real Stock Price Index	53.01%	55.95%	61.10%
SH 11	Stock Price Index	40.04%	43.72%	49.34%
SH 12	US National Debt	32.21%	37.53%	49.40%

The percentage of variation described, explained and predicted by *K'ul* Resonance was substantially greater than that explained by either GTW Compression or K'ul Response Effect. The results, which are significant at the .001 level, hold for all the economic indexes (SH_1 through SH_{12}). Hypothesis H_{22} is retained for SH_1 through SH_{12} and $Null_{22}$ is summarily rejected.

On average, the *K'ul* Resonance predictor improved the percentage of variance explained in economic measures (SH_1 through SH_{13}) by 9.36%. On average it also explained 4.87% more of the variance than the *K'ul* Response Effect predictor. *K'ul* Resonance is not a perfect universal predictor. It is suspected, however, that it would have pleased an ancient Maya Shaman. What's more, *K'ul* Resonance predicted the majority of the variation in all but the SH_{11} Stock Price and SH_{12} US National Debt indexes. Not too shabby for a single measure of economic performance, even by today's standards. This is especially the case when you consider the fact that *K'ul* Resonance predicts all but 1.11 percent of the variation in the smoothed trend in Human Innovation over the last 512 years.

Scary Smart Maya

By the birth of Jesus the Maya were an empire. By the time of the Roman occupation of the British Isles, Maya shaman were describing, explaining, and predicting World Age sub-cycle driving Precession Epochs, and Earth's Obliquity, orbital Inclination, and eccentricity. Using their Long Count Calendar they managed predictions of their climate and human behavior to rival anything we possess today. That is, until now.

The secrets of the Maya's success are written in the shape of their pyramids. The broad bell-shape of the pyramids constructed by the Olmec and earlier civilizations, accurately represent the distribution of the great Galactic force we now recognize as a Gravitational Torque Wave. As demonstrated, GTW Compression is a strong predictor of the behavior of the visible Galaxy, our solar system, this planet, and human behavior. It is believed that Olmec shaman first became aware of the World Age Precession cycle around 1200 BC.[194]

By 200 BC the Maya had built their World Age Long Count Calendar system. By then they were precisely mapping the past, describing the present, and predicting the future.[195] What ever measure of *K'ul* the Maya used, they were associating it with the sub-cycles of the Long Count by then. When and if the Maya multiplied that *K'ul* response effect times the values of the sub-cycles remains unknown. If they did, they would have found it a near universally valid, reliable, practical, and economic predictor.

The ancient Maya were consummate scientists and mathematicians. As repeatedly demonstrated, they possessed the basic means to describe the past, explain the present, and predict the future with unprecedented accuracy.

All these cited accomplishments make the Maya prediction of the "end of space and time" on the Winter Solstice of Friday December 21, 2012 seriously disturbing. What could their Long Count divination system have possibly revealed to prompt them to make such a bizarre assertion? The answer is as scientific as it is misconstrued.

Before that genuinely disconcerting business is taken up, though, a really challenging question requires an answer. That question is: 'How the heck does GTW Compression get into human beings to evoke the *K'ul* response?' The next three chapters provide the surprisingly obvious yet previously overlooked answer.

11

WAVES OF HUMMING EARTH

Earth's Gravitational field is machine measured 24 hours a day 7 days a week around the globe. Earth's more subtle gravitational field variations are often eclipsed by headline grabbing geological and meteorological events, but they're perpetually present. Until two decades ago the Earth's weaker seismically transported signals were ignored as irrelevant noise. Similarly, Human responses to those short slow waves of gravitational compression are frequently measured, but go totally unrecognized for what they truly are.

This chapter and the next bring together recent breakthroughs in geology and neurophysiology. The purpose of this meta-conjoining is to reveal how Galactic GTW compression evokes a relentless stream of responses from this planet and the people that inhabit it. A third chapter tests those effects to expose the driving source of the fundamental human sense of self.

Measuring Earth's Gravity

On Earth, gravity is measured using a gravimeter. "A gravimeter is an instrument ... [used] to measure [oscillations in] the local gravitational field of the Earth. A gravimeter is a type of accelerometer specialized for measuring the constant downward acceleration of gravity."[196]

"A high-grade calibrated spring gravimeter – such as the portable LaCoste-Romberg gravimeter – can measure the Earth's gravitational field to 0.1 nanometer/s²"[197] (an amplitude sensitivity of one tenth of one billionth of a meter per second, per second of acceleration). "The most accurate relative gravimeters are superconducting gravimeters ... The superconducting gravimeter [or SG] achieves the extraordinary sensitivities of one thousandth of one billionth of the Earth surface gravity."[198]

The downward acceleration exerted by Earth's gravitational field fluctuates in amplitude, wavelength, frequency, speed, and direction. While the gravitational field varies geographically with every tick of the clock, it also shifts seasonally. Such variations encode a plethora of things like: the spin of the Earth's core, seismic events (earthquakes, volcanic eruptions, tsunami ...), variations in Earth's hydrology, ocean tides, and the tugs of the moon, planets, sun, and other nearby stars. Embedded deepest within those signals is the excruciatingly slow compression of space-time density imposed by the Galactic Gravitational Torque Wave now creeping through our solar neighborhood.

In a 2006 report to the Finnish Geodetic Institute, Heikki Virtanen described the best known sources of variation in the dynamics of Earth's gravitational field.

Sources tracked by the Superconducting Gravimeter (GWR T020) at Metsähovi in Finland.

The temporal variation in [Earth's] gravity consists of numerous phenomena with different periods and amplitudes [which are grouped into what are called *wave bands, wave modes,* or *wave spectra*].

[The] <u>Earth tide</u> [mode] has the strongest effect, measured at Metsähovi station, at about 2250×10^{-9} ms^{-2} peak to peak. *[where $10^{-9}m$ is one billionth of a meter, and ms^{-2} symbolizes square meters per second per second of acceleration, in accordance with Isaac Newton's famous formula equating force, mass, and acceleration to gravitation.]*

The next largest [mode] is the variable gravity effect of the <u>atmosphere</u>. The range of atmospheric pressure at Metsähovi station ... corresponds to 300×10^{-9}ms^{-2} in gravity. The <u>pole tide</u> at the Metsähovi station is 80×10^{-9}ms^{-2} (peak to peak).

<u>Hydrological phenomena</u> such as variation in soil moisture content, groundwater level, and snow cover have an influence up to 80×10^{-9}ms^{-2}. The loading effect of the Baltic Sea varies about 40×10^{-9}ms^{-2} (peak to peak). ...

The ground acceleration due to strong <u>earthquakes</u> exceeds all magnitudes above. The effect of a strong microseism [seismic tremor effect] can be about 1000×10^{-9}ms^{-2}.

The weakest observable periodical phenomena, such as <u>free oscillations of the Earth</u> [called Earth's Hum] are 0.01×10^{-9}ms^{-2} in amplitude.[199]

The T020 SG at Metsähovi, Finland is part of the Global Geodynamics Project network. Figure 68 depicts the locations of the Global Geodynamics stations that existed when Virtanen made his report to the Finish Geodetic Institute. By 2009 the number of global superconducting Earth gravity monitoring stations in the network had increased to 24.[200] In addition, the CHAMP (CHAllenging Mini-satellite Payload) satellite placed in orbit in 2000, and the two satellites of the GRACE (Gravity Recovery and Climate Experiment) mission launched in 2002, contribute data to the Global Geodynamics Project network.[201]

The vast majority of things that happen on Earth are encoded in the variations in Earth's gravitational field. This concept was anticipated very early on in the history of western science. The notion was first publicly expressed by the Reverend Neil Maskelyne, 18th Century Astronomer Royal of Great Britain, in 1772.

> If the attraction of gravity be exerted, as Sir Isaac Newton supposes, not only between the large bodies of the universe, but between the minutest particles of which these bodies are composed ... it will necessarily follow, that every hill must, by its attraction, alter the direction of gravitation in heavy bodies in its neighborhood ... [202]

And what a "neighborhood" it turned out to be. For distant Earthquakes in Peru and Tsunamis in Sumatra paint large on the superconducting gravimeter (SG) in Finland. The same can be said for every other gravimeter sitting on this world or orbiting Earth for weeks following a large geologic event. From the decoded gravitational waveforms, geologists tease out what they call gravitational modes. These gravitational wave modes reveal tectonic, volcanic, oceanic, atmospheric, cryospheric (Earth surface), and even human activity event signatures all over the planet.

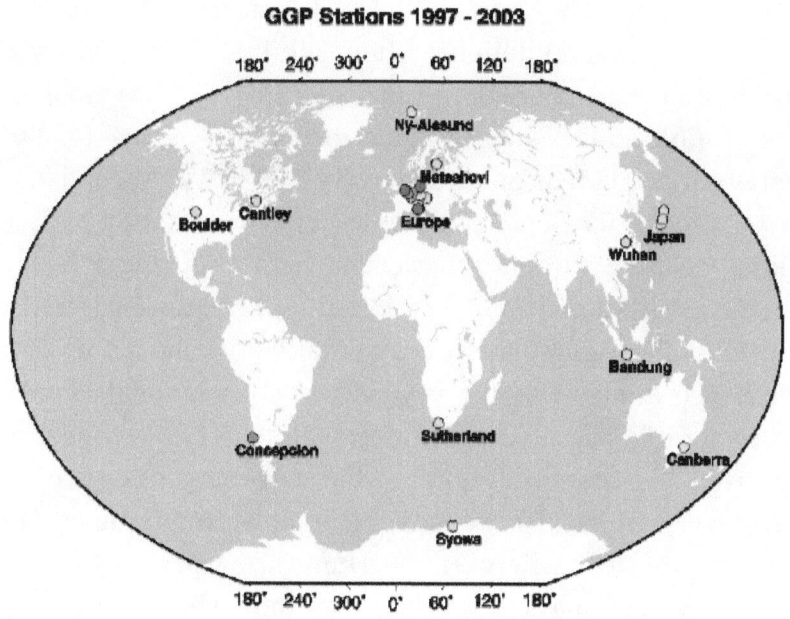

Figure 68 Global Network of SG stations on June 2005
New stations in Taiwan and South Korea not shown
Source: Virtanen, Heikki, 2006, Studies of Earth Dynamics with the Superconducting Gravimeter, Academic dissertation in Geophysics, University of Helsinki, Finland, Finish Geodetic Institute ISBN-13: 978-952-10-3057-4 (PDF) and ISBN-10: 952-10-3057-7 (PDF) p.12

Of greatest importance here, today's geologists monitor the extremely subtle peaks and valleys in the Earth's continuous free oscillation waveforms. What geologists call "the Earth's Hum."[203] This Hum is the subtlest and slowest waves written in Earth's gravitational field. All of the SGs regularly report this seasonally varying gravitational "Hum of the Earth," which was first described in 1993 by J. Peterson.[204] In the lowest mode, these are the "free oscillations of the Earth at $0.01 \times 10^{-9} ms^{-2}$ in amplitude."[205] That persistent gravitational Hum exhibits as "modal peaks within the frequency range from 2 to 7 mHz",[206] (symbolizing 2 to 7 millihertz, which equates to 0.002 cycles to 0.007 cycles per second, or one wave every 500 to 143 seconds).

Figure 69 depicts the two persistent modes of Earth "Hum" relative to atmospheric and micro seismic effects in Earth's gravitational field, as plotted by Peterson in 1993. As shown in the figure, there are actually two modes of Hum. These two modes are: the very slow "Gravitational Hum" in the range of one wave cycle every 500 to 143 seconds, and the faster "Microseism Hum" in the wavelength range of one wave every 143 to 10 seconds. The slow Gravitational Hum and the faster Microseism Hum are persistent, vary seasonally, are masked by large earthquakes, and are best sensed by Superconducting Gravimeters.

The "Microseism Hum" is also mapped by vast networks of seismographic stations that track the waves between stations. Theoretic models have traditionally attributed the Gravitational and Microseism Hum modes to the effects of ocean tides and atmospheric pressure. In other words, Earth bound sources. Recent discoveries have cast serious doubt on the completeness of such assertions, as will be explained shortly.

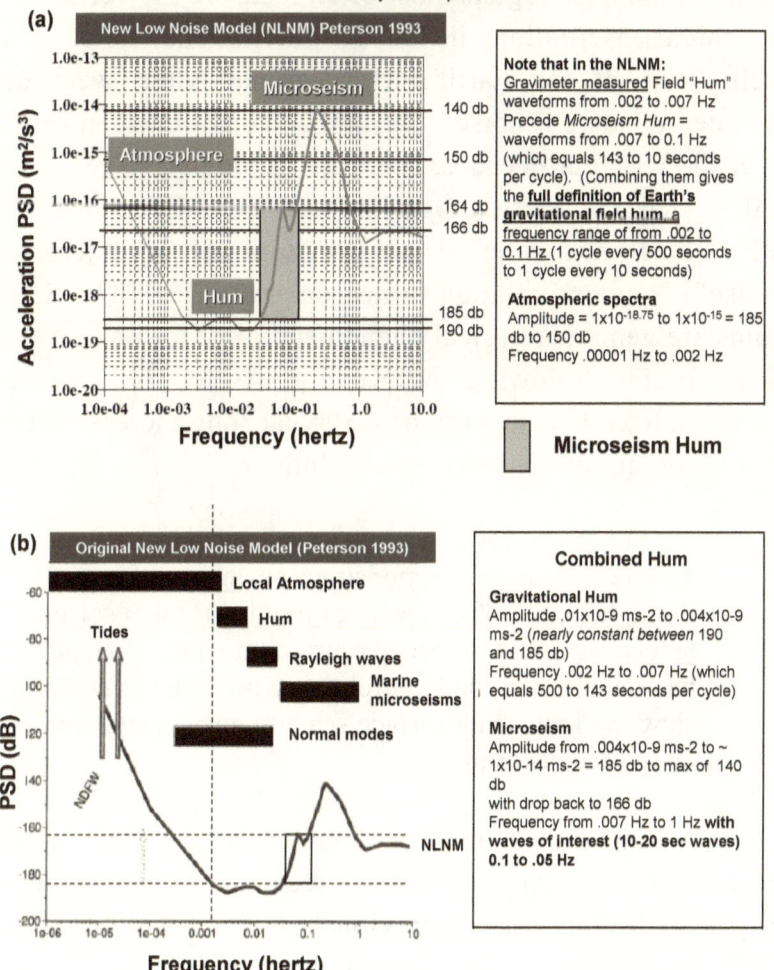

Figure 69 Earth's Gravitational Hum Relative to Atmospheric and Micro Seismic Events

Source: T. Tanimoto1 and Juliette Artru-Lambin2, 2006, Interaction of Solid Earth, Atmosphere & Ionosphere, Department of Earth Science, University of California, Santa Barbara, California, 93106, October 30, 2006, p.53

253

Using data collected from gravimeters and seismographs, geologists most often measure the vertical (the spheroidal) oscillations in Earth's gravitational field. These oscillations of the Earth's Gravitational Hum were once considered mere "noise," meaning "instrument noise."[207] Over the last four decades, however, geologists have come to realize that the Earth's Gravitational and Microseism Hum (still sometimes referred to gravitational or microseism "noise") "have no relation to instrument noise." The Earth's Hums are genuine physical phenomena.

In the following, McNamara, Aster, Hutt, and Gee report their 2007 discovery of a 30 year long increasing trend in the strength of the Microseism Hum.

> For the period 1972-2007 we detect clear micro-seismic signal and <u>demonstrate an increasing trend through time. The positive trend in microseism levels</u> suggests that oceanic wave maximum heights have increased, and correlates with other metrics showing increasing surface sea and air temperatures and oceanic storminess."[208]

McNamara et al have demonstrated a clear positive increase in Gravitational and Microseism Hum over the last 30 years (the full span of previously collected usable seismic data). In doing so, they point to but one of the theoretical sources ascribed to both types of vertical oscillation born Hum – the ocean tides. Other authors attribute the *vertical* Gravitational and Microseism Hum to fluctuations in the Earth's atmospheric pressure.[209, 210 & 211] Other researchers cite combinations of ocean tides and atmospheric pressure.[212]

The most recent breakthrough in the investigation of the Earth's Hum came in 2008. That shocking discovery was reported by D. Kurrle and R. Widmer-Schnidrig. These investigators were among the first to mine the twisting horizontal (torodial) variations in accumulated multi-station Gravitational and Microseism Hum data. At the American Geophysical Union's Fall Meeting of 2008 they reported that:

> Despite the difference in noise levels between vertical and horizontal components, we analyzed horizontal component seismic data from the quietest stations and found indications for the continuous excitation of fundamental toroidal [horizontal] modes between 3 and 7 millihertz. We could identify numerous peaks at both fundamental spheroidal and toroidal mode frequencies in seismic noise spectra. Both kinds of modes exhibit similar amplitudes. ... This finding confirms that the horizontal hum of the Earth is composed of both fundamental spheroidal and toroidal modes.
>
> Regardless of whether the spheroidal and the toroidal modes are excited together or not, <u>new theoretical models will be necessary to explain the torsional hum of the Earth,</u> since [ocean and atmospheric] <u>pressure forces usually invoked for the excitation of the spheroidal hum cannot explain the excitation of toroidal modes at the observed level.</u>[213]

It has been suggested that the long term positive increase in the Gravitational Hum, as well as the unexplained horizontal variation in Earth's Hum, point to an extraterrestrial source. Charles Q. Choi (2008) commented:

This discovery [the D. Kurrle et al results] should force researchers to significantly rethink what causes Earth's hum. While the spheroidal [vertical] oscillations *might* be caused by forces squeezing down on the planet — say pressure from ocean or atmospheric waves — the twisting ring-like phenomena might be caused by forces shearing across the world's surface, from the oceans, atmosphere or possibly even the sun.[214]

Echoing Choi, Larry O'Hanlon of Discovery News, states: "It [the toroidal, meaning horizontal, Hum] could even be caused by the sun …. Oscillations in the sun may be picked up by earth's geomagnetic field and cause earth to hum a solar tune."[215]

Catherine Brahic of the New Scientist (2008) adds:

Listen closely, and you'll hear the Earth humming - in not just one note, but two. The source of this second signal [that described by D. Kurrle et al] is a mystery.

For around a decade we've known about Earth's quiet "vertical" hum, probably caused by the steady thumping of deep waves on the ocean floor. Now a team in Germany has discovered a second "horizontal" note, too, and nobody knows what's causing this new signal.[216]

Scientists have known for some time about the seasonal variations in the Microseism Hum. Stehly, L., Campillo, M., and Shapiro, N.M, (2006) conducted a "Study of the Seismic Noise from its Long Range Correlation Properties,"[217 & 218] Figure 70 is taken from that study of seasonal variations in the direction, wavelength, and number of wave cycles comprising the Microseism Hum.

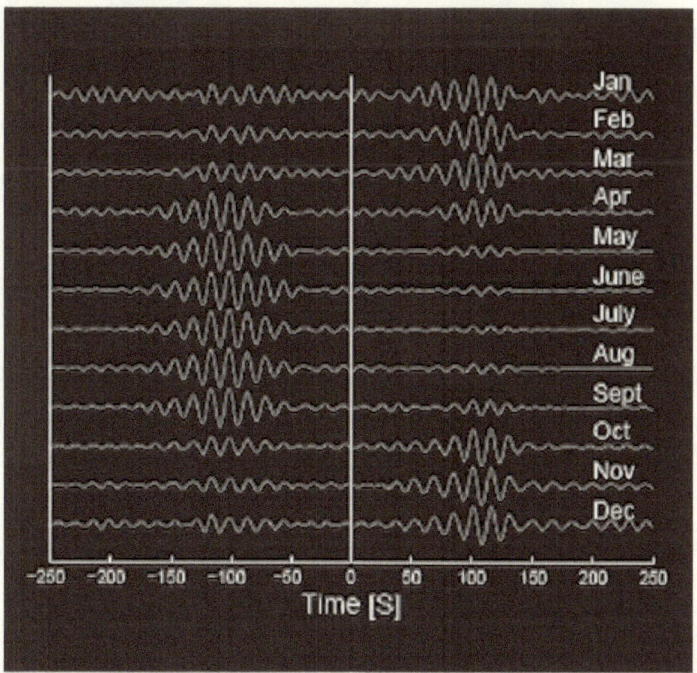

Figure 70 Seasonal Variations in Microseism Hum
Source: Stehly, L., Campillo, M., and Shapiro, N.M, (2006), A Study of the Seismic Noise from its Long Range Correlation Properties, Draft: December 21, 2005, p.22

While the Earth's Hum may appear to come from the Sun, we assert that the true source of the two Hum modes is a Galactic Gravitational Torque Wave that entered our solar neighborhood around 1500 AD. The *local* source of the Hum may indeed be atmospheric pressure and the ocean tide activity, but the originating source is most likely that GTW.

For earth bound creatures, Earth's gravitational field has provided a valid, reliable, practical, and economic external reference. Really early on, primitive life forms took notice. Over 570 million years ago the most primordial force in the universe – gravity – gave rise to the "first organ to appear in the animal kingdom, the statocyst."[219] Ever since the statocycst sensory organ has provided invertebrate creatures with an orienting sense of Earth's gravitational field and referenced their movement and balance relative to it.

Human Sensing of Earth's Hum

At least 400 million years ago the statocyst evolved into the vertebrate vestibular system. For the last 200 million years that system has been the first system to emerge and mature in every viable mammalian fetus. The 85 million year old primate vestibular system is remarkably similar to the early versions that persisted throughout the 2.5 million year old genus *Homo*. The *Homo sapiens* who lived 200,000 years ago had essentially the same vestibular system we have today. The latest adaptation in the vestibular system has been a substantial increase in the input and output connections of the main neural relay component of the system. These connections were needed to fully integrate the two thirds larger brains of emergent *Homo sapiens*.

The vestibular system consists of three main parts. These are the:

1) Two *otolith* organs (the utricle and saccule) of the inner ear that sense linear acceleration and its gravitational equivalent.

2) Three *semicircular canals* of the inner ear sensing rotational (roll, pitch, and yaw) movements in space.

3) Four *vestibular nuclei* in the older part of the brain, making up the system's various input and output connections.

Figure 71 depicts the anatomical locations of the otolith organs (the utricle and saccule) and the semicircular canals. Figure 72 depicts the four vestibular nuclei that continuously associate and relay a surprising variety of brain functions. These range from the highest levels of consciousness to the most automatic of reflexes.

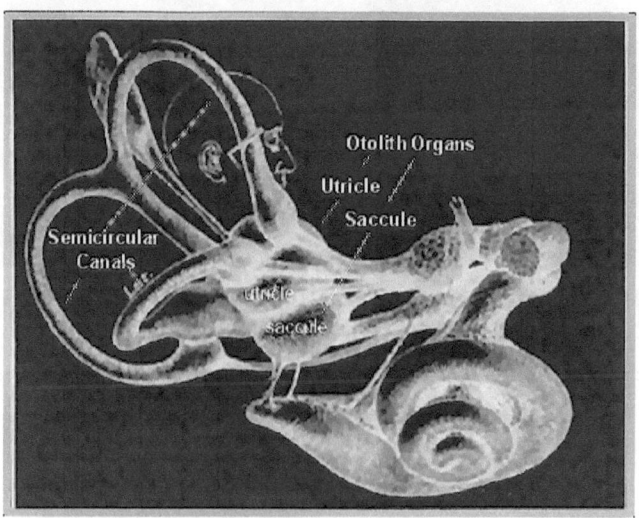

Figure 71 Vestibular Sensory Apparatus
Source: http://www.neuro-kinetics.com/whitepapers.aspx, Revisited October 8, 2009

The study of the human vestibular system has sustained two central focuses. The first is the investigation of how the vestibular system influences the autonomic nervous system. The second is the study of the effects of vestibular system impairment on human health and behavior. As you will soon see, there is a third seriously neglected focus.

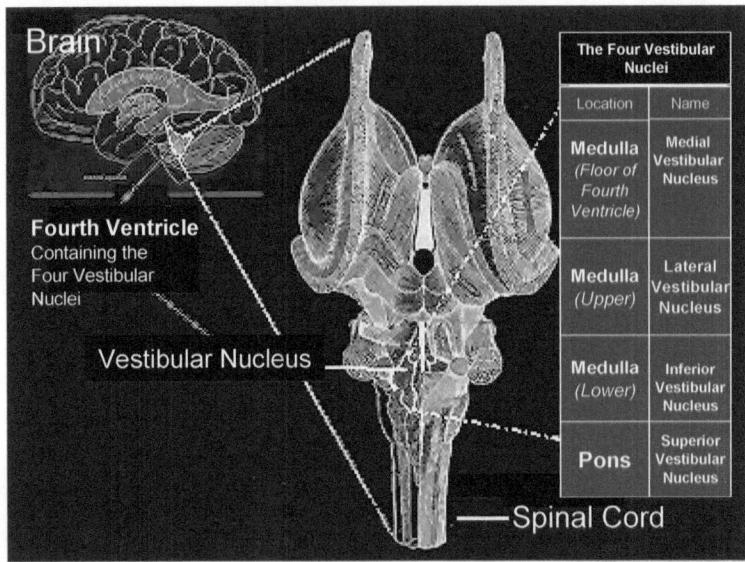

Figure 72 the Location of the Four Vestibular Nuclei of the Vestibular Nucleus

Source: http://en.wikipedia.org/wiki/Vestibular_nuclei, Revisited December 27, 2009

Neural physiological researchers – to include those investigating the significant effects of microgravity on astronauts – treat gravity as a predominantly constant environmentally specific force. In 2005, B.L. Day and R.C. Fitzpatrick noted:

Small, intricately formed and locked in the skull, the vestibular organs continuously bombard the brain with messages. The messages are quite unlike any others. They tell of accelerations, how the head is rotating and translating and its orientation in space. The messages never stop and cannot be turned off. Even when we are completely motionless, they signal the relentless pull of gravity. Perhaps because of their constant monologue, the vestibular sensation is different from the other senses. There is no overt, readily recognizable, localizable, conscious sensation from these organs. They provide a silent sense.[220]

Sensitivity of the Vestibular System

The best criteria for evaluating the sensitivity of the human vestibular system are the precise measurement sensitivities of modern gravimeters. The state of the craft in machine captured vertical gravitational wave modes range in amplitude from "$1000 \times 10^{-9} ms^{-2}$ to $0.01 \times 10^{-9} ms^{-2}$" and in frequency from "0.01 millihertz,(mHz) to 1000 mHz."[221] Machine sensitivity to Earth's Hum (the weakest gravitational mode) ranges in amplitude from "$0.01 \times 10^{-9} ms^{-2}$"[222] to "0.04 ngal," which equals $0.004 \times 10^{-9} ms^{-2}$"[223] (a difference of only six thousands of a billionth of a meter, which is assumed by geologists to be "rather constant in time").[224]

Within the *"microseism modes"* of the machine measured Earth's Hum, gravitational waves exhibit "modal peaks" that vary in frequency from approximately 50 millihertz (mHz) with a wavelength of 20 seconds, to 100 mHz with a wavelength of 10 seconds. That is why these waves are referred to as "10 to 20 second waves".[225]

The most frequently observed lower frequency of the vertical Earth Hum varies only slightly in amplitude, but ranges in frequency from "2 to 7 millihertz" (a wavelength of .002 to .007 hertz, or 1 cycle every 500 to 143 seconds).[226] The rarely analyzed 3 to 7 millihertz horizontal modes of the Earth's Hum reverse the direction of their twisting torque of the Earth's surface about once every three minutes.[227]

Hypothesis H₂₃ asserts that the human vestibular system is sensitive enough to detect and respond to the variations in the Gravitational Hum and Microseism Hum modes. **Hypotheses Null₂₃** posits that the human vestibular system is not sensitive enough to detect and respond to the variations in either the Gravitational Hum or the Microseism Hum modes.

The available literature makes the case that H_{23} is true and $Null_{23}$ false. The gravitational field sensors of the autonomic vestibular system are outrageously small and exquisitely sensitive. The sensory apparatus of the human vestibular organs are nearly as sensitive to the dynamic variations of the gravitational field as the best available spring gravimeters. A typical spring gravimeter detects changes in gravitational wave amplitude of one tenth of one billionth of a meter. The otolith organs of the vestibular system can detect wave amplitude oscillations down to at least three tenths of a billionth of a meter. But, as we are about to reveal, the vestibular organs are in fact even more sensitive than that.

The human vestibular sensors are much-much smaller and much more finely resolving than is required to maintain basic body-balance and motion functions. Neurologists describe the sensory apparatus of the vestibular system this way.

The vestibular organs form two functional units. The two otolith organs sense linear acceleration and its gravitational equivalent, and the three semicircular canals sense rotational movement in space.

The hair cells of the utricle and saccule form a two-dimensional array with their cilia [hairs] embedded in a membrane of dense calcium crystals known as otoliths ("ear stones"). Movement of the membrane by gravitational or inertial forces maximally activates [by displacement of the endofluid] those hair cells that are aligned with the movement.

With the two organs oriented at right angles to each other, the direction of linear acceleration is spatially encoded in three dimensions and the magnitude of the acceleration is encoded by the [neuron] firing rate. As the head rotates, the inertial force of the fluid [called the endofluid] in the semicircular canals deflects the cilia of hair cells aligned with the canals, modulating the firing of the afferent nerves. With the three semicircular canals aligned at right angles to each other, rotation in any direction can be resolved.[228]

The vestibular resolution of human head and body motion is accurate to at best "0.5 degree." Specifically, "these sensors can detect a change in orientation of the head of 0.5° from the upright position, a change of 5° from the horizontal position, or a change of 15° from the upside down position."[229] So why is the sensitivity of the Otolith organ membrane approximately equivalent to that of a high resolution spring gravimeter? Figure 73 depicts the sensory apparatus of the Vestibular System.

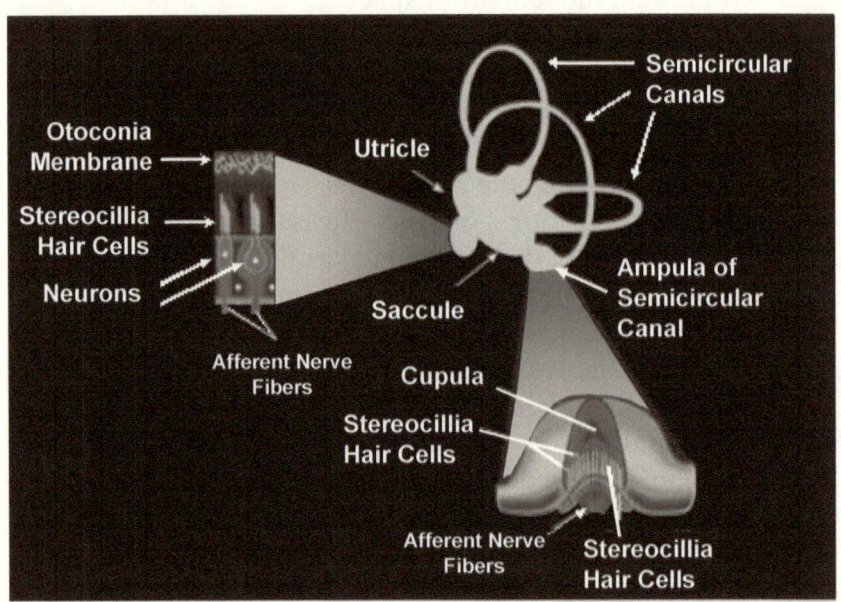

Figure 73 Sensory Apparatus of the Vestibular System

Source: "http://www.scholarpedia.org/article/Vestibular_system, Revisited October 6, 2009

It would appear that the vestibular system must serve a far more demanding purpose than basic motion and balance control. Still, it is apparent that present day neurophysiologists are fully fixated on the ability of the human vestibular system to resolve and coordinate physical movements relative to the gravitational field. Researchers ignore any human capability to sense and decode the subtle variations in the Earth's gravitational field. Consequently, researchers overlook the fundamental effects gravitational variations might be having on the central nervous system. That oversight has led to a blinding bias unknowingly exemplified in the following statement by T. A. and S. M. Jones, and L. F. Hoffman in 2008.

> The consensus hypothesis states that such [resting] discharge patterns [the spontaneous activity of primary vestibular neurons] are independent of stimulation and depend instead on excitation by vestibular hair cells due to background release of synaptic neurotransmitter.[230]

What fueled that seriously misleading view is: what was studied, how, and at what level of precision. Here's why. Each calcium carbonate crystal on the Otoconia membrane of the otolith organs and the Semicircular canals covers an average area of 3.2 to 4.5 square billionths of a meter. These genuinely tiny crystals (or ear stones) have seriously misled many physiologists.

For example, Jones et al 2008 found that "the afferent neuron firing rate of vestibular neurons varied only slightly in calcium carbonate crystal deficient mice when compared to intact mice (mice with carbonate crystals)."[231] The widely held assumption being that the crystals alone determine the "mean discharge rate" of the vestibular neurons. That incorrect assumption caused Jones et al to falsely conclude:

> These results confirm the hypothesis that resting activity in macular primary afferents occurs in the absence of ambient stimulation. [232]

What is fundamentally wrong with that Jones et al conclusion is that the calcium carbonate crystals are not the source of vestibular fluid displacement. The crystals do not produce neuronal hair cell deformation and gravitational field sensing in the vestibular system. That is what the Jones team actually demonstrated.

The Jones team genetically removed the crystals from the vestibular membranes of experimental mice. Then, they found no difference in the firing rate between the altered mice and the control mice. What they actually studied was the importance of otoconial membrane deformation, with or without the presence of dampening carbonate crystals.

In 2007 J.L. Davis, J. Xue, E.H. Peterson, and J.W. Grant constructed "finite element [software based three dimensional dynamic] models of the otoconia membrane" from the Otolith organ of the inner ear. They did that, "to investigate the effects of three geometric variables on static and modal response of the otoconia membrane."[233] That work makes it clear that Jones' et al had pursued an incomplete notion. Jones assumed that the movement of the two dimensional arrays of calcium carbonate crystals maximally activates the hair cells. Davis et al demonstrated that patently false. It just isn't what actually happens in the fluid filled organs of the vestibular sensory apparatus.

Davis demonstrated that the carbonate crystals are merely the load-dampening weights of the deformable membrane. It is the complex deformations (mechanical gain) of the three ultra thin flexible layers of the otoconia membrane itself that respond to gravitational variations. The three dimensional deformations of the highly complex membrane displaces the gelatinous fluid (i.e., the endolymph fluid) that fill the otolith organs and the semicircular canals. Taken in sequence, gravitational field driven membrane deformations displace the fluid, bend the neuronal hair cells, and thereby alter the neuronal firing rates within the vestibular system.

According to the Davis team (and many others):

The otoconia membrane ... is comprised of three layers: an otoconial layer, a compact gel layer, and a column filament layer. Accelerations and ambient gravity act on the higher density otoconial layer, shearing the column filament layer and gel layer with a magnitude of displacement that is proportional to the component of the acceleration vector in the plane of the otoconial organ. Shearing of the compact gel layer and a column filament layer displaces the hair cell bundles."[224]

How sensitive is the otoconial layer to gravitational field variations? According to Davis et al:

There were changes in static gain over the surface [of the membrane] when the stimulus acceleration was in a specific direction, and the <u>maximum</u> variation [displacement in nanometers (nm)] observed in the model was 15nm/G" [i.e., ranging from 0.0 to 15 nanometers (billionths of a meter) per unit gravity].[235]

Davis also makes it clear that the natural frequencies responded to by the Otoconial membrane (from flat to variable curved surface) deformation, range from "1,396.84 cycles per second" to "1,530.19 cycles per second."[236] That equates to a frequency range nearly two orders of magnitude (100 times) more sensitive than that needed to detect the 2 to 7 mHz frequency of the Earth's Microseism Hum. Davis et al note:

> The amplitude of displacement of the Otoconical Layer is inversely proportional to the square of the frequency. Low frequency displacement modes [like those just cited] are easily excited (they require less energy to excite) and are therefore most frequently excited. Secondly, in our otoconical layer models these low frequency displacement modes represent shear displacement of the Otoconical layer, i.e., the characteristic natural motion of the Otoconical membrane."[237]

Photo 2 depicts a microscopic photograph of a Vestibular Hair Cell bundle sitting atop the head of a single vestibular neuron. In point of fact, the Otolith organ sensors should be sufficiently sensitive to completely sense all of the gravitational modes (from earthquakes to the free oscillations of the Earth's Hum) captured by spring gravimeters and satellite sensors of the Global Gravitational Project and seismic monitoring network.

In the human vestibular organs, from 50 to 100 sensory stereocilia hairs populate each 200 micrometer (200 millionths of a meter) diameter neuronal cell in the vestibular sensory apparatus. That would imply that each stereocilia hair has a diameter equal to or less than 0.2 millionths of a meter. Observation bears this out.

As reported in 1983 by Tinly et al: The stereocilia vary from a maximum of 5.5 millionths of a meter in length to a minimum of 1.5 millionths of a meter in length, and from 0.12 millionths of a meter in diameter to 0.2 millionths of a meter in diameter. [238]

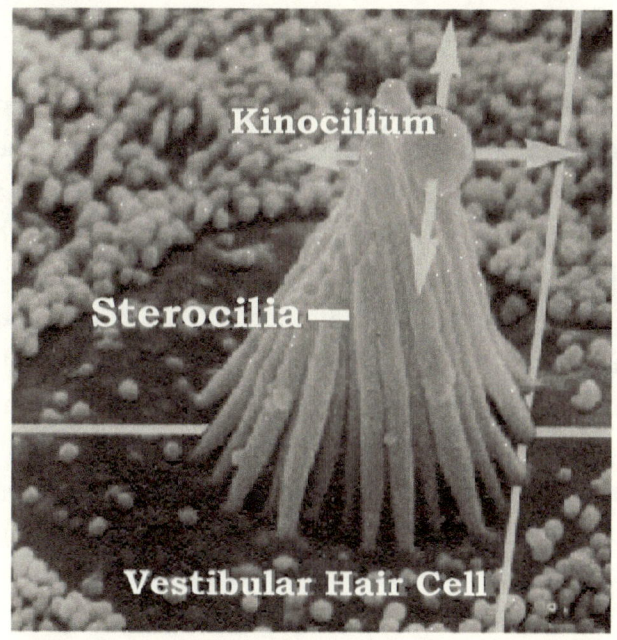

Photo 2 Vestibular Hair Cell Bundle
Source:
http://www.medicine.mcgill.ca/physio/cullenlab/introtovest3.html,
Revisited October 11, 2009

The sensory protein chains that tie vestibular stereocilia together are woven of actins (individual molecules less than 1 billionth of a meter in diameter). Tugs on these infinitesimal protein chains bend stereocilia and open potassium ion channels just over 1 billionth of a meter in diameter. Ions flowing through these channels between the stereocilia cause each vestibular neuron to depolarize and fire. All this microstructure suggests that the downward and lateral acceleration of the Earth's weak Gravitational and Microseism "Hum" could easily deform the vestibular membranes (e.g., the otoconial membrane), displace the endofluid, bend the stereocilia, activate the protein filaments and ion channels, and fire the vestibular neurons.

At rest, vestibular neurons fire with a frequency ranging from 70 to 100 cycles per second. Hence, the vestibular system likely *over samples* the Earth's persistent Gravitational Hum at a ratio of 100,000 to 1 and the stronger seismic event frequencies (e.g., that of substantial earthquakes) at a ratio of 100 to 1.

Hypothesis H_{23} is retained. The human vestibular system is sensitive enough to detect and respond to the variations of the Gravitational and Microseism Hum modes. The question becomes, does the human autonomic nervous system actually attend to any of the Earth's horizontal and vertical gravitational variations it is capable of sensing?

The answer appears to be yes. In fact, your literal autonomic sense of self may well depend upon it. To understand why this is true, we must transcend the more woolly-minded descriptions of the sense of self afforded by psychology, sociology, and social-psychology. In the next chapter it is the literal physical source of the autonomic human sense of self we are after.

12

WAVES OF HUMMING IDENTITY

Neurophysiologists often hint at the role of vestibular sensed gravity in defining and sustaining the human sense of self. What they have failed to do is pursue the notion. The following quote by B.L. Day provides a good example of the prevailing perspective.

> The value of the vestibular sensory system to brain functions include perception of self and non-self motion, spatial orientation, navigation, voluntary movement, oculomotor control, and autonomic control, comes from their unique and complete description of head motion and orientation in three dimensions.[239]

Hypothesis H$_{24}$ asserts that the neuronal signals of the human vestibular system are used by the brain to discriminate the internal human frame of reference (the self) from the external frame of reference, the non-self (i.e. the environment). **Hypotheses Null$_{24a}$** posits that no such discrimination takes place. **Hypotheses Null$_{24b}$** asserts that the source of self and non-self discriminations in the brain is other than vestibular.

A comprehensive review of the literature provides evidence supporting the neural physiological discrimination of "body-centric and "geo-object-centric frames of reference." That review was provided by Christophe Lopez and Olaf Blanke (2007) in their paper, "Neuropsychology and Neurophysiology of Self-Consciousness; Multisensory and Vestibular Mechanisms."[240] Results related by Lopez and Blanke staunchly support Hypothesis H_{24} (which has long been *a priori* accepted by prevailing neurophysiologists), and demonstrates Hypothesis Null$_{24a}$ and Null $_{24b}$ to be patently false. That same body of research implicates multi-sensory information in the discrimination of the human self from the object-environment. That integration is believed to occur primarily through the vestibular nuclei, with "the integration and discrimination of the frames of reference occurring mainly in the cerebellum."[241]

The Default Mode Network

In the human body "the sense of self" is believed to be the product of the "slow synchronous waves" of the "default mode network" in the brain, "one of the most mysterious and well connected networks of all."[242] **Hypothesis H_{25}** argues that the autonomic human sense of self is the product of the Default Mode Network in the healthy adult brain. **Hypothesis Null$_{25}$** posits that the discrimination of the autonomic self from the non-self (the environment) is not related to the cycles of the Default Mode Network in the brain.

There is mounting evidence that Hypothesis H_{25} is true, but the relation between the Default Mode Network and the discrimination of the self from the non-self remains hypothetical. Hypothesis H_{25} is strongly supported by Research summarized at key points in this chapter.

"Despite the laid-back-name, which Marcus Raichle coined in a 2001 paper, the default mode network is one of the hardest-working systems in the brain."[243] "Using [mostly] PET" (Positron Emission Tomography) and "fMRI" (functional magnetic resonance imaging) scanners, the default mode network was "discovered accidentally by researchers watching the activity of brains at work on various tasks."[244]

Figure 74 depicts the paths and brain site tracings of the default mode network cycles, overlaid on the brain as seen from above. The synchronous cycles of the Default Mode Network trace through the brain between "the two major hubs of the network, the Posterior cingulate cortex (*Precuneus*) and the Medial prefrontal cortex."[245] In the resting brain, these cycles have a wavelength (measured between peaks, zero crossings, or troughs) of one cycle every 10 to 20 seconds (which is 0.1 to .05 cycles per second or Hertz).

The Default Mode Network is called "the network that never sleeps."[247] According to "Peter Franson, a neuroscientist at the Karolinska Institute in Stockholm," and "Peter Williamson, a psychiatrist at the University of

Western Ontario," 'You don't even have to be conscious for it [the Default Mode Network] to be apparent.'"[248]

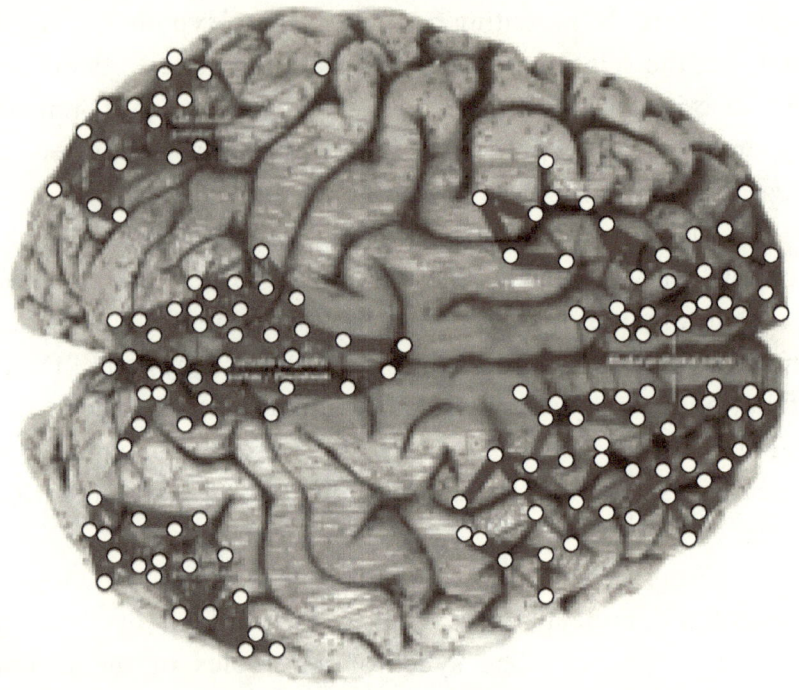

Figure 74 Default Mode Network Map
Source Saey, Tina, H, Science News, "You Are Who You Are by Default," Vol. 176, No. 2, July 18, 2009, page 16-17]

Figure 75 plots this Default Mode Network variation in terms of neural tissue blood flow (percent Blood Oxygen Level Dependence, or "BOLD, change"[246]) as revealed and measured using fMRI scans. Saey (2009) contrasts the relentless autonomic persistence of the Default Mode Network with conscious processing.

Slow yet continuous fluctuations in activity bind the [default mode] network together. The syncopations continue even while people are asleep, under anesthesia, or in coma. But it is unlikely that such activity reflects ongoing conscious processing ... The fluctuations that move through the network are incredibly slow, one cycle every 15 to 20 seconds [with the brain at rest this is actually one cycle every 10 to 20 seconds]. Most conscious thought happens in split seconds [e.g., 100 to 500 milliseconds]...[249]

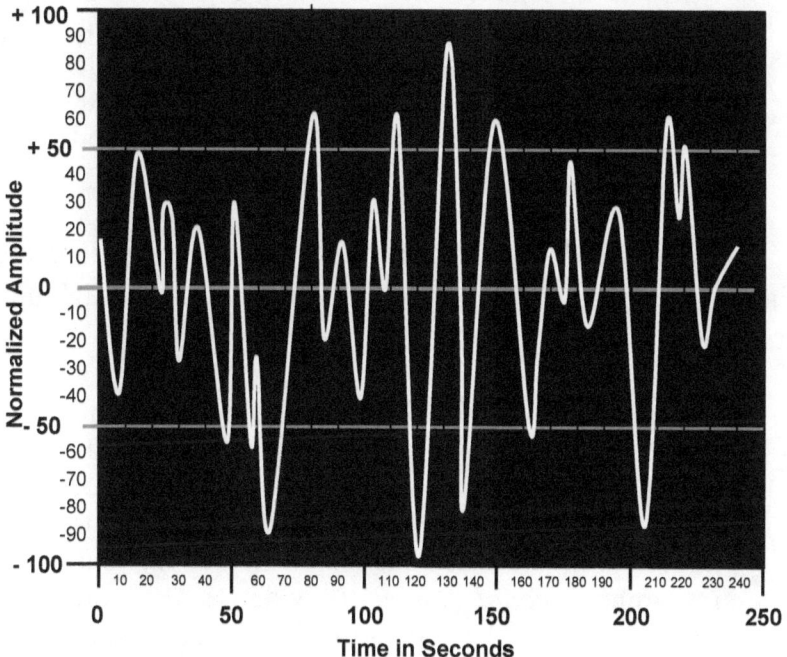

Figure 75 Resting Default Mode Network Waves,
Source: Harrison et al, 2008, Ben J. Harrison, Consistency and Functional Specialization in the Default Mode Brain Network, PNAS July 15, 2008, Vol. 105, No. 28, 9781–9786: Http://www.pnas.orgcgidoi10.1073pnas.0711791105

It has long been recognized that the sense of self (one's unique isolated identity) first emerges between the ages of 3 and 4 years of age. The classic child development psychology argument is put as follows by mental health practitioners.

> During early childhood, children start to develop a 'self-concept,' the attributes, abilities, attitudes and values that they believe define them. By age 3, (between 18 and 30 months), children have developed their Categorical Self, which is a concrete way of viewing themselves in 'this or that' labels. For example, young children label themselves in terms of age 'child or adult', gender 'boy or girl', physical characteristics 'short or tall', and value, 'good or bad.'[250]

Now, "new ways of analyzing neural connections are bringing the picture [of the development of the self] into better focus."[251]

> Franson and his colleagues used fMRI to scan the brains of sleeping premature infants who had reached the equivalent of 40 weeks of gestation to see whether the default network is already in place when babies are first born. The researchers *could not find* evidence that the default mode is operational in newborns, although five other brain networks are already online, the team reported in 2007.
> ...
> But a recent study by Weili Lin, a neuroscientist at the University of North Carolina at Chapel Hill, and colleagues shows

that infants as young as 2 weeks have rudimentary, incomplete default mode networks. The study published in the April 21 *Proceedings of the National Academy of Sciences*, tracks development of the network from shortly after birth into toddler-hood. Newborns' default networks connect six brain regions, Lin's group found.

It doesn't take long for the brain to develop a default mode, Lin showed. By age 1, babies link 13 brain regions in their default network. In 2-year olds, the default network is even bigger, comprising 19 regions, 13 consistent with the network in adults. But bigger networks can also be inefficient, Lin says, noting that adult default networks have been pruned of extraneous connections.

...

Preliminary data from 4-year olds indicate that extra connections are severed as the brain ages ...[252]

The early childhood development of the Default Mode Network tracks well with the established milestones of developmental psychology. In her article on the Default Mode Network, Saey points out another dead-on match.

A group of researchers at Washington University ... are piecing together the network's development from age 7 into early adulthood.

Brain connections in 7-year olds are organized differently than in adults. Children have more short-range connections among neighboring brain regions and fewer long-range connections, particularly among the parts of the default network in the back and the front [of the brain] ... As children age, the connections are rewired. Adolescents have a network structure somewhere between that of elementary-age children and adults.[253]

Saey is quick to point out that "kids default networks aren't clunky." [Ibid] She states:

While children's connections are structured differently, they have enough shortcuts to make information transfer in the networks just as efficient as in adults, the scientists reported online May 1 in *PLoS Computational Biology*.[254]

One of the most interesting things about the Default Mode Network is the small amount of variation in its cycles and tracings among healthy adults. Saey relates that:

Once people reach adulthood, activity in the network is fairly consistent from person to person, with some slight differences between the sexes and older versus younger people, Williams and his colleagues wrote in a 2008 paper in *NeuroReport*.

This consistency in the networks from person to person is remarkable, especially considering what its function is supposed to be. 'Everyone's brain is thinking different thoughts while in the default mode,' Fair says, 'and yet all healthy brains in default mode, look essentially alike.'"[255]

While each person's sense of self identity is unique, the rhythms and source of it form a kind of common denominator across child development and adult populations. Although strongly persistent in all waking and unconscious states, the Default Mode Network and its product sense of self can be seriously disrupted by trauma and disease. Much of that work on abnormal subjects supports Hypothesis H_{25}, which argues that the Default Mode Network is the source of the fundamental human sense of self.

In March 2010 Marcus E. Raichle and his team at the Washington University School of Medicine reported:

> Analysis indicated that performing a particular task increases the brain's energy consumption by less than 5 percent of the underlying baseline [Default Mode Network] activity. A large fraction of the overall activity – from 60 to 80 percent of all energy used by the brain – occurs in circuits [of the Default Mode Network] unrelated to any external event.[256]

Calhoun of the University of New Mexico and his colleagues studied Default Mode Network activity in a large sample of people with schizophrenia and a healthy control group. They found that, "some sub-networks in the default mode network had trouble disengaging in people with schizophrenia, impairing their ability to focus on the task"[257] Citing the 2007 Calhoun study, Saey (2009) relates that:

> People with schizophrenia also have faster cycles of activity in their default networks during a resting state than normal people do.[258]

Saey further notes that "Williamson and colleagues have shown that the default network's connections with other parts of the brain may be important in determining who develops Post Traumatic Stress Disorder (PTSD) after a traumatic event."[259]

People who have been traumatized can become numb and loose their sense of self, Williamson says. The researchers examined the default networks in women who developed PTSD after trauma in childhood. The study found altered levels of connectivity among parts of the default network as well as between the network and other parts of the brain. The findings published in May [2009] in *the Journal of Psychiatry & Neuroscience*, could indicate that trauma creates disturbances in the network's ability to create a sense of self.[260]

Greicus and his colleagues reported online in June 2008 at *PLoS Computational Biology* that "activity in the default network is affected by Alzheimer's disease."[261] "At least one study suggests that the default network may be vulnerable to Alzheimer's disease decades before any symptoms or plaques show up."[262] "Young people who carry the genetic risk factor for the disease have more activity in the default network, particularly in the hippocampus, than young people who don't have the genetic risk ... "[263]

A 2009 study found:

A natural, sleep-induced reduction of consciousness is reflected in an altered correlation between default mode network components, most notably a reduced involvement of the frontal cortex. This suggests that the default mode network may play an important role in the sustenance of conscious awareness."[264]

Yvette Sheline, professor of psychiatry at the Washington University in St. Louis, recently reported online that, "People with depression may not be able to "lose themselves" in work, music, exercises, or other activities that enable most healthy people to get 'outside' of themselves."[265] In a study using fMRI, Sheline demonstrated that:

> Brain regions of the default mode network behave differently in depressed people. The default network typically is active when the mind wanders. It shuts down when an individual focuses on the job at hand. But the researchers found the network stays active in people who are depressed, even when they are concentrating on specific tasks.[266]

Sheline is now looking at the brains of depressed people following treatment with antidepressant drugs. "Preliminary results suggest that their default networks function more normally."[267]

Evidence suggests that a malfunctioning Default Mode Network is likely involved in diseases and disorders as diverse as Alzheimer's, autism, depression, post-traumatic stress disorder, Tourette syndrome, amyotrophic lateral sclerosis, schizophrenia and attention-deficit/hyperactivity disorder.[268]

fMRI and PET scans have predominated in the discovery and subsequent study of the Default Mode Network. In 2007 Andrew C.N. Chen et al succeeded in recording the default network using EEG (Electroencephalograph) recording instruments.[269]

Source of the Default Mode Net Rhythm

A question looms large. What is the *source* driving the clearly autonomic and involuntary Default Mode Network? Particularly as observed in human brains at rest, anesthetized, or in coma? Nothing else within the human body exhibits the slow cyclic frequency of the default mode network. No autonomic cardiovascular, neurovascular, or respiratory function consistently counts in the 2.5 cycles (waves) every 50 seconds of the Default Mode Network. The Default Mode Network cycles as follows:

Cycles per second ≈	0.05
Cycles per minute ≈	3
Cycles per hour ≈	180
Cycles per day ≈	4,320
Cycles per 20 days ≈	86,400

Note: *It is an extremely curious sort of coincidence that the number of cycles of the Default Mode Network in a 20 day period appears to exactly match the number of standard seconds per 24 hour rotation of the Earth about its vertical axis.*

Definitely Neuronal

Hypothesis H$_{26}$ posits that the source of the Default Mode Network as evidenced by fMRI (functional Magnetic Resonance Imaging) BOLD (Blood Oxygen Level–

Dependent) fluctuations is predominantly neuronal. **Hypothesis Null$_{26}$** asserts that fMRI BOLD measurements of functional networks like the Default Mode Network are generated by artifacts, such as aliasing of cardiac or respiratory pulsations in the low-frequency range, vasomotor oscillations, or effects of attention.

On July 17, 2008, J.S. Anderson of the Department of Neuroradiology at the University of Utah published online a literature survey and the results of a controlled experiment.[270] As described earlier, fMRI (functional Magnetic Resonance Imaging) and PET (Positron Emission Tomography) scans rely on percent Blood Oxygen Level Dependent (BOLD) fluctuations in neuronal tissues. In his pre-experiment survey of the literature, Anderson notes:

> Early speculation about the technique [i.e., fMRI BOLD measurements] questioned the possibility that fluctuations might be generated by artifacts such as aliasing of cardiac or respiratory pulsations in the low-frequency range, vasomotor oscillations, or effects of attention. Subsequent investigations now provide strong evidence that synchronous low-frequency fluctuations have a neural source. [271]

Anderson's own experiment reinforced those earlier findings. Those findings indicate that the source of fMRI measurements used to discover and test the Default Mode Network are the result of neural activation. Anderson concludes:

> Therefore, although low-pass filtering from neurovascular coupling likely makes a large

contribution to the frequency components observed in BOLD fluctuations, it is insufficient to explain the observed frequencies by itself. [272]

Assessments like Anderson's established that the Default Mode Network fMRI, PET, and related EEG measurements are a result of neural activity. Thus, Hypothesis H_{26} is retained.

Still, such studies only vindicate measurement methods and affirm the validity of what is actually measured. The question remains; what is the *source* of the waves of the Default Mode Network in awake, resting, anesthetized, and comatose human brains?

Case for an External Stimulus

The Default Mode Network must be the composite product of the brain and some external wave force. Clearly, the primary senses are all externally sourced. Those senses are not sufficiently transmitted and processed – particularly when a subject is asleep, anesthetized, or comatose – to initiate and sustain the Default Mode Network. The waves of the Default Mode Network persist in all brain states. Further, the five basic senses account for too little of the variation in neural processing to be considered viable candidate sources of the Default Mode Network. Marcus Raichle (2010) put it this way:

> Researchers have known for some time that only a trickle of information from the virtually infinite flood in the surrounding environment reaches the brain's processing centers.

Although six million bits [out of the equivalent of 10-billion bits per second arriving at the eye] are transmitted through the optic nerve, for instance, only 10,000 bits make it to the brain's visual processing area, and only a few hundred are involved in formulating a conscious perception – too little to generate a meaningful perception on their own. The finding suggested that the brain probably makes constant predictions about the outside environment in anticipation of paltry sensory inputs reaching it from the outside world. [273]

There are other candidate sources of the Default Mode Network. A substantial case in point is the interaction of the suprachiasmatic nuclei and Pineal gland of the brain. It is the source of the biological clock that underlies the majority of human waking activity and sleep cycles.

Hypothesis H$_{27}$ asserts that the genetically controlled and externally cued rhythms of the suprachiasmatic nuclei are the source of the persistent low frequency rhythms of the Default Mode Network. **Hypothesis Null$_{27}$** argues that the externally cued rhythms of the suprachiasmatic nuclei are *not* the source of the persistent low frequency rhythms of the Default Mode Network.

The Canadian Institute of Neurosciences and the McGill University website describes that candidate source as follows:

Most human bodily functions and behaviors are not 'steady-state'. Instead, they fluctuate in 24-hour cycles, such as the sleeping and waking cycle, and the cycles for body temperature, hunger, and the secretion of various hormones.

The central clock that regulates all of these circadian cycles is located in two tiny structures in the brain, at the base of the left and right hypothalamus. Each of these structures is no larger than a pencil tip and contains several tens of thousands of neurons. These structures are called the suprachiasmatic nuclei because they are located just above the optic chiasma, where the left and right optic nerves cross paths.

This strategic position enables the suprachiasmatic nuclei to receive projections from the optic nerve that tell them the intensity of the ambient light entering the eyes. The neurons of these nuclei use this information to resynchronize themselves with daylight every day, because like any clock, the human biological clock is not perfect and does need to be reset periodically.[274]

The Mcgill neuroscientists make the following physiological case for the genetic basis of this rhythmic source of human timing.

Despite this need to resynchronize with an external cue, it has been shown that the suprachiasmatic nuclei do in fact constitute a biological clock with its own independent rhythm. First, many experiments have shown that the fluctuations of the human circadian cycle persist even when individuals are cut off from the light of day.

Second, in experiments where the suprachiasmatic nuclei were destroyed in animals such as hamsters, their cyclical behaviors, such as their sleep/wake cycles, become completely disorganized. When suprachiasmatic nuclei were transplanted from hamster fetuses into these animals, their biological rhythms returned but with the properties of the donors.[275]

The genetically controlled and externally cued rhythms of the suprachiasmatic nuclei *cannot* be the source of the persistent low frequency rhythms of the Default Mode Network. The suprachiasmatic nuclei oscillators reset at 3 to 4 hour intervals, peak at approximately 8-12 hours, and vary over a 24 hour period with light changes to produce sleep cycle and waking state behavioral variations. As stated, the waves of the Default Mode Network persist 24 hours a day, 7 days a week, at a nearly constant rate of one cycle every 10 to 20 seconds. Default Mode Network rates endure in healthy mature: waking, sleeping, anesthetized, and even comatose brains. Such differences guarantee that the multi-hour wavelength of the sub-groups of suprachiasmatic nuclei oscillators cannot possibly bear any relation to the wavelengths of the Default Mode Network waveforms.

Hence, **Hypothesis Null$_{27}$** is retained. The externally cued rhythms of the suprachiasmatic nuclei are *not* the source of the persistent low frequency rhythms of the Default Mode Network. Some other environmentally cued rhythmic neural oscillator must be responsible for the relatively constant waveforms of the Default Mode Network.

Locator Candidates

It seemed obvious that the isolating sense of self might have begun as a primitive sense of location. One such candidate is Magnetoception. **Hypothesis H$_{28}$** posits that

Magnetoception is the source of the persistent low frequency rhythms of the Default Mode Network. **Hypothesis Null$_{28}$** argues that Magnetoception is not the source of the persistent low frequency rhythms of the Default Mode Network in the brain.

> Magnetoception (or magneto-reception) is the ability to detect a magnetic field to perceive direction, altitude or location. This sense plays a role in the navigational abilities of several animal species and has been postulated as a method for animals to develop regional maps.
>
> Magnetoception is most commonly observed in birds, where sensing of the Earth's [electro] magnetic field is important to navigational abilities during migration; it has also been observed in many other animals including fruit flies, honeybees, turtles, bacteria, fungi, lobsters, sharks and stingrays.[276]

Magnetoception has also been demonstrated to exist in human beings, though this line of research had humble beginnings.

> Bones in the human nose, specifically the ethmoid bone, contain magnetic deposits of ferric iron.[277] Beginning in the late 1970s, the group led by Robin Baker at the University of Manchester began to conduct experiments that purported to exhibit magnetoception in humans: people were disoriented and then asked about certain directions; their answers were more accurate if there was no magnet attached to their head.[278] These results could not be reproduced

by other groups and the evidence remains ambiguous[279]

Recently some other evidence for human magnetoception has been put forward: [indicating that] low-frequency magnetic fields can produce an evoked response in the brains of human subjects.[280]

Indeed, in 2007 Carrubba, Frilot II, Chesson and Marino of the Department of Orthopedic Surgery at Louisiana State University, found that "Human subjects respond to low-intensity electric and magnetic fields."[281] These researchers determined that "if the ability were a form of sensory transduction, one would expect that fields could trigger evoked potentials, as do other sensory stimuli."[282] They tested this hypothesis by examining electroencephalograms from 17 subjects for the presence of evoked potentials caused by the onset and by the offset of a field-strength comparable to that in the general environment. They reported that:

Using the [nonlinear] method of recurrence analysis, magnetosensory evoked potentials (MEPs) in the signals from occipital [brain region] derivations were found in 16 of the subjects ($P<0.05$ for each subject). The potentials occurred 109–454 milliseconds after stimulus application, depending on the subject, and were triggered by onset of the field, offset of the field, or both. ... MEPs in the signals from the central and parietal electrodes were found in most subjects using [nonlinear] recurrence analysis. ... The occurrence of MEPs in response to a weak magnetic field suggested the existence of a human magnetic sense.[283]

A few basic facts quickly eliminate electromagnetic fields and "Magnetoception" as candidate sources of the Default Mode Network. First of all, if the Default Mode Network were driven by the electromagnetic field, most modern humans living in cities, under power lines, standing in microwave transmissions, talking on cell phones, working on computers, and being treated in electromagnetic spectra filled hospitals, would have seriously disrupted Default Mode Networks – and they don't.

Secondly, the wavelengths of the electromagnetic spectrum vary from radio waves at 10^3 Hertz (i.e., 1,000 cycles per second) to Gamma rays at 10^{-12} Hertz.[284] These wavelengths can and do affect individual particles, atoms, cells and molecules of atomic and subatomic quantum physics. The collective signal patterns of the Default Mode Network, however, – at persistent wavelengths of about once every 10 to 20 seconds (0.1 to 0.05 Hertz) – are totally insensitive to the entire electromagnetic spectrum. If they weren't, the wave of the Default Mode Network would be broad-band-jammed right out of existence.

Not only do we live in a world chocked full of man-made electromagnetic signals, but Earth generates its own immense electromagnetic field, which sits in the even greater primordial electromagnetic field that pervades *all* space-time. Figure 76 illustrates just how great the electromagnetic field current sheet is in our local solar neighborhood. The tiny sphere in the central region of the figure is our Sun.

**Figure 76 Heliospheric Current-Sheet about the
Sun Out to the Orbit of Jupiter**
*Source: http://en.wikipedia.org/wiki/Heliosphere, Revisited January 8,
2010*

Hypothesis Null₂₈ is retained. Magnetoception can
not be the source of the persistent low frequency rhythms of
the Default Mode Network in the brain. The slow rate and
persistence of the Default Mode Network strappingly
indicates that Magnetoception simply cannot be the source of
the Default Mode Network driven sense of self.

The Most Likely Candidate Source

The sole candidate source of the waves of the Default
Mode Network is Earth's gravitational Hum – specifically the

waves of the Microseism Hum. It seems only logical that the source ultimately giving rise to the human sense of self should be universal, primordial, and emerge from the basic need to orient relative to the gravitational field. As related by Melvin J. Cohen back in 1955:

> The great majority of animals normally maintain a definite attitude within the gravitational field which Fraenkel & Gunn (1940) [*The Orientation of Animals,* Oxford: Clarendon Press] termed the primary orientation. It is the characteristic position an animal assumes when at rest and from which all active movements originate. Such a normal position is generally not defined by the organism's physical centre of gravity, but is actively achieved and maintained. An orientation of this type is possible only if receptors are available which are sensitive to the magnitude and relative direction of gravitational force.[285]

As detailed earlier, the sensory organs of the vestibular system detect the subtlest variations in the Earth's gravitational field. In discussing the *Statocyst organ* – an early precursor of the vestibular system – Cohen notes:

> The organ classically associated with geo-orientation is the statocyst, a term which has been applied to the vertebrate utriculus and sacculus as well as to the invertebrate equilibrium organ [the otolith] (Fraenkel & Gunn, 1940; von Buddenbrock, 1952).
>
> ...

The statocyst can be described as an ectodermal sac, fluid-filled and lined by hairs which are in contact with a relatively dense mass, the statolith. Such structures occur in all major metazoan phyla (Plate, 1924; Kolmer, 1926) and their appearance in the Hydromedusae, which are still at a tissue level of organization, qualify them in a real sense as the first organ to appear in the animal kingdom.[286]

Cohen emphasizes the surprisingly universal neuronal language employed by the diverse range of gravitational field sensors that have evolved throughout the animal kingdom.

Throughout the animal kingdom actively mobile animals differing greatly in phylogeny have independently evolved a similar means of signaling orientation within the gravitational field. The common functional solution is arrived at by using fundamentally different anatomical accessories. The lobster and the ray both have receptors capable of detecting rate and direction of angular displacement; but the significant point is that in animals widely separated in evolution this information is presented to the central nervous system in a similar 'language'. The similarities are not superficial; they extend from the broad plane of identical response patterns to minute details of threshold. Convergent evolution has occurred in this instance with respect to the coding of afferent messages.[287]

The 10 to 20 second waves of the Earth's Microseism Hum stand as a candidate match to the 10 to 20 second waves of the Default Mode Network. Bridging this candidate cause and effect is the vestibular sensory and signal distribution apparatus. The human vestibular system transmits signals to all of the key areas of the brain directly associated with those of the Default Mode Network. Figure 77 illustrates this physiological fact.

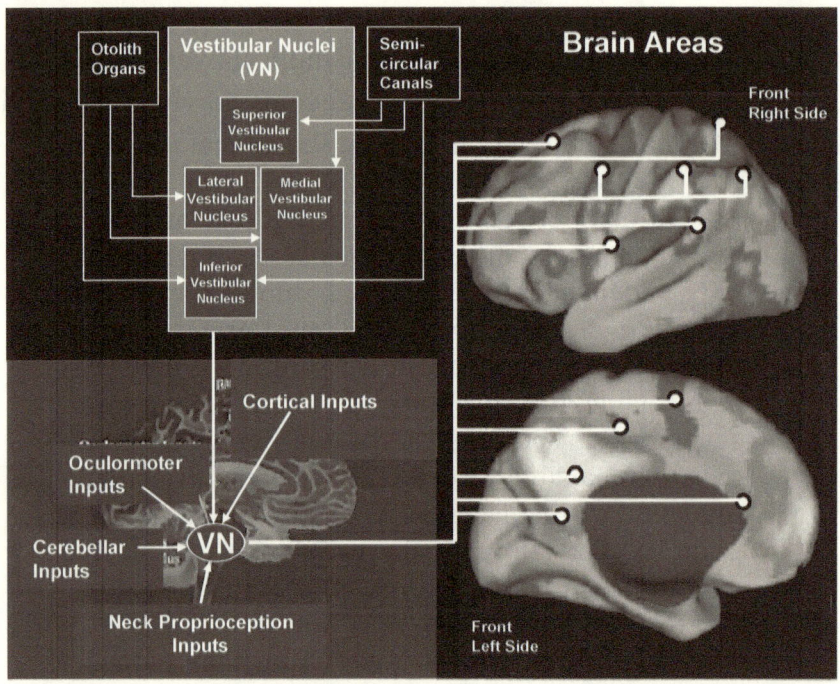

Figure 77 Correspondence of Vestibular System Outputs to the Brain Sites of the Default Mode Network

Source:http://www.scholarpedia.org/article/Vestibular_system, Revisited October 6, 2009

Only the vestibular system integrates and contributes all of the sensory information needed to define and differentiate an internally integrated sense of self. A self discriminated from the gravitational field environment. Given these facts, the next chapter relates a test of the notion that the waves of Earth's Microseism Hum are the source of the waves of the human Default Mode Network.

13

WAVES OF HUMAN INFLUENCE

Question

Is the *source* of the waves of the Default Mode Network the Earth's Microseism Hum? If the Microseism Hum does drive the Default Mode Network, it could explain how the density compression imposed by a Gravitational Torque Wave exerts its demonstrated influence on fundamental human behavior.

The case has been made in this book that Gravitational Torque Wave density compression describes, explains, and predicts the majority of variation in the diffusion of human innovation and change. What remains to be seen is how GTW Compression stimulates the human physiology to evoke that causal relationship. We assert that the waves of the Microseism Hum are that means of GTW stimulation. We further posit that the otolith organs of the human vestibular system constitute the primary transducer of that GTW driven environmental stimulus.

The human vestibular sensory apparatus is clearly capable of sensing the waves of the Microseism Hum. The discharge rates of vestibular hair cell neurons of the otolith and semicircular canal organs way over-samples the variations in such gravitational wave pressures. Moreover, vestibular afferent nerve fibers distribute those signals to the very brain sites implicated in the cycles of the Default Mode Network. Most importantly, the 14.7 second peak-to-peak mean-cycle-times of the waves of the Microseism Hum are a strong match to the 14.3 second average wave cycle rates of the resting Default Mode Network. If the variances of the respective wave parameters are equal – how could the waves of the Microseism Hum *not* be the primary source of the resting Default Mode Network?

A rigorous test of the posited causal assertion constitutes a genuinely large order. It requires the collection of Microseism Hum waves, in parallel with the collection of Default Mode Network wave data from a representative sample of subjects, at a single geographic location under precisely controlled conditions, over an extended period of time. Such an experiment would be demanding, expensive, and mandate the use of a host of medical and gravitational field measurement instruments. Heretofore, no such experiment has been proposed, let alone performed.

A gross approximation of such a test is enabled. Using previously collected Microseism Hum and Default Mode Network wave data, we realized a crude comparison of the waves of the Microseism Hum and the resting Default Mode Network. This chapter reports the means, methods, and results of that study. That preliminary investigation strongly indicates that a more ambitious experiment is clearly warranted.

Data

The data sets selected were collected five years apart, on two different continents, by scientists not likely to have knowledge of each other's experiments. Hence, any possible correspondence between the waves of the resting Default Mode Network and seasonally varying waves of the Microseism Hum would have to be genuinely strong to emerge.

The Microseism Hum data used were reported by Stehly, Campillo, and Shapiro, (2006), in "A Study of the Seismic Noise from its Long Range Correlation Properties," December 21, 2005, p.22.[288] Microseism Hum wave data were collected at the California "MLAC and PHL" seismographic station sites "for the months of January thru December of 2002."[289]

The Default Mode Network functional magnetic Resonance Image (fMRI) data used were collected in 2007 in Spain. These data were reported by Ben J. Harrison, Jesus Pujol, Marina Lo´ pez-Sola`, Rosa Herna´ ndez-Ribas, Joan Deus, Hector Ortiz, Carles Soriano-Mas, Murat Yu¨ cel, Christos Pantelis, and Narcı´s Cardoner in 2008 in a paper titled "Consistency and functional specialization in the default mode brain network," PNAS (National Academy of Sciences), July 15, 2008 vol. 105 no. 28, 9781–9786.[290]

The Microseism Hum (MSH) has exhibited an increasing long term trend over the last 30 years.[291] Hence the five years that elapsed between the collection of the Microseism Hum wave data in 2002, and the collection of

Default Mode Network (DMN) waves in 2007, further handicapped the tests. Any correspondence between the waves of the MSH and the DMN would have to emerge despite the total absence of that five year long trend in the Microseism Hum data.

Measures

Figure 78 depicts definitions of the measures used to describe and compare the DMN and MSH waveforms. The example figure contrasts one respective MSH and one DMN wave cycle. Each wave in the figure has a Wavelength (λ) of 14 seconds (360 degrees), and varies in Amplitude (γ) from a normalized value of + 50 to -50 units. Each of the two example waves depicted has a Velocity (V) of 25.7 degrees per second, and a Frequency (n) of 1.8 times (vibrations) per second. The Greek symbol θ (Omega) is used in the figure to represent: the Phase (θ_1), Time (θ_2), and Amplitude (θ_3) shifts (offsets) between the two waves per quarter cycle (90 degrees). The formula for the calculated seismic Wave Number (k) used in the study is also defined in the figure.

Hypotheses

Hypotheses (**H$_{29}$ through H$_{100}$**) under test take the form: $\mu_1 = \mu_2$, or in words, that the variances of two compared samples are equal and hence from the same population at the .05 level of confidence. In each hypothesis the two samples compared are: 1) the characteristic waveform of the resting DMN and 2) each of the MSH characteristic waveforms for the twelve months from January through December. The null hypothesis for each test conducted is that $\mu_1 \neq \mu_2$, or in words, that the variances of the two samples are *not* equal.

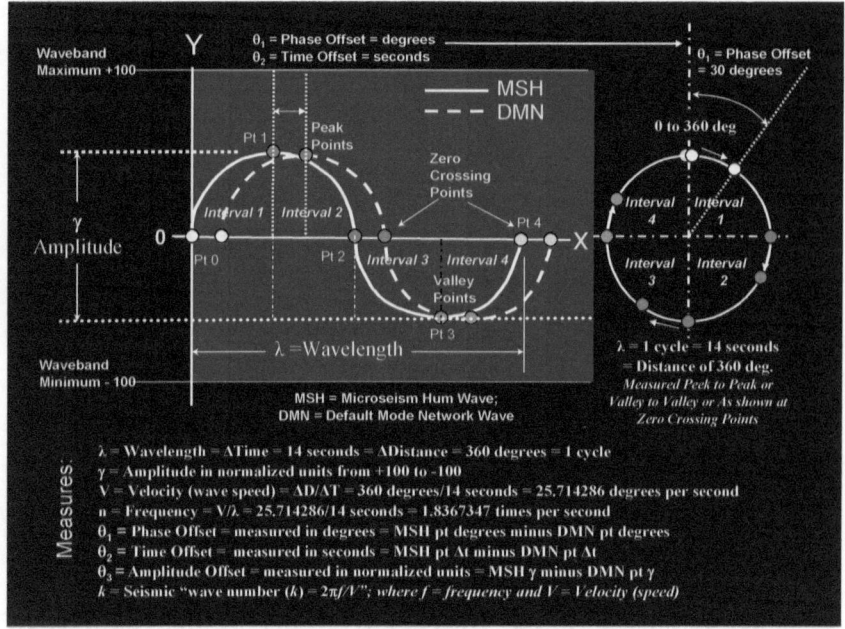

Figure 78 Study Metrics

Hypotheses H_{29} to H_{40} compares the resting DMN waveform variances in *Wave Number* (k) with the paired wave number variances of the MSH waveforms. The twelve subscripted hypotheses match the months of MSH data collection from January through December of 2002. Each *Wave Number* (k) is calculated using the seismic wave number formula, $k = 2\pi f/V$; where f = the wave frequency in vibrations per second and V = wave Velocity in degrees per second. That formula was defined by Sivaji, C., Nishizawa, O., Kitagawa, G., and Fukushima, Y., (2002) in, *"A Physical-Model Study of the Statistics of Seismic Waveform Fluctuations in Random Heterogeneous*

Media," as reported in the Geophysics Journal International, 2002, 148, 575-595 on page 588.[292]

Hypotheses H_{41} through H_{52} compares the DMN waveform variances in peak, zero-crossing, and valley point *Arrival Times* with matching Arrival Time points of the MSH waveforms for each of the months from January through December.

Hypotheses H_{53} through H_{64} associates the DMN waveform variances in peak and valley point *Amplitudes* (in normalized units) with the paired points of the MSH waveforms for the months of January through December.

Hypotheses H_{65} through H_{76} compares the DMN waveform variances in *Wavelength* (in seconds) with the paired Wavelengths of the MSH waveforms for the months of January through December.

Hypotheses H_{77} through H_{88} contrasts the variances in DMN waveform *Velocities* (in degrees per second) with the paired Velocity variances of the MSH waveforms for the months of January through December.

Hypotheses H_{89} through H_{100} compares the variances in DMN waveform *Frequencies* (in vibrations per second) with the paired Frequency variances of the MSH waveforms for the months of January through December.

Hypothesis H_{101} asserts that in the majority of cases where the variances of the DMN waveform parameters tend to equal the variances of the MSH, the *Velocity* (V) of the Default Mode Network waves will equal or lag the *Velocity* of the Microseism Hum waves. Hypothesis H_{101} thus posits a cause and effect relationship between the MSH and DMN. The logic being that the *cause* must equal or precede the time of the *effect*. Hypothesis Null$_{101}$ counters that the waves of the Microseism Hum tend to be slower in Velocity than the waves of the Default Mode Network, thus precluding any causal occurrence and relation.

Results

The twelve Figures which follow (79 through 90) depict the single characteristic resting DMN waveform, overlaid on each calendar month specific characteristic MSH waveform. The figures progress from the month of January (Figure 79) through December (Figure 90) with each unique MSH wave being graphically contrasted with the single characteristic resting DNM wave.

The DMN waveform corresponds best to the MSH waveforms from the months of January through March. This first quarter match up is hardly a surprise. Seasonal MSH waves exhibit changes in direction in April and reduced numbers of wave cycles starting in June. The MSH waves change direction again in October, and reach the annual minimum of wave cycles in December. This seasonal grouping of sets of MSH waveforms makes the isolation of the associated DMN waveform highly probable.

What was surprising was that even though the DMN wave data were collected five years after the MSH data, not only the calendar quarter but the specific month in which the DMN data were collected appears to be discriminated by the corresponding MSH waveforms. Even a visual comparison of the overlaid waveforms indicates that the DMN wave data were collected in the first quarter of 2007, and most likely in the month of February. The question became, would subsequent quantitative analyses validate that casual observation.

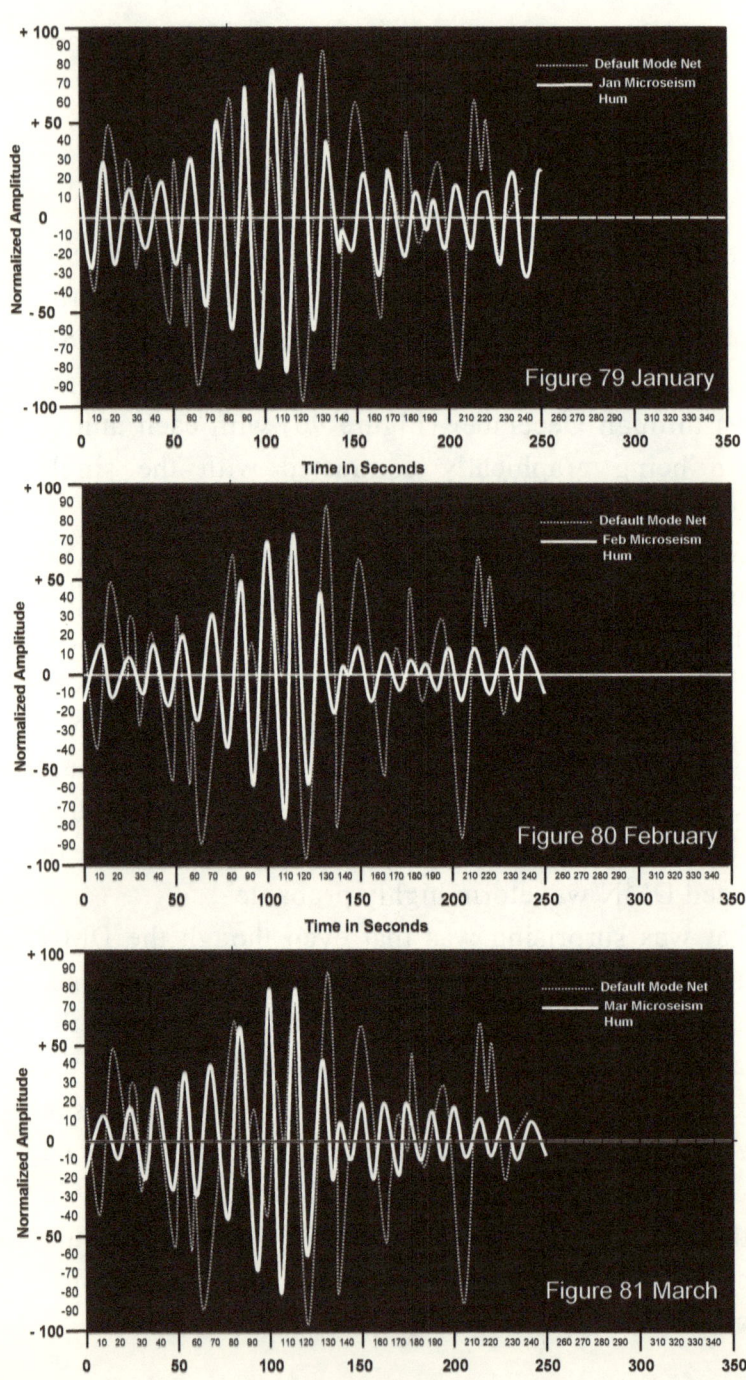

Figure 79 January

Figure 80 February

Figure 81 March

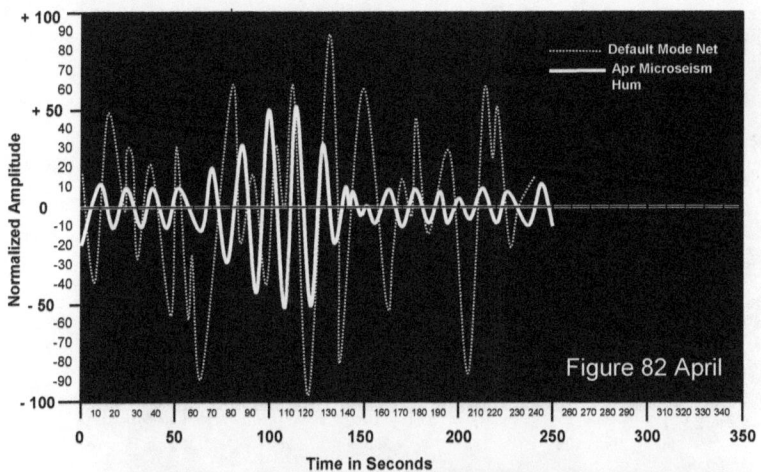

Figure 82 April

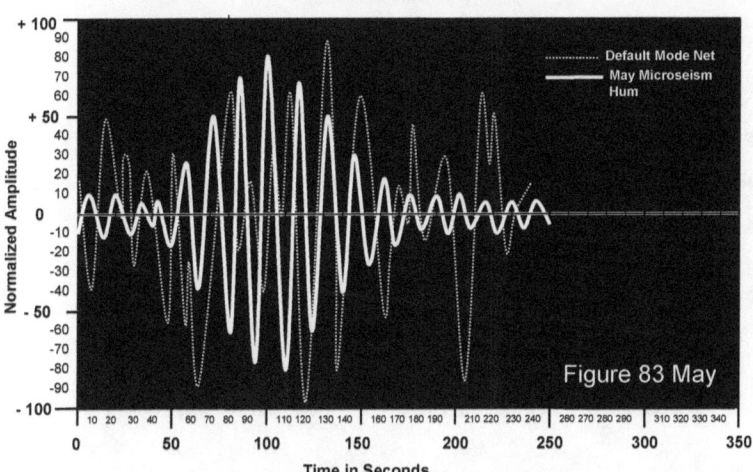

Figure 83 May

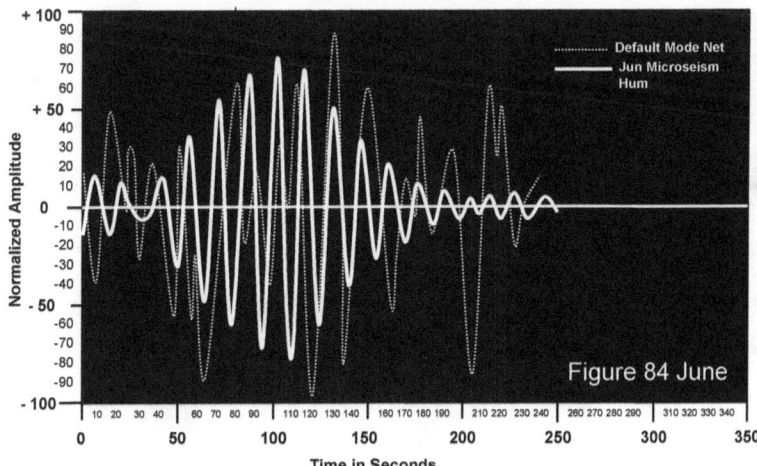

Figure 84 June

305

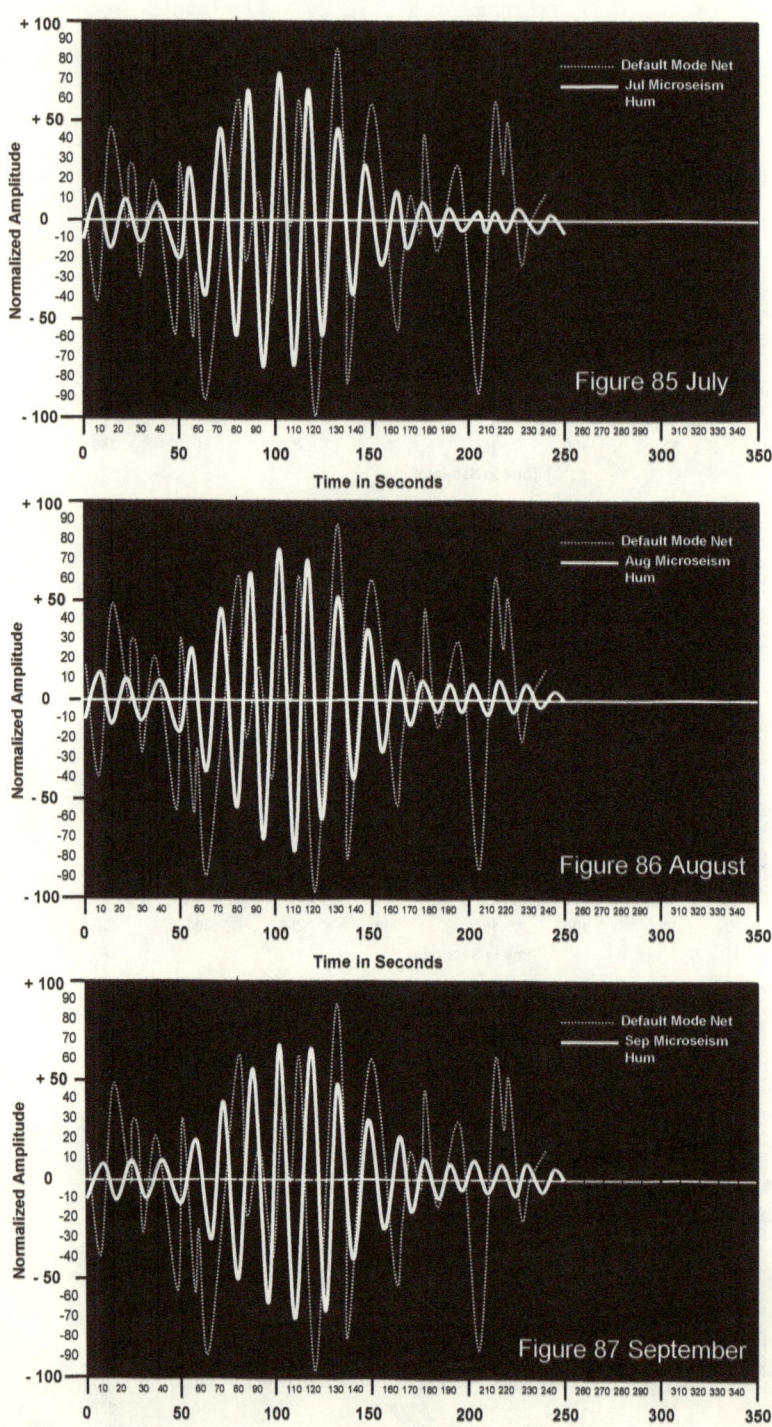

Figure 85 July

Figure 86 August

Figure 87 September

306

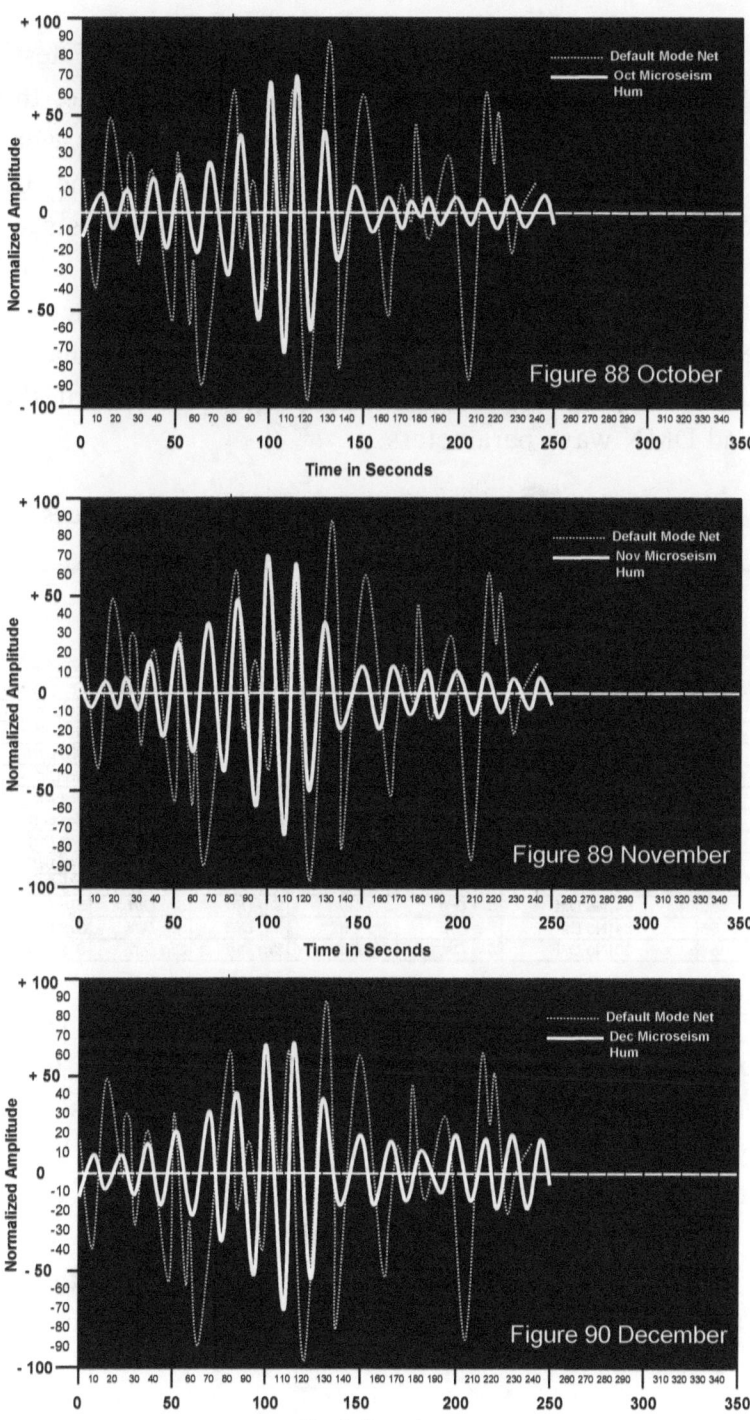

307

Table 19 summarizes the results of statistical F-tests conducted to test **Hypotheses H$_{29}$ to H$_{100}$**. In the table, the words 'No Diff' indicate that no significant differences were found between the MSH and the DMN variances at the .05 level of confidence. In such cases the null hypothesis of significant differences is rejected. The term *'Sig Diff'* indicates that the null hypothesis of significant differences is retained. To aid the examination of the results, gray shaded cells denote significant differences in the variances of the MSH and DMN wave parameters.

Table 19 Two Sample F-Tests for Equality of Variances

MSH Months	No. of Cases	Deg. of Freedom	Spatial Ratio of Freq to Velocity / Wavenumber / W.No (k)	(In Secs) Peeks, Vallies, & Zero Crossing / Pt Times	Amplitude Offsets Normalized / Amplitudes	Time Offsets (In Secs) / W.lengths	Phase Offsets (Deg/Sec) / Velocities	(Vib/Sec) / Frequency
Jan	67	66	No Diff	No Diff	*Sig Diff*	No Diff	*Sig Diff*	No Diff
Feb	67	66	No Diff	No Diff	No Diff	*Sig Diff*	*Sig Diff*	No Diff
Mar	67	66	No Diff	No Diff	*Sig Diff*	No Diff	*Sig Diff*	*Sig Diff*
Apr	67	66	No Diff	No Diff	*Sig Diff*	No Diff	*Sig Diff*	No Diff
May	67	66	No Diff	*Sig Diff*	*Sig Diff*	No Diff	*Sig Diff*	*Sig Diff*
Jun	65	64	No Diff	*Sig Diff*	*Sig Diff*	*Sig Diff*	*Sig Diff*	*Sig Diff*
Jul	65	64	No Diff	*Sig Diff*	*Sig Diff*	No Diff	*Sig Diff*	*Sig Diff*
Aug	65	64	No Diff	*Sig Diff*	*Sig Diff*	No Diff	*Sig Diff*	*Sig Diff*
Sep	65	64	No Diff	*Sig Diff*	*Sig Diff*	*Sig Diff*	*Sig Diff*	*Sig Diff*
Oct	65	64	No Diff	*Sig Diff*	*Sig Diff*	*Sig Diff*	*Sig Diff*	*Sig Diff*
Nov	65	64	No Diff	*Sig Diff*	*Sig Diff*	*Sig Diff*	*Sig Diff*	*Sig Diff*
Dec	61	60	No Diff	*Sig Diff*	*Sig Diff*	*Sig Diff*	*Sig Diff*	*Sig Diff*

Hypotheses H$_{29}$ through H$_{40}$ compared the resting DMN waveform variances in *Wave Number* (k) with the paired wave number variances of the MSH waveforms for the months from January through December. The DMN variances in Wave Number (k) are *not* significantly different from the MSH Wave Number variances for any of the 12 months of MSH waves at the .05 level of confidence. Therefore Hypotheses **Null$_{29}$ through Null$_{40}$** are rejected.

Hypotheses H_{41} **through** H_{52} compared the DMN waveform variances in wave peak, zero crossing, and valley point *Arrival Times,* with the matching arrival time points of the MSH waveforms for the months of January through December. DMN variances in Arrival Times are not significantly different from the MSH wave point arrival times for the months of January through April. The variances in Arrival Times are significantly different for the months of May through December. Therefore, **Hypotheses H_{41} through H_{44}** are *retained* at the .05 level of confidence and the associated null hypotheses are rejected. Correspondently, **Hypotheses H_{45} through H_{52}** are *rejected* and their associated null hypotheses retained.

Hypotheses H_{53} **through** H_{64} compared the DMN waveform variances in peak and valley point *Amplitudes* with the paired points of the MSH waveforms for the months of January through December. With only one exception, the variances in wave peak and valley Amplitudes for all other months (**H_{53}, and H_{55} through H_{64}**) of MSH are significantly different from the variances in resting DMN Amplitudes. The February MSH Amplitude variances (**H_{54}**) were *not* significantly different from those of the DMN at the .05 level. Hence, only the February **Hypothesis, H_{54}** is retained.

Hypotheses H_{65} **through** H_{76} compared the DMN waveform variances in *Wavelength* (in seconds) and the paired Wavelengths of the MSH waveforms for the months of January through December. The DMN variances in Wavelength are found *not* significantly different at the .05 level of confidence from those of the MSH for the months of January (**H_{65}**), March through May (**H_{67} to H_{69}**), and July (**H_{71}**) through August (**H_{72}**). For all the other months (February **H_{66}**, June **H_{70}**, and September **H_{73}** through

December H_{76}) the null hypotheses of unequal variances are retained.

Hypotheses H_{77} through H_{88} compared the variances in DMN waveform *Velocities* (in degrees per second) with the paired Velocity variances of the MSH waveforms for the months of January through December. The DMN variances in Velocity (speed) are all significantly different from those of the MSH for all twelve months. Therefore, Hypotheses Null$_{77}$ through Null$_{88}$ are retained.

Related Hypothesis H_{101} asserted that in the majority of cases where the variances of the DMN waveform parameters tend to equal those of the MSH, the *Velocity* (V) of the DMN waves will equal or lag those of the MSH waves. Hypothesis H_{101} thus posits a cause and effect (lead-lag) relationship between the waves of the MSH and DMN. Table 20 summarizes the categorical velocity differences found.

Table 20 Velocity Percentages

| | | Percent of Cases | | | % DMN |
MSH Months	No of Cases	DMN Faster	DMN = MSH	DMN Slower	Equal or Slower
Jan	67	45%	22%	33%	55%
Feb	67	37%	21%	42%	63%
Mar	67	46%	18%	36%	54%
Apr	67	37%	19%	43%	63%
May	67	46%	10%	43%	54%
Jun	65	46%	14%	40%	54%
Jul	65	48%	15%	37%	52%
Aug	65	52%	11%	37%	48%
Sep	65	45%	25%	31%	55%
Oct	65	52%	11%	37%	48%
Nov	65	45%	23%	32%	55%
Dec	61	51%	13%	36%	49%

The shaded cells of the table indicate those months in which the MSH and DMN wave parameter variances tend to be equal. In the months from January to April, the hypothesized DMN and MSH Velocity (speed and lead-lag) relation is found. In fact, overall, the wave cycle velocities of the DMN tend to equal or lag those of the MSH waves. Thus, **Hypothesis H_{101}** is supported for all months save August, October, and December. MSH waves tend to precede or coincide with those of the DMN. While a cause and effect relationship cannot be confirmed, one is at least implied and appears to warrant further investigation.

Hypotheses H_{89} through H_{100} compared the variances in DMN waveform *Frequencies* (in vibrations per second) with the paired Frequency variances of the MSH waveforms for the months of January through December. Hypotheses of no significant differences were supported for the months of January (**H_{89}**), February (**H_{90}**), and April (**H_{92}**). For those months the variances in the Frequencies of DMN waves equaled those of the MSH at the .05 level of confidence. **Hypotheses** of significant differences, however, are retained for the months of March (**$Null_{91}$**) and May through December (**$Null_{93}$ through $Null_{100}$**).

Table 21 provides the squares of the correlation coefficients indicating the proportion of variance in DMN explained by the MSH. As before, shaded cells mark those cases in which the variances of the DMN wave parameters do *not* equal those of the MSH wave parameter. Shaded cell squared correlations must be considered suspect. The formula for the calculation of correlation coefficients assumes the equality of sample variances.

Table 21 Squared Correlation Coefficients

	N	Wave Number 2πf/V W. No (k)	Peak, Valley, & Zero Crossing Pt Times	Amplitude Offsets Amplitude	Time Offsets W.Length	Phase Offsets Velocity	Freq
Jan	67	98.62%	98.56%	35.24%	2.21%	0.04%	32.29%
Feb	67	97.81%	98.65%	30.44%	2.08%	0.47%	9.01%
Mar	67	98.62%	98.82%	36.58%	4.21%	4.94%	43.10%
Apr	67	98.62%	97.47%	30.17%	0.18%	7.06%	4.95%
May	67	97.81%	99.49%	37.58%	0.30%	7.06%	30.82%
Jun	65	98.52%	98.95%	37.04%	0.79%	0.99%	39.99%
Jul	65	98.52%	98.85%	33.86%	0.88%	0.00%	36.65%
Aug	65	98.52%	98.96%	34.84%	0.36%	0.13%	39.10%
Sep	65	98.52%	99.01%	37.87%	5.37%	4.76%	46.08%
Oct	65	97.66%	98.80%	30.22%	0.00%	0.45%	17.45%
Nov	65	98.52%	99.66%	33.40%	0.59%	10.71%	54.01%
Dec	61	98.44%	99.40%	39.57%	2.31%	0.59%	27.77%

The Wave Numbers of the Microseism Hum almost perfectly describe, explain and predict the corresponding Wave Numbers of the Default Mode Network for all 12 months of the MSH. This is the case despite the five years and substantial geography that separated the two samplings.

The variances of the MSH and DMN wave peak, valley, and zero-crossing point Arrival Times are *not* equal for the months of May through December. For those months, the fundamental assumptions of the correlation calculation are violated. That inequality of variance violation renders those squared correlation coefficients in the shaded cells suspect and unusable.

The squared correlations between the peak, zero-crossing, and valley point Arrival Times for the months from January through April are valid. Those values indicate that the MSH waveform Arrival Times near perfectly predict the majority of the variation in associated DMN waveform Arrival Times.

Figure 91 depicts that predictive relation between the MSH and DMN wave point Arrival Times for the month of February. The plot is indicative of the predictive relations between MSH and DMN arrival times for the months of January through April. Since the Arrival time variances are equal for those particular months, the near perfect prediction of the DMN wave point Arrival Times by the MSH are valid and reliable.

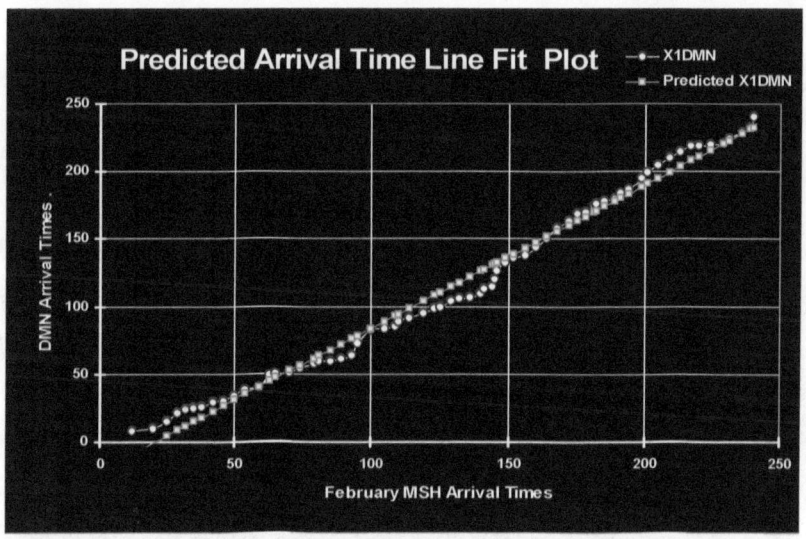

Figure 91 February Microseism Hum Arrival Times Prediction of Default Mode Network Wave Arrival Times

The variances of MSH wave *Amplitudes* equal those of the DMN waves solely during the month of February. February MSH Amplitudes describe, explain, and predict just over 30 percent of the variation in DMN wave amplitudes. For all other months, the Amplitude variances are *not* equal, and the squared correlation results suspect and unusable.

The variances of the MSH Wavelengths are not significantly different from the Wavelengths of the DMN for the months of January, March, April, May, July, and August. Yet the MSH and DMN Wavelengths are uncorrelated for all the cases analyzed. The lack of correlation appears to result from the nonlinear nature of the wavelengths, and the lead-lag relationship between associated data points. Figure 92 illustrates the nonlinear tendencies and lead-lag relations of the Wavelength data for the month of February.

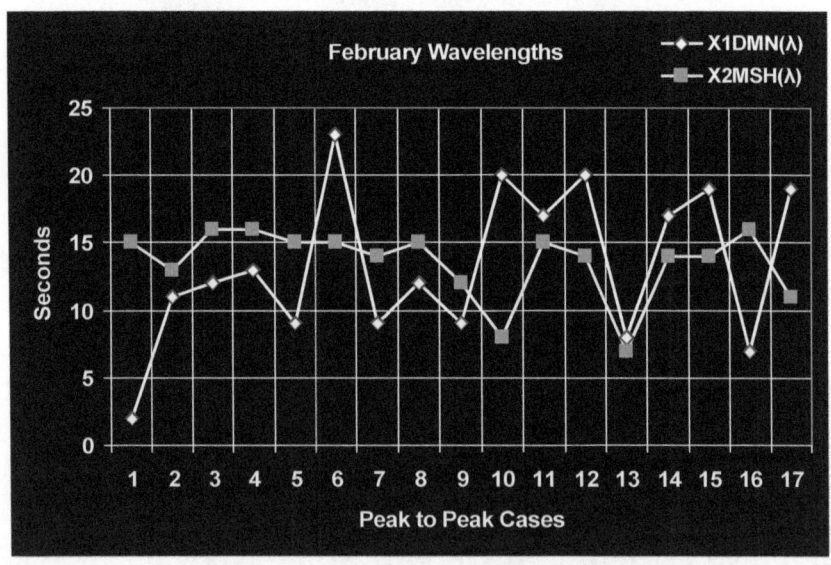

Figure 92 February MSH and DMN Wavelengths

It should be noted that two *perfectly correlated* parabolas (arcs) will return a product moment correlation coefficient of zero. The false result – the indicated absence of a correlation – is due to the nonlinear nature of parabolic data. The formula of the product moment correlation assumes a straight line linear relationship, and a parabola is a curved arch. That explains the lack of correlation between the MSH and DMN wavelengths and frequencies for the months of February and March.

For the majority (9 out of 12) of the months, DMN wave velocities tend to be equal to or slower than those of the MSH waves. That offset, and the nonlinear nature of the velocity data, explain the inequality of variance in velocity and the poor correlations

Figure 93 depicts a pie chart contrasting the percentages of DMN wave velocities that are Faster than, slower than, and equal to the February MSH Velocities. Figure 93 is reasonably representative of nine of the total twelve months sampled.

Only the January MSH wave Frequencies predicts a significant portion of the variation in the Frequencies of the DMN waves. Figure 94 charts the January MSH and DMN Frequencies. The presence in the figure of extreme Frequency values in both the MSH and DMN waveforms reveals an interesting feature of each.

The Microseism Hum waveband is described in the prevailing literature as ranging in frequency from 0.1 hertz (10 second waves) to 0.5 hertz (20 second waves). Yet, as shown in Figure 94, in two of the January quarter cycles sampled the Microseism waves reached 1.0 vibration per second. Microseism frequencies ranging from 0.5 hertz to

1.0 hertz are those signals ascribed to seismic events (e.g., tremors) in the greater Microseism waveband. Clearly, these anomalous spikes extend beyond the range of the Earth's Microseism Hum.

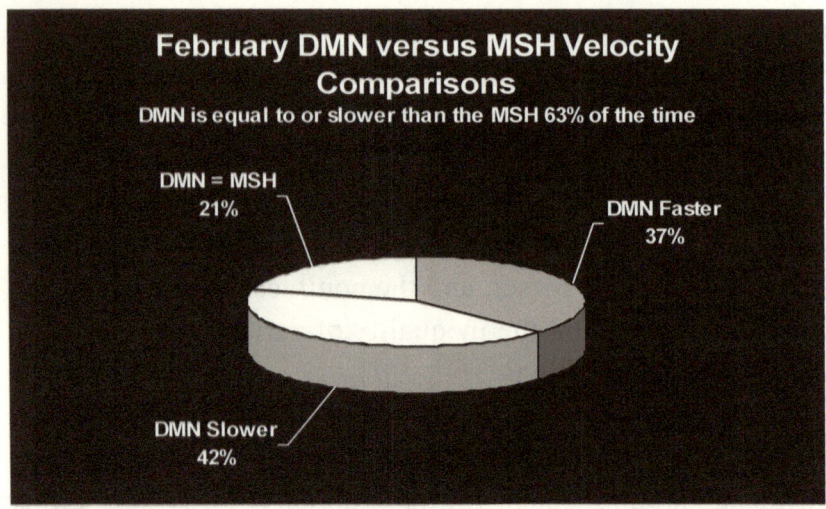

Figure 93 Categorical Comparison of February Velocities

Also shown in Figure 94 are three DMN wave spikes in frequency. These spikes indicate neurophysiologic events recorded by fMRI measures across subjects. Such signals indicate either an anomalously large externally imposed stimulus, or measurement errors that occurred during data collection.

Only one of the MSH extreme values is paired with a matching extreme DMN frequency value. Five years elapsed from MSH to DMN data collection. Hence, these spikes cannot possibly represent any DMN and MSH time correlated phenomena. The dramatic spikes in both data sets must be considered unrepresentative events that occurred outside of the normal range of either the DMN or the MSH wave bands of interest.

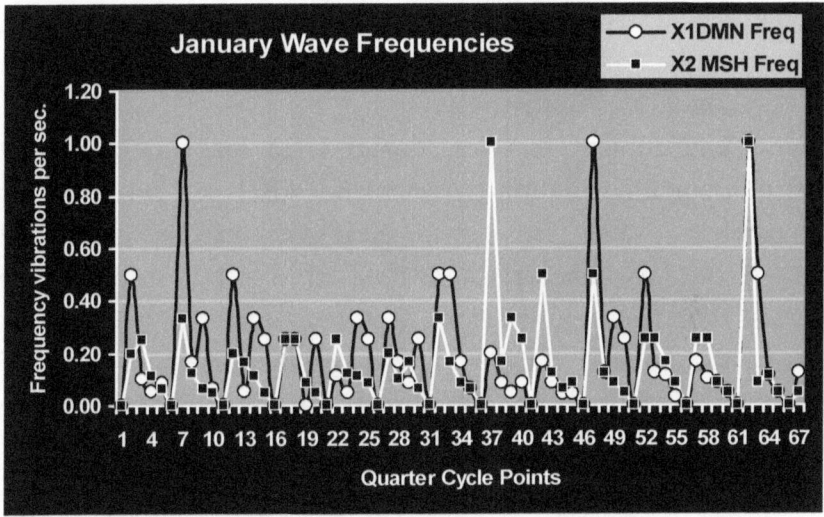

Figure 94 January MSH and DMN wave Frequencies

It is suspected that all the valid correlations between the February MSH and DMN amplitudes, wavelengths, frequencies, and velocities are substantially greater than those reported here. The strength of the correlations is most likely suppressed by the time and geography differences between data collection events, the extreme outlier cases included in both samples, and the less than linear nature of certain of the measures (wavelengths, frequencies, and velocities).

Interpretation

The preliminary test results just reported support but one conclusion. The source of the waves of the resting Default Mode Network could very well be the gravitational waves of the Microseism Hum. Waves sensed and decoded by the human vestibular system.

The cursory test was designed to make it extremely difficult for any relation between the MSH and DMN waves to emerge. Yet, significant equalities of variances and descriptive, explanatory and predictive relationships were found. The test results strongly suggest that a more rigorous test of the posited causal relation between the waves of the Microseism Hum and the waves of the Default Mode Network is warranted. Though such an experiment mandates the once-a-month use of expensive medical and gravitational field measurement instruments for a full year of testing, the investment appears justified.

Theoretic Implications

This research points to the fundamental mechanism by which Gravitational Torque Waves most likely influence human behavior. Once every 240 seconds, the 2.5 modal wave peaks of the very slow Gravitational Hum are associated with 16 to 18 modal peaks of the Microseism Hum. The slow waves of the Gravitational Hum are the likely result of transverse density compression waves of vertical and horizontal (twisting torque) imposed on this planet by the current GTW. The waves of the Gravitational Hum and Microseism Hum appear to represent this planets' momentary, daily, and seasonally varying responses to that GTW. The GTW stimulus and the resulting Hum modes of response appears to echo the GTW Compression cause and *K'ul* response effects demonstrated in earlier chapter test results. The fact that the Microseism Hum has been steadily increasing over the last 30 years is highly consistent with that interpretation.

The demonstrated variation in the slow waves of the Gravitational Hum and planetary Microseism Hum appears to be consistent with the Maya Kin (day), Uinal (20 day), Tun (360 day), Katun (7,200 day) and Baktum (144,000 day) cycles of the Long Count calendar. As explained earlier, those Cycles are but scaled subdivisions of Maya World Age Precession. All of the sub-cycle scaling subdivisions have been shown to describe, explain, and predict the majority of variation in the celestial motion of our solar system, this planet, and Earth's climate.

The waves of the Microseism Hum near perfectly predict the wave numbers and Arrival Times of the waves of the resting Default Mode Network. The waves of the Microseism Hum appear to partition nearly one third of the variation in DMN wave amplitude for the month of data collection. Moreover, MSH waves tend to precede or equal DMN waves, suggesting a possible cause and effect relation between the two. The reported equality of variances and predictive relations between the MSH and resting DMN waves tend to indicate that the cyclic MSH wave responses of this planet are the originating source of the cyclic DMN waves in the human brain.

Most importantly, the relation between the waves of the MSH and the DMN provide a viable explanation as to how and why GTW Compression multiplied by the *K'ul* response (*K'ul* Resonance) so strongly predicts the economic behavior of human beings over the last half millennia. Such a casual relation also sheds light on why GTW Density Compression near perfectly predicts 280,000 years of human innovation and change events.

Hypothesis H$_{24}$ asserted that the neuronal signals of the human vestibular system are used by the brain to discriminate the internal human frame of reference (the self) from the external environmental non-self. **Hypotheses Null$_{24}$** posited that either: no such discrimination takes place, or that the source of self and non-self discriminations in the brain is other than vestibular.

Results reported in this and the prior chapter support **Hypothesis H$_{24}$** and seriously diminish the likelihood that **Hypotheses Null$_{24}$** is true. Substantial prior research implicates the vestibular system and the persistent waves of the DMN in the physiological "discrimination of the body-centric human self from the external-object-centric non-self."[293] The human internal frame of reference is widely believed to be "subtracted out of the sensed gravitational field environment frame of reference by the cerebellum."[294] The discrimination of the physiological self from the non-self is considered "a fundamental requirement for unconscious and conscious neural processing." The discrimination of the physical self from the environment is considered to be one of "the oldest neurological developments in the animal kingdom."[295]

Such findings might inadvertently lead to a false conclusion. That being that specific human behaviors – what is specifically done as opposed to how much is done how frequently -- result from the slow gradual influence of Galactic Gravitational Torque waves, the Microseism Hum response, and the resulting waves of the human Default Mode Network. That interpretation is incorrect in the extreme.

GTW compress the density of space-time. The effect of that compression is to multiply – as oppose to reduce – the number of proximate options available for free will to operate on. Rather than deterministically limiting specific human choices and actions, the progressive effect of GTW Density Compression is to vastly expand the number of options available, the number of choices made, and the number of actions taken. Physicist Markopoulou Fotani (1998) explains the scientific basis for this mathematical fact in excruciating detail in her paper, "Internal Description of a Causal Set: What the universe looks like from the inside."[296]

If you haven't found the motivation to expand your conceptual horizons yet, stay tuned. The next three chapters dare to investigate the possibility of a rapidly approaching GTW driven transformation of our entire solar neighborhood in under a minute. The very Transformation predicted by the ancient Maya.

14

TSUNAMI TRAINS OF TRANSITION

Maya Prophecy of Transformation

T he consensus among historians and archaeologists is that the ancient Maya constructed their Long Count Calendar by working backwards from a specific distant future date.[297] That now famous date equates on our Gregorian calendar to the winter solstice of Friday, December 21, 2012 AD.[298] The Maya believed that on that date (and near the end of all World Age Precession cycles) a two fold change will occur: "the destruction of the old age, and the *transformation* into a new, better age."[299 & 300] "The Maya world view," Say Synthia and Colin Andrews, held humans as "an integral part of an interactive living galaxy."[301] The Andrews describe the importance of this view to the Maya.

> The ancient Maya believed the center of the Galaxy, birthplace of the stars, was the home of the creator god,

Hunab K'u. Hunab K'u sends emissions from the center of the Galaxy that pulse through all the celestial alignments, bringing enlightenment. They believed that the alignment of the sacred cross with the center of the Galaxy will create a portal; a direct connection to *Hunab K'u*, accelerating the transmissions from the center of the Galaxy.[302]

"According to both the Maya and Toltec myth, humanity saves itself at these critical nexus points only by transforming, by mutating into something totally unrecognizable, a new being altogether."[303] Although the phrasing implies some sort of intentional transformation on the part of the Maya, the concept is actually more one of a naturally imposed transformation. According to Maya day keeper shaman, historian, and anthropologist Carlos Barrios (2002), both ancient and surviving Maya shaman and priests claim that the end of the World Age in 2012 is *"the end of space and time,"* the "Great alignment," and "The Transformation."[304]

Now don't go galloping off to the all too familiar and totally false conclusion being promulgated in the popular media. The shaman *did not say* that the end of the current World Age brings about the end of the solar system, this planet, or humanity.[305] After all, World Age Precession cycles have ended hundreds of thousands of times before. Over the last 4.6 billion years, our solar system, this planet, and more recently people, have continued; though often exhibiting significant changes.

The Maya stated that *during* any World Age transformation linear space-time measurements and predictions no longer apply. There is every indication that this is why the Maya elected to end (roll over the count of) their Long Count calendar on Friday, December 21, 2012. They simply felt they could not accurately predict Long Count cycle effects beyond that date. This chapter explains the astoundingly physical reasons why the Maya may have been smack on right about that.

Taken literally, the Maya were insisting that any World Age transformation is *"non-analytic."* That very modern term means that a post transformation state cannot be quantitatively predicted in advance of the transformation event. In theory, prior to the transformation of 2012, mathematical functions can be formulated to describe the progressing states of most things. Inside the transformation, however, no math functions apply. Once the transformation is complete an entirely new set of functions is required to describe, explain, and predict the states of things; functions that cannot be determined in advance of the transformation.

These kinds of pre and post transformation distinctions are described in detail by our current science. Not surprisingly, those very modern scientific details are of paramount importance in any attempt to equate the Maya concept of World Age Transformation to the ideas of our prevailing science. That is what is done here, but not to revive some spooky old Amerindian prediction. It is done to solve several irritating enigmas that the present research kicked up.

Wave Train and Caboose Quandary

Although delightfully successful, the relating of Maya
World Age Precession to GTW Compression raised some
seriously troubling questions. First off, how can the GTW
compression and the associated *K'ul* response to a *single*
GTW predict the majority of variation in the Rate of Human
Innovation from 1615 to the present? After all, each World
Age Precession cycle is the apparent product of 2.5 GTW.
Earth's Obliquity cycle emerges from 4 GTW. Earth's
orbital Inclination cycle results from 7 GTW. Earth's orbital
Eccentricity cycle results from 10 GTW. So why aren't those
cumulative wave-train effects needed to predict the variation
in the Rate of Human Innovation from 1615 to the present?

What makes all that even more worrisome is that it
takes cumulative GTW *wave trains* counted out in Degrees of
Maya World Age Precession to predict 280,000 years of
Human changes, Global Temperature change, and variations
in Long Lived Green House Gases. Yet, the compression
effects of a *single* GTW predict all those very same measures
from 1500 AD to the present, the last even Epoch in an entire
World Age. What gives? How can one GTW do in the short
term what it takes many GTW to explain in the longer term?

Hold on, it gets a whole lot worse. How can a World
Age Precession Cycle consist of just 2.5 GTW? How can a
World Age *cycle* end in a wave peak with no matching
trough? To cut straight to it, how can a *continuous* cycle end
in a *discontinuous* point? Lest we forget, those oddly ending
cycles perfectly predict Hominid speciation, extinction,
genetic bottlenecks, major mutations, and mass migration
events. If the Maya had somehow bungled the World Age

Epoch count, those pristine correlations and predictions could not exist – and they do.

Oddly, those glaring wave train and cycle end point anomalies weren't the least anomalous to the ancient Maya. Oh no, those long-gone folk saw such bizarre World Age endings as *signposts*. Milestones that mark the arrival of each major make-over of pretty much everything – events they eerily termed *the end of space and time*.

The fringy few prophesying the December 2012 end of the world, unanimously fail to cite a source of our wholesale destruction. They merely give us an end date that flies straight into the mountains of evidence proving our world has not been totally destroyed once every 25,626.8 years. Nevertheless, these persistent purveyors of doom continue to extol various versions of impending hell, in print, film, and on-line. Choosing to ignore these ever-present woolly minded peddlers of smog, we concentrate instead on the perplexing puzzles left behind by the clever head-haunting-shaman. Maya wise-men we'd yet to prove wrong.

Resonant Swing Solution

There had to be some decidedly physical explanation for the cumulative build up of GTW Compression effects. It had to be a model of an analogous energy storage process that could:

1) Explain the short term predictive success of single GTW

2) Explain the long term predictive success of World Age GTW wave trains

3) Solve the paradox of a continuous cycle that ends in a discontinuous point.

Strangely enough, part of the answer was found in the pendulum analogy of Galileo's physics of 1602.[306] Picture yourself sitting in a child's playground swing. Initially, the swing hangs straight, like a pendulum at rest. You now exercise the swing by leaning back and kicking out. That sends the swing gliding forward and then back again. In a way, this action grossly approximates the build up and storage of energy by the first GTW in a World Age cycle. The major exception being that in the swing example the build up and storage of energy is very visible. In the case of a World Age GTW, the build up and storage of energy remains invisible. That is, until it is abruptly released in a burst of work.

The swing example is different. The motion (energy amplitude) gained from the pumping action of a single swinging motion, visibly represents each gain imposed. The swing example is depicted on the left side of Figure 95.

Let's put a bit more detail into the swing analogy. Your swing action follows the depicted arching motion from zero (the rest state) to a positive forward motion of say 16.2 degrees, a positive gain of 18 percent of a possible 90 positive degrees of arch. Then, it glides back past the resting state to a negatively signed motion of 16.2 degrees. Your total gain is then 32.4 degrees of motion. The total 32.4 degrees of motion here symbolizes the cumulative amount of hidden compression imposed by a single GTW over the course of 10,250.7 years. Yes, we know, it is a really slow swing, but then our part of the Galactic playground is outrageously large as well.

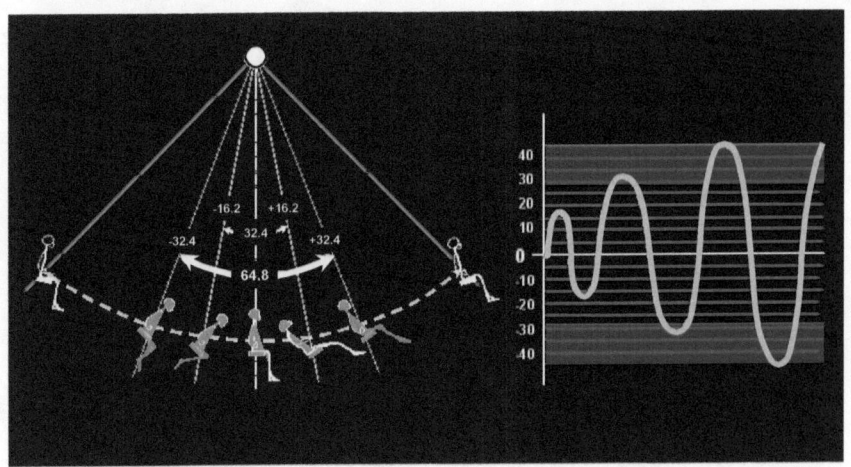

Figure 95 Swing Energy and Resonance

Now, let's say you exert about the same amount of energy as you did before and pump the still moving swing yet again. This time your body builds on the prior gain in altitude (i.e., the stored energy amplitude). Now you've accumulated 32.4 more degrees of total forward sweep and 32.4 degrees more in the backward direction. Your total gain is now 64.8 degrees of motion. What's more, your initial 18 percent energy gain in each direction (36 percent overall) just doubled to 36 x 2 or 80 percent. Your total swing is now roughly equivalent to the *cumulative* compression effect of 2 passing GTW of a World Age Precession Cycle; albeit, after the excruciatingly long period of some 20,501.4 years.

At this point something special happens, something called *resonance*. Our playground swing has acted as a pendulum. Pumping the swing in time with the natural interval of the swing (called *its resonance frequency*) makes the swing go higher and higher, until it reaches its maximum energy amplitude. Warning: any attempt to pump the swing at a faster or slower tempo results in smaller arcs. That is because the energy the swing *absorbs* is maximized when the pumps or pushes are *in phase* with the swing's oscillations. So, some of the swing's energy will actually be extracted by any pumps or pushes that are out of phase with the swing.

Let's assume for the sake of discussion that you are a real no-kidding swinger, you haven't buggered the analogy, and you're approaching genuine resonance. "At resonant frequencies [the gray areas of the amplitude graph in the Figure 95], even really *small* periodic driving forces can produce really *large* amplitude gains, because the system *stores* energy."[307] That means that if your next tiny pump is in time with the natural interval of the swing, with relatively little effort on your part, the swing will achieve its' maximum energy amplitude.

The very small effort needed to achieve maximum effective swinging amplitude at resonance explains a lot. It suggests how the last GTW in the current World Age predicts all the variation in the last 508 years of change, while the effects of multiple wave trains are needed to explain changes in the far distant past. The resonant nudge of the current GTW is in phase with the GTW train pushing our swing of change. The compression effects of each GTW, brings all *the systems* it encounters into resonance with it. Hence, the energy effects of 2.0 GTW accumulate in the systems compressed, just like the swinging energy motion imposed on the playground swing.

The systems a GTW encounters and compresses must absorb and store all the gravitational torque energy imposed, which they seem to readily accommodate. The GTW compression energy absorbed and reacted to by each absorbing system, however, remains hidden within those systems. But, after just one more *Baktun* cycle (394.25846 Gregorian calendar years) of GTW compression, the effected systems achieve World Age cycle resonance and attain their maximum effective energy absorption limit – the critical point. In theory at least, something has to give, and according to the Maya, it does.

Unfortunately, our quaint seventeenth century swing analogy only hints at what happens at optimum resonance and maximum energy absorption. It tells us nothing of what happens when that threshold is exceeded. It does not even begin to explain how one World Age Precession Cycle abruptly ends in a discontinuous transformation. Nor does it tell us why or how the effected systems might abruptly reset to their resting states of equilibrium at the start of a new World Age Cycle. Those parts of the problem form the crux of the yet to be addressed *"Transformation"* the Maya of 2,100 years ago set as the rollover date of their Long Count Calendar.

Transformation Precedence

We diligently searched the available literature for examples of sudden changes from one phase of energy and matter to another. The type of changes sought had to be *non-analytic,* as previously defined. They had to be readily apparent and *conceptually simple*. The sorts of transformations an ancient Maya shaman could have readily observed and measured, but that are *difficult to precisely define*. We looked for physical transformations that could develop over an extended period of time, then suddenly manifest to dramatic effect. Of dominant interest were transformations that could be viewed as fundamentally *universal*, yet largely independent of the idiosyncratic microscopic physics of things. What that means is that any candidate transformation has to allow the same sort of changes to arise in a wide variety of systems (e.g., planets, climate, people, and sub-atomic particles). The real differences in the transformations across systems had to reduce to matters of scale.

The kind of transformations sought also had to be able to produce both *stable and unstable* transitions and outcomes. For example, a candidate transformation of people had to allow species like *Homo erectus, Homo neanderthalensis* (Neanderthal man), and *Homo sapiens* to be transformed in parallel. The transformation had to also allow *Homo erectus* and the Neanderthal species to go unstable and extinct before the end of the subsequent World Age cycle.

What seemed a very tall order quickly proved not at all difficult to fill. Every search retrieved examples of *universal* processes that we rely on every day. These transformations are as *conceptually simple* and common as: the freezing of liquid water to make ice; the boiling of water into a gaseous vapor; and the melting of solids into liquids. Such readily observable changes are: *non-analytic, difficult to precisely describe,* and involve *stable and unstable* forms. The unstable stages and outcomes of such transformations are known as *metastable phases*.[308]

In modern scientific disciplines transformations in the phases of *energy and matter* are known as *"phase transitions."*[309 & 310] In the world of atomic and sub-atomic (i.e., quantum) physics, phase transitions address particle fluctuations in energy absorption, emission, and effective mass. That is, as opposed to transitions induced by the thermodynamic properties of changing temperature and pressure.[311 & 312]

There is no question that the ancient Maya observed and measured thermodynamic phase transitions. Phase transitions like the: melting (solid to liquid), freezing (liquid to solid), boiling (liquid to gas), and condensation (gas to liquid). Due to the so called *non-analyticity* (discontinuous nature) of all such phase transitions, the free energy on either side of a transition can only be described by phase separate mathematical functions (e.g., one set for the old phase and a different set for the new phase).[313] Regardless of whether a phase transition takes the form of macro thermodynamic phase changes or as sub-atomic particle energy absorption or emission, objects in the new phase will behave very differently from objects in the old phase.[314]

While examples of phase transitions in basic biology and physics abound, there is another domain in which the Maya likely observed phase transitions – *the heavens.* Phase transitions are readily observed, measured, and reported by astronomers, astrophysicists, and cosmologists.[315, 316, 317, 318 & 319] The Maya were without question the greatest astronomers and cosmologists to occupy the ancient world.

Back in Earth's kitchens, the melting, freezing, boiling, and condensation examples of phase transitions at conventional pressures, most commonly deal with the property of *heat capacity*.[320] As shown in Figure 96, during a phase transition, the heat capacity of the phase of matter being transformed may become *infinite*, *jump* abruptly to a different value, or exhibit a *kink* or discontinuity.[321]

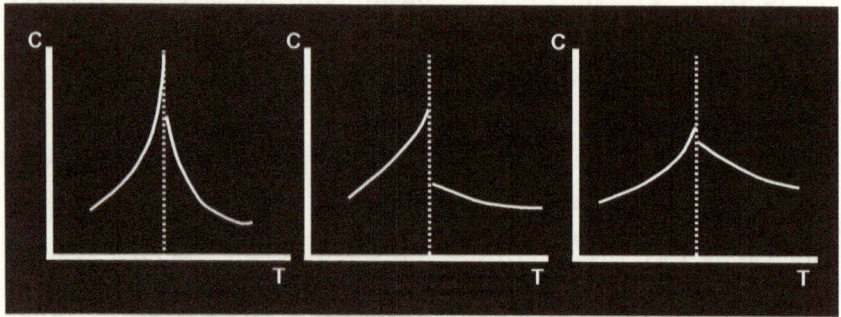

Figure 96 Graphs of Heat Capacity (C) Against Temperature (T) at a Phase Transition
Source: http://www.nationmaster.com/encyclopedia/Phases-of-matter, Revisited February 8, 2008

As shown in Figure 96, an existing phase will build slowly to the critical transition point and then suddenly change to the new phase. For example, at severe pressures carbon may turn to graphite or even be transformed into diamond.[322] Of course the change to the diamond phase of matter is extremely slow, like that predicted by the Maya. What's more, diamond is an *unstable phase of matter*, i.e., *metastable*.[323] The reason for this is that if you heat a diamond sufficiently, it returns to the graphite phase of matter.[324] Stating the obvious, the thermodynamic variables most often used to define elementary phase transitions are Temperature and Pressure.[325]

The phase transitions (Transformation) predicted by the Maya for 2012 are in the main, non-thermodynamic.

Still, the thermodynamic cases of phase transition are highly instructive. That is why the classic Phase Diagram, Figure 97, is inevitably included in nearly every physical chemistry textbook.

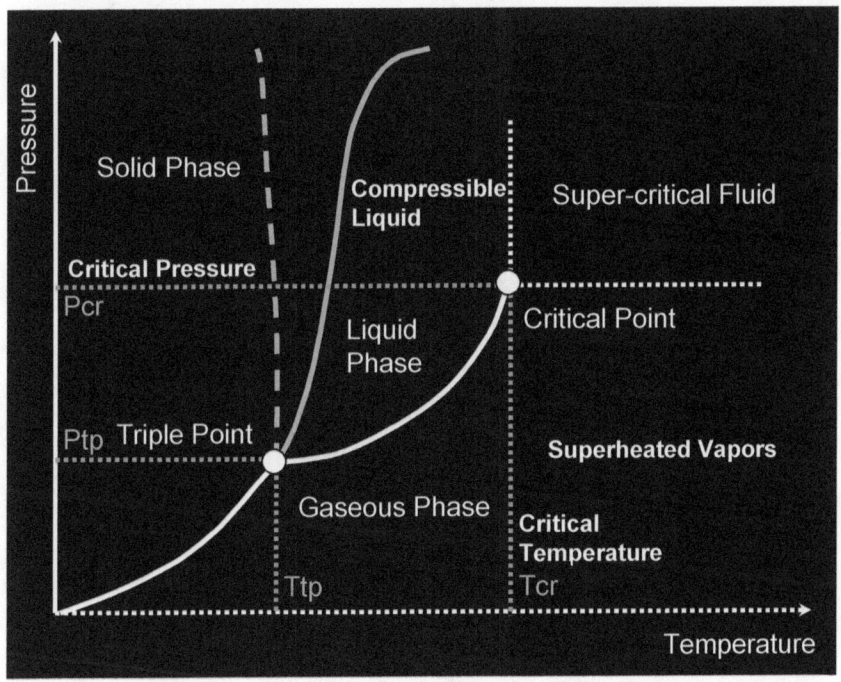

Figure 97 Typical Phase Diagram

Source: http://en.wikipedia.org/wiki/Phase_transition, Revisited February 27, 2009

The Phase Diagram describes a typical material exhibiting thermodynamic solid, liquid, and gaseous phases.[326] The lines in Figure 97 show the boundaries where the free energy becomes *non-analytic*.[327] The labeled open spaces depict the areas where the free energy is analytic (can be described by a continuous function) and correspond to the common phases of matter.[328] The lines called *Phase Boundaries* are where phase transitions occur.[329]

Importantly, the properties of material and energy phases are largely independent of the underlying microscopic physics.[330] Hence, as per our search requirements, these types of phases arise in a wide variety of systems.[331] As if further anticipating our every need, this fact is called *the principal of universality*.[332] That means that "the only real differences in this great variety of phase transitions reduce to matters of scale."[333]

Physical systems under stress develop dynamic fluctuations.[334.] While a system is still far from achieving a phase transition, the fluctuations start increasing in spatial extent[335] on the old phase side. At the ideal transition point the spatial extent of the fluctuations becomes as large as the system itself, and everything abruptly shifts to the new phase side. When that happens, our science is unable to predict the behavior of the system[336] until the new phase is complete.

Curious isn't it that the unpredictability of the outcome of more complex phase transitions was the very reason the ancient Maya supposedly gave for ending their Long Count calendar on Friday, December 21, 2012. It seems our so called modern *universality principle* of phase transitions is only passing for *new*. It was immediately obvious that the Maya shaman had adopted this universality principle over 2,000 years ago, and then proceeded to generalize it to every *seen and unseen* thing in the universe.

Tiny Echoes of the 2.5 Wave-Anomaly

To understand why Maya World Age Precession Cycles end in discontinuous half wavelengths, we need to take a close-up look at the world of quantum physics.

In the sub-atomic world, electrons orbit the nucleus of atoms to form the shell of the atom. Protons orbit their own axis way-way down in the central nucleus of those atoms. Both electrons and protons change energy states by absorbing or radiating energy *photons* (the unit quanta of the electromagnetic spectrum). When electrons or protons absorb or radiate photons, they change their orbit of spin, from one discrete angle and level to another. Specifically, the electrons change the angle of their spin orbit, and with it, the angle of "precession" (the wobble called "Larmor *Precession*") of the vertical axis.[337]

Yes, you read that correctly. This is the exact same kind of precession that defines Maya World Age Epochs. Only in this case, precession occurs in the polar magnetic moment (axis) of particles at the sub-atomic level, as opposed to the axis of our entire solar system. And, it is another charming example of *universality*.

Most important here is the electron or proton (particle) energy waveform generated when one of these particles changes its spin axis as a result of a discrete energy (photon) absorption or emission. Those waveforms look a whole lot like miniature versions of the GTW that drive World Age Precession. Figure 98 depicts the discrete energies and associated analog wave functions of a particle.[338] Of particular note is the fact that each energy wave function must possess one more *half-wavelength* than it did at the energy level below it to achieve and sustain that discrete energy level. Just as each Maya World Age Precession cycle must end in a *half-wavelength* to complete a World Age, and according to the Maya to achieve *Transformation*.

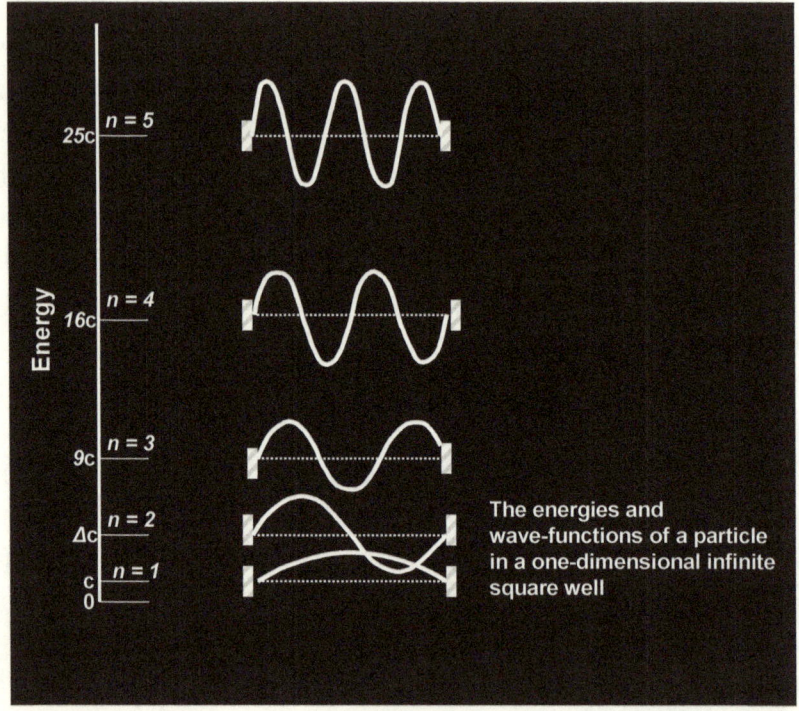

The energies and wave-functions of a particle in a one-dimensional infinite square well

Figure 98 Energies and Wave Functions of a Particle

Source: Atkins, P.W., 1978, Physical Chemistry, W.H. Freemanp & Co, San Francisco, p.401

Increasing the number of half-wavelengths implies a sharpening of the curvature of the wave function. This further implies that the kinetic energy of the particle is increased and stored in each energy level by additional half wavelengths (what the Maya called *Epochs*). When a photon is absorbed by a sub-atomic particle, the energy level jumps up by a half wavelength to the next level. Whenever a photon is radiated (emitted) by a particle, the energy level jumps down by a half wavelength. The overall tendency is for the particle to shed excess energy, and thereby return to its lowest possible energy state (i.e., equilibrium). Interestingly, each absorption and emission change is a discrete mostly non-thermodynamic *phase transition* that occurs in steps of *half-wavelengths*.

Apparently, there exists substantial precedence in the world of quantum physics for cycles that abruptly end in phase transitions occurring at half wavelengths (continuous wave cycles ending in a discontinuous point). It would seem that the Maya phase transition *universality principle* is substantially more *universal* than our own. Clearly, the error is ours ... and not the Maya's.

Is it just a remarkable coincidence? Do microscopic phase transitions of quantum physics just happen to coincide near perfectly with the Transformation predicted for the macroscopic domains of World Age Precession and Galactic GTW? It seems highly unlikely.

Fortunately, the ancient Maya recently gained the company of a number of leading physicists and cosmologists. They include, among others: Nassim Haramein (Director of Research at the Resonance Project Foundation), Michael Hypson (Research Director at the Sirius Institute), and E. A. Rauscher (of the Tecnic Research Laboratory). Those named scientists introduced in the 2008 Proceedings of The Unified Theories Conference (Budapest, Hungary), a paper entitled

"Scale Unification – A Universal Scaling Law for Organized Matter." In it they assert.

> For observational data and theoretical analysis, we demonstrate that a scaling law can be written for all organized matter utilizing the Schwarzschild [as in Karl Schwarzschild of 1916 General Relativity Theory fame] condition, describing cosmological [macro scale] to sub-atomic [micro scale] structures. Of interest are solutions involving *torque* [as in particle and solar system precession] and *Coriolis effects* [an apparent deflection of moving objects when they are viewed from a rotating reference frame] in the field equations. ...[339]

Whew – talk about wordy-weird. Well, all that boils down to something that can be said a bit more simply. The idea that from the realm of the most microscopic wave-particles to the most macroscopic aggregates of ordinary matter (like our Galaxy and the entire universe) these folks believe they have discovered the irreducible first principals of scaling that can tie everything together. In short, they've discovered "a scaling method to support the formulation of a theory that will at long last combine all of the forces in nature."[340]

In a similar vein, in 2005 Researcher G. Van Hooydonk (Ghent University) in Belgium demonstrated that:

> ...a universal equation of state exists and is valid for any phase transition in any system. ... We show quantitatively that his [referring to Johannes Diderik van der Waals'] standing wave equation

leads to *the equivalence of macro-and microscopic phase transitions*, due to natural chiral symmetry breaking."[341]

Okay, let's stand back a bit and skip over the complex notion of symmetry braking. A Grand Unification theory (GUT) refers to any of several very similar unified field theories. Specifically, it refers to models in physics that predict that at extremely high energies, the electromagnetic, weak-nuclear and strong-nuclear forces fuse into a single unified field.[342 & 343] Scientists tell us that:

"Thus far, [physicists pursuing such goals] have been able to merge electromagnetism and the weak nuclear force into the electroweak force. Work is now being done to merge electroweak and quantum chromodynamics into a QCD-electroweak interaction, sometimes called the electro-strong force. Beyond these so called grand unifications, there is also speculation that it may be possible to merge gravity into a theory of everything."[344]

My earlier book 'Waters of Creation and Reality' (2006),[345] demonstrated that our best modern efforts to achieve a Theory of Everything pale in comparison to a much earlier attempt. Specifically, that documented in a 2,000 year old Aramaic book called the *Sefer Yetzirah* (meaning the book of Creation or Formation). Please note that the *Sefer Yetzirah* was most likely based on a 4,000 year old Mesopotamian oral tradition. The Maya, like the Mesopotamians, simply took it on faith that the unseen part of the universe behaves at all scales like the seen world.

More specifically, that physical phase transitions are universal, non-analytic, and largely matters of scale. What is astonishing about that notion is that the more we learn about the big picture of the previously unseen micro and macro scale constituents of this universe, the more we seem to validate that antediluvian world view. Yeah ... talk about whiplash.

Well, it's time to pull all this together into a theoretic structure. That theory is 'WARPS,' which spells out to 'World Age Resonant Phase Synchronization', and it's the subject of the next Chapter.

15

WORLD AGE RESONANT PHASE SYNCHRONIZATION (WARPS)

A Composite Theory; WARPS

Given the perspectives related in the prior chapter, this chapter makes a bold speculative move. To cut right to it, it combines our less than universal scientific knowledge of phase transitions and transformations with the World Age notions espoused by the Maya. The result is the falsifiable theory called 'World Age Resonant Phase Synchronization, or 'WARPS.'

WARPS goes like this. The Galactic plane and the extent of each GTW are divided into *two* well defined and widely accepted parts: 1) the part that resides *inside* the Co-rotation Radius, and 2) the part which exists *outside* the Co-rotation Radius.[346] Yes, that is the same Co-rotation radius defined in Chapter 5 and Figure 99.

Inside the Co-rotation Radius, each sweeping GTW *pulls* toward the Galactic Center the angular momentum and effective energy-mass it extracts from objects (the gas, dust,

343

and stars) rotating clockwise about the inner Galactic plane.[347]

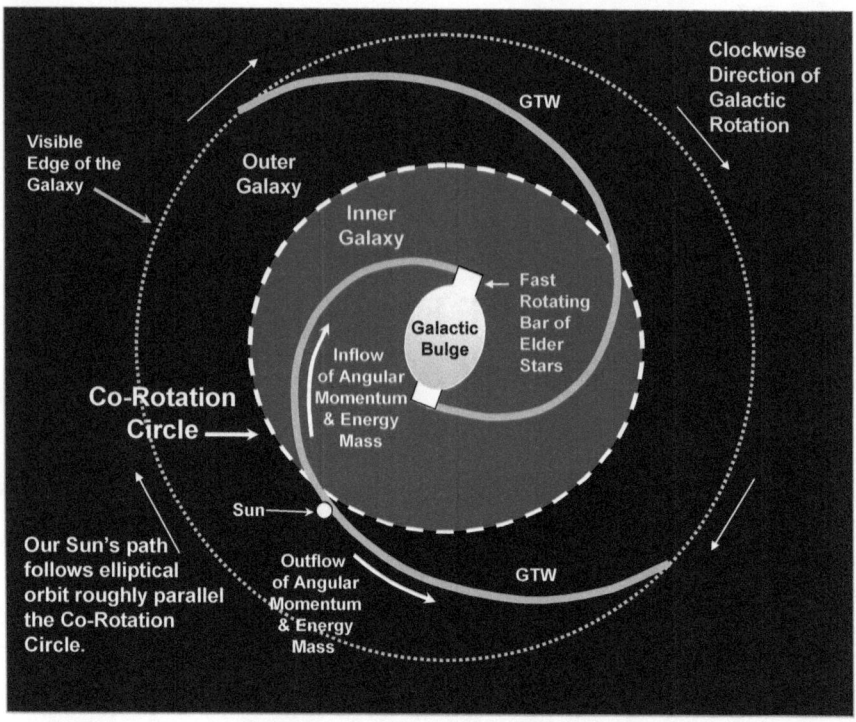

Figure 99 Galactic Co-Rotation Radius

This tugging transfer produces visible processions of gas and dust that trace out the torques of each GTW. These massive matter streams are called *"dust lanes,"* and they have been shown to extend all the way from the Co-rotation radius into the Galactic Bulge that enshrouds the Galactic Center.[348]

Outside the Co-rotation Radius, where our Sun and solar system orbit the Galaxy, each sweeping GTW abruptly reverses its imposed effects.[349] At these radial distances the torques of each GTW *push* angular momentum and effective energy-masses extracted from objects orbiting closer to Co-

344

rotation outward to well beyond the furthest edges of the visible Galactic plane.[350] Along the way, the GTW deposit a great deal of that angular momentum and effective density compressing energy into the stuff of our solar neighborhood. The objects of our solar system absorb and store that angular momentum and energy over the course of the first two GTW of each World Age Precession cycle.

Both inside and outside the Co-rotation radius, each GTW is communicated and propagated within all levels of the visible and invisible energy and material masses (hereafter referred to as the '*systems*') it encounters. Working our way up the size scale and down the energy scale, these wave-particles '*systems*' are:

- *Vacuum Dark Energy (DE)* which pervades all space [351] and is believed to consist of vast numbers of ten types of super-partner wave-particles (five positively signed bosons and 5 negatively signed fermions existing in superposition, which means all possible energy states simultaneously).[352] These infinitesimal super-partner wave-particles are believed to range in mass from the energy equivalent of 500 to 100 billion electron volts, each.[353] These super-partners are said to interact (exchange one for another) to form what is believed to be 32 supercharges that radiate enormous amounts of energy. Enough energy in fact to explain the accelerating expansion of our entire universe.[354]

- *Dark Matter (DM)* is believed to be composed of hypothetical wave-particles of resting energy-masses of around 100 to 200 billion electron volts.[355] DM is believed to be the parent of ordinary matter (the stuff we're made of), and pass straight through ordinary matter and cluster (clump), often wherever ordinary matter is found.[356]

- *The Electromagnetic Field (EMF)* which pervades all space [357] and is composed primarily of photons (which are described as mass-less particles), protons of a resting energy-mass of approximately 1 billion electron volts, and electrons (and their antimatter-twins positrons). Some scientists argue that the free electrons of the EMF are simply photons trapped in a circular orbit.

- *Micro level Ordinary Matter (OM $_{micro}$)* which behaves in accordance with the Standard Model of quantum physics and comprises all of the atoms of our Periodic Table of the Elements.[358] The atomic shell of each atom is made up of orbiting electrons with energy masses of approximately 500,000 electron volts.[359] The nucleus of each atom is composed of protons and neutrons.[360] Protons and neutrons are in turn composed of gluon bound quarks[361] that possess energy masses estimated at from 3 million to 4.3 billion electron volts).[362]

- *Macro level Ordinary Matter (OM $_{macro}$)* which behaves in accordance with Einstein's Theory of General Relativity, and is comprised of vast aggregations of atoms that make up all the objects we directly experience from molecules like DNA, to great clusters of galaxies.[363]

Figure 100 depicts the breakdown of the fabric of space-time and the energy mass density of the fabric of the universe.

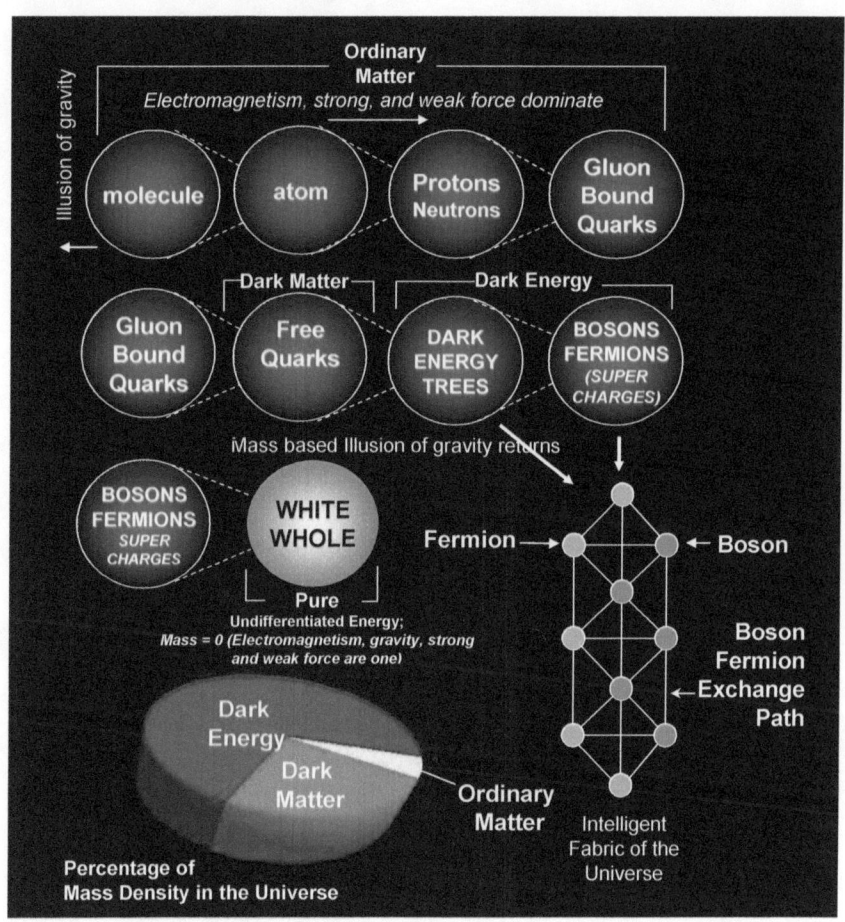

Figure 100 Breakdown of the Energy Mass Density of the Universe

Scientists are pretty sure that approximately 70 percent of all the energy-mass density of everything (e.g. our: universe, Galaxy, Sun, planet, selves, and every molecule, atom, proton, neutron, electron, and quark) is comprised of invisible vacuum dark energy (DE).[364]   About 26 percent of the energy-mass density of everything, and 85 percent of all matter, is believed to be comprised of invisible dark matter (DM).[365]   It is argued that only about 4 percent of all the mass density of things in this universe are comprised of microscopic and macroscopic ordinary matter (OM).[366]

Dark Energy (DE) and the Electromagnetic Field (EMF) pervade all space. Dark matter tends to cluster, often about collections of ordinary matter. Given that, all things (wave-particles) of all types, sizes, and energies must indeed be connected and interdependent.[367] Hence, all particles regardless of type, size, energy mass, and distance apart can become *entangled* (capable of collective resonant synchronization of behavior and energy state exchange, irrespective of the distances and scaling differences between them).[368]

Gravity (and hence gravitational torques) is the only known force that interacts with all levels of all systems (i.e., DE, DM, the EMF, and OM micro and macro). While the direct effects of Gravity become infinitesimally small at atomic and sub-atomic scales (i.e., the OM micro realm) the effects of gravitational torques (AKA *density waves*) are observable, measurable, and significant. At that scale, the strong nuclear force, weak nuclear force, and Electromagnetic force are orders of magnitude stronger than gravity. But, those forces exist and operate solely in the presence of ordinary matter, and ordinary matter comprises only 4 percent of the stuff of this universe. Gravity, in stark contrast, operates at every level of the universe. What's more, gravity is the literal source of space-time curvature. Hence, in terms of scope, Gravity (and thus gravitational torques) is the queen mother of all forces and is as old as the universe itself.

Whomp of WARPS

Our solar neighborhood occupies a tiny cube measuring about 3 light years on a side. Our entire solar system occupies a volume of space about three thousandths the size of that neighborhood. The solar neighborhood traces a clockwise orbit around the Galaxy just outside the Co-rotation radius. Thus, each of the GTW comprising each Maya World Age Precession Cycle extracts angular momentum and effective energy mass from stars (like our Sun) closer to the Co-rotation radius, and deposits some of that angular momentum and effective energy in the celestial objects of our solar system.

Over the course of a World Age Precession cycle, each passing GTW imposes a cumulative density compression effect on our solar neighborhood (much like the pumping of the playground swing described in the prior chapter). Each passing wave thus increases the particle excitation (energy absorptions and emissions) at all levels of all of the effected systems (i.e., DE, DM, EMF, OM $_{micro}$, and OM $_{macro}$) in our neighborhood.

During the first two GTW of a World Age, the respective phases of these systems remain in relative equilibrium, doing their level best to absorb and accommodate the GTW compression imposed density and energy changes. This buildup (without a phase transformation) is depicted in Figure 101 (repeated from Chapter 5).

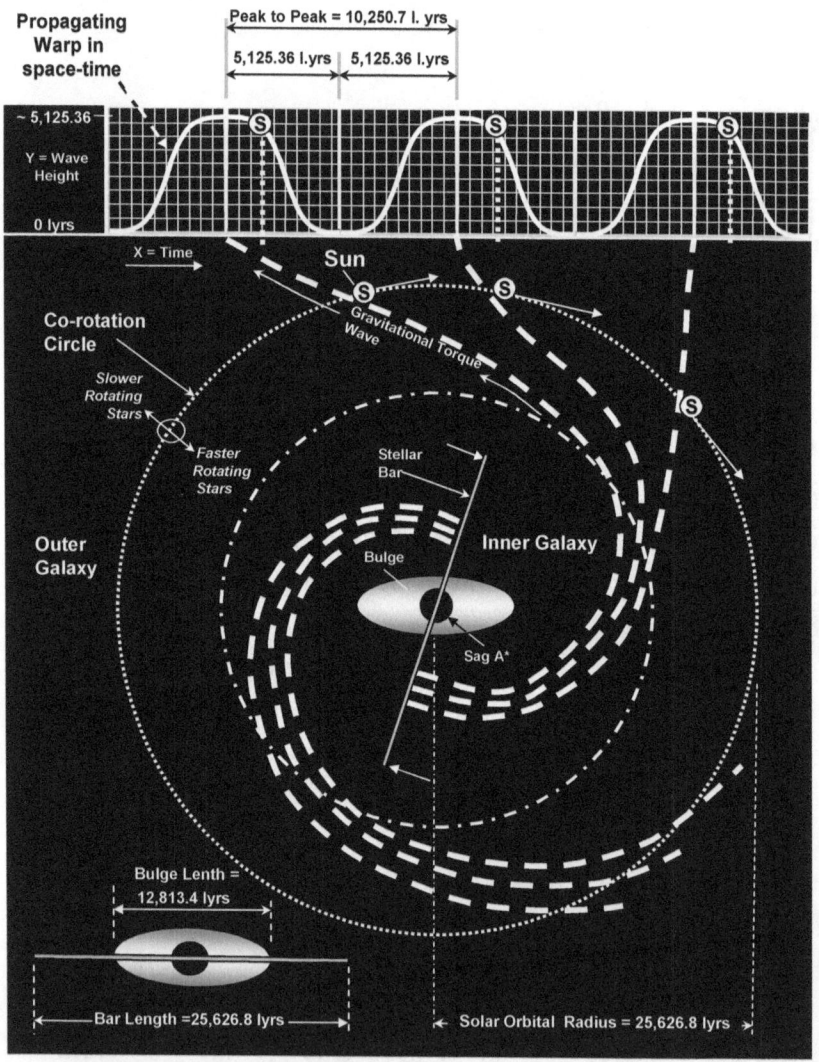

Figure 101 GTW Propagation and the Sun's Galactic Orbit about the Co-rotation Circle

At the beginning of the third GTW of each World Age (our current situation), the cumulative build up of angular momentum and energy reaches a level of maximum resonance, *the critical point,* within all respective systems. Exceeding these critical points will trigger a cascade of phase transitions within all the systems. The phase transitions result in the resonant synchronization of energy emission, absorption, and exchange within and between the systems. Now the system specific changes that have been developing for thousands of years in existing phases, abruptly transition to a new synchronous phase. The new phase constitutes an *optimum* energy and information distribution which harmonizes the particle properties, states, and behaviors of all the systems in an immeasurably short instant.

During that exceedingly short transition interval, the radically compressed conventional dimensions of time and space smear out into a wave of un-manifest probabilities. As a result of the cascading phase transitions, the wave-particles of the systems become temporarily entangled (i.e., precisely synchronized such that they operate as if they were one great big wave-particle). Since each GTW stretches from the Galactic Center to the edge of the Galactic Plane, it is possible that this temporal state of entanglement extends all the way to the Galactic Center as well. If that is the case – and WARPS asserts it is – then everything within that immense gravitational wave will be temporally entangled with the Galactic Center, just as the Maya world view espoused. Figure 102 Symbolizes the in Transformation Process.

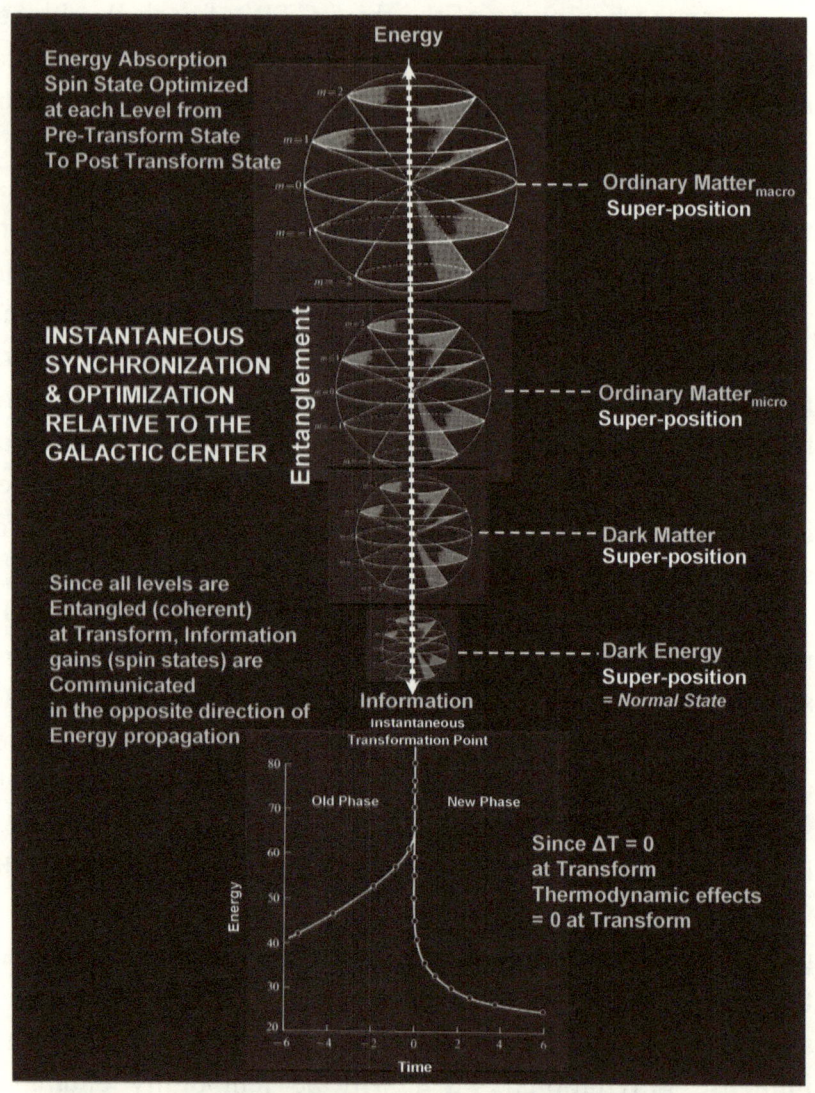

Figure 102 Cascade of Phase Transitions at WARPS Transformation

The state of solar neighborhood wide entanglement, if it in fact occurs, will be very short lived. Rapidly ensuing local system variations will instantly force the collapse of the entangled wave state, and degrade the multi-system resonant synchronization. More specifically, the *non-analytic* synchronous systems abruptly collapse into the *analytic* phase of a new World Age of the solar system. Over time, the systems diverge into progressively more asynchronous local behaviors. This gradual drift from a state of optimum Galactic synchronization will continue to increase until the onset of the next World Age Transformation.

When the excess energy that triggered the phase transition of the systems has been effectively radiated away, the new World Age Precession cycle (GTW Compression effects) will equal zero. At this point, the wave particles of the systems will be at equilibrium (their lowest stable energy state). In this new phase, clocks, calendars, and rulers will once again apply, and the new World Age Precession cycle will once again begin to count anew in a predictable way.

Wave Warp Drive

Now we come to the real reason for naming the theory 'WARPS.' The idea of the *warp-drive* originally made famous by science fiction writers, is *not* an engine that propels a spacecraft faster than light speed.[369] In theoretical physics, at least, a warp-drive is a hypothetical device that generates an enormous – and enormously powerful – gravitational field.[370] That field then compresses the space-time ahead of the space-craft and decompresses the space-time aft of the craft.[371] Applying just a little conventional rocket thrust, the craft effectively travels immense distances in very short intervals of craft relative time.

Theoretically, at least, the warp driven craft could conceivably arrive at its destination quicker than a beam of light traveling the full uncompressed distance[372] – hence the seriously misleading Sci-Fi reference to faster than light travel. In truth, the spacecraft never comes anywhere near the speed of light, and being made of matter can never exceed it.[373]

The WARPS theory hypothesizes that the cumulative effect of World Age Gravitational Torque Waves accomplishes the same thing as the hypothetical warp-drive. Only in the case of WARPS, the space-craft is replaced by our entire solar neighborhood, and the compression and decompression of space-time is accomplished by a compressing train of 2.05 GTW. If the WARPS theory and Maya predicted date of Transformation is correct, on Friday December 21, 2012 two rather amazing things will happen really-really fast. First, our entire solar neighborhood will be moved *the functional equivalent* of 4.6128224 light years along our Galactic orbit in an immeasurably short instant. Second, our entire solar neighborhood and everything in it will undergo 4,612.824 years worth of evolutionary change in less than a minute.

WARPS Hypothesis H_{102} becomes: Our Solar System (and hence everything within it) experiences a hyper-jump (the functional equivalent of a sudden abrupt acceleration) of 4.6128224 light years and an equivalent amount of evolutionary change in under a minute, once every 21,013.9760 years. **Null Hypothesis H_{102}** states that: The Solar system has not exhibited, nor will it exhibit, any solar neighborhood wide fixed-time-interval evolutionary transformations.

WARPS Hypothesis H_{103} follows, asserting that: Any residual evidence of the sudden (less than 60 seconds long) compression and decompression of space-time imposed on the solar system by 2.05 GTW once every 21,013.9760 years, will appear to have occurred at an interval of 25,626.8 years; a chronological offset of 4,612.824 years (or 90 percent of one Maya Epoch) more than actual. **Null Hypothesis H_{103}** is the same as **Null Hypothesis H_{102}**.

The two WARPS Hypotheses are illustrated in Figure 103.

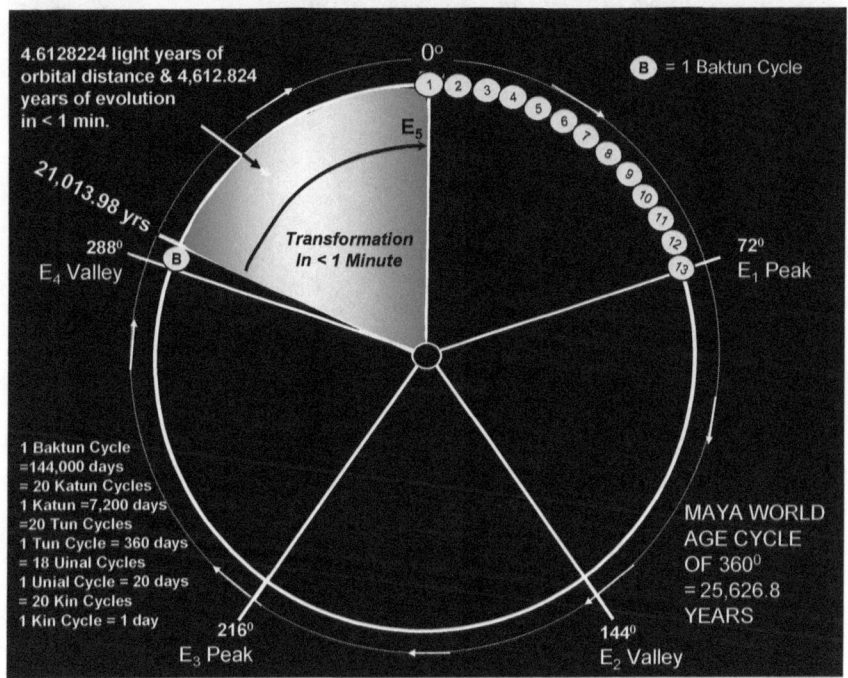

Figure 103 World Age Cycle with Transformation

The next figure, Figure 104, Depicts the two GTW between the prior Transformation (A) and pending World Age Transformation (B). In Figure 104 the Greek letter lambda (λ) is used to symbolize the cascade of Phase Transitions and the resulting Transformation.

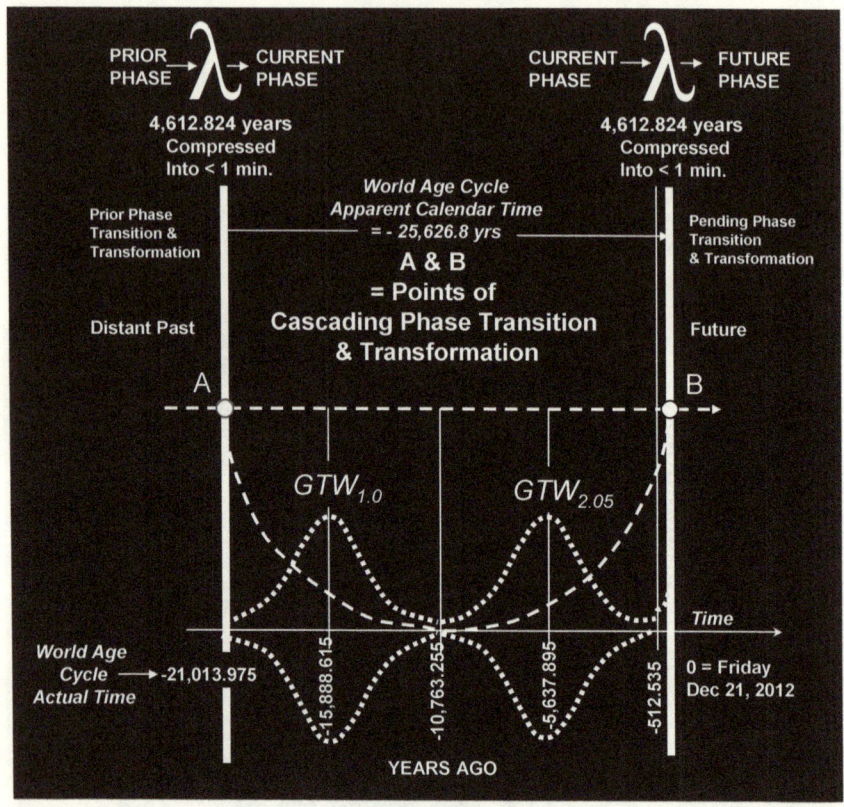

Figure 104 Prior and Pending World Age Transformations

It is important to note that in an application of the WARP drive of science fiction fame, the spacecraft and its occupants remain unaltered. That is because the spacecraft is unaffected by the WARP drive gravitational field compression ahead of the craft, and the decompression behind it. The spacecraft is not compressed, accelerated,

356

entangled with space-time, or decompressed in the process of achieving what only appears to be faster than light travel. In WARPS theory, however, the entire solar neighborhood and everything in it is compressed, accelerated, entangled, transformed, and decompressed. Hence, any object in any of the constituent systems in the solar neighborhood *is* altered by the Transformation.

The extent to which an object is changed by the Transformation is the difference between the pre and post Transformation states of that energy-mass object. That pre-post difference of any such object is the difference of its pre-Transformation state from its optimum post Transformation state. Strangely, if an object's pre-transformation state is too different from post transformation optimum, all trace of its' prior existence will vanish at the instant of Transformation. Only energy-masses that can be optimized relative to the conditions of the New World Age will survive Transformation.

If the WARPS predictions are correct, it might solve a paradox that has long befuddled evolutionary biologists. This is a dilemma posited by scholars and scientists for centuries. Notable among them, the 5th century BC Greek philosopher Anaxagoras, Benoît de Maillet (1743), Jöns Jacob Berzelius (1834), Kelvin (1871), Hermann von Helmholtz (1879), Svante Arrhenius (1903), more recently Sir Fred Hoyle and Chandra Wickramasinghe, and most recently the late great Nobel Laureate Francis Crick and Leslie Orgel (1973), and Thomas Dehel (2006). These eminent sources assert that an Earth based origin of organic life appears to have occurred in an inadequate if not impossibly short period of time.[374] It is suggested that that time reduces to "A window of at most 1 billion years (from 4.55 billion to 3.5 billion years ago), most plausibly 400 million years (from 3.9 billion to 3.5 billion years ago), and possibly less than 100 million years ago (from 3.9 billion to 3.85 billion years ago). ... We know that the Earth has provided a life-friendly environment for at least 3.5 billion years,"[375], but the speed at which life forms have emerged and diversified into known species on Earth appears paradoxically rapid and impossibly short.

The WARPS theory hypothesizes a hyper-jump of 4,612.824 years worth of evolutionary change transpiring in under a minute every 21,014 years. That would suggest that *each World Age* affords a sudden abrupt hyper-leap of species perfecting optimization. Such hyper-boosts of evolutionary change would exceed by orders of magnitude that possible had organic life on Earth emerged and evolved linearly over 3.5 billion Earth years.

A Falsifiable Theory

We arrive at the fully falsifiable theory of World Age Resonant Phase Synchronization (WARPS). By Saturday, December 22, 2012 we will know with certainty if the WARPS theory is *false*. If it is, all of the WARPS assertions will simply fail to emerge. Our position relative to the stellar constellations will look much the same as it did the day before. Solar System Precession will not have leaped some 64.8 degrees to abruptly reset to zero. Earth's obliquity will not have advanced by 0.54 degrees, Earth's orbital Inclination will not have increased by 0.00000518 degrees, and Earth's orbital Eccentricity will not have jumped 0.0046 units. No discernable increase of 4.6 light years in the state of the solar system Galactic Orbit will be apparent, and this planet will continue to experience an increasing rate of global warming. Most importantly, plant and animal species, including our own, will not have been significantly altered in an incalculably short period of time.

The converse, however, may not be equally true. If the WARPS theory is *correct*, we will most likely never know it. The reason why this is the case is just plain daunting. If the WARPS theory is correct, all record and memory of the WARPS theory will likely be erased by the very events that could demonstrate it correct.

Why? Well, it seems that the audit trail of events leading up to the critical point of each phase transition and Transformation is recorded in the familiar space-time dimensions of X, Y, Z, and T. At the critical point of Transformation, those dimensions will be smeared out into the multi-dimensional configuration of an exceedingly narrow vertical waveform. As a result, conventional space-time geometry will become a shifting mass of yet to be manifest probabilities. Once the phase transitions and Transformation are complete, that waveform will abruptly collapse into newly defined phases of energy and matter, with a history to match. Only then will the terms of conventional space-time geometry once again apply.

If the late Physicist Richard Feynman is to be believed[376] each and every particle instanced by the collapse of the Transformation waveform will bring its own history right along with it; a history perfectly tailored to the new World Age.[377] It is extremely unlikely that those transformed particles will convey a history inclusive of the WARPS theory. Make no mistake, no energy or information will have been lost in Transformation. It is just that the Maya Prophecy and WARPS theory may no longer be accessible to the transformed people of the next World Age.

Hence, if WARPS is *wrong* we will know it bright and early Saturday, December 22, 2012. If WARPS is *right*, we will probably be irreversibly changed into something new, with an equally different history. How different might we be? That is something that WARPS aptly declares to be *unknown and unknowable* in advance of its occurrence.

What is certain is that if the Transformation does occur, the survivors will be as different from our present selves as we are from our predecessors of 25,626.8 years ago. One cannot help but wonder. Might those transformed Hominids of the next World Age one day hypothesize a WARPS theory of their own? Will those transformed – like the ancient Maya – ultimately detect the invisible gravitational rhythms that drive Galactic change?

There is little point in dwelling on those sorts of questions, or even the probability of a WARPS transformation occurring. Those are things well beyond the reach of current humanity. What this book does offer is a considerable amount of evidence that there is a firm fixed cyclic rhythm to our Galaxy, solar system, planet, and selves. A rhythm our ancient ancestors knew well, set clocks by, and used to divine the general trends of the future. They tried their level best to leave us monuments and artifacts to communicate their wisdom, knowledge, and understanding of that cyclic cause and its effects. But, like all immature children, we had to learn those lessons for ourselves, the hard way. As our grandfathers used to say as they watched us toddle off to make more foolish mistakes – "too soon old and too late smart."   So, by way of wising up, the next three chapters explain how a non-destructive Transformation works and offer evidence that some really big changes are in the works.

16

WAVES OF CASCADING TRANSFORMATION

Non-Destructive

How can a World Age Resonant Phase Synchronization (WARPS) possibly alter everything from dark energy to vast aggregations of ordinary matter without literally destroying the entire solar system, this planet, and every living thing on Earth? It seems only logical to assume that the instantaneous release of 21 millennia worth of pent up gravitational compression would annihilate all the matter in its path. Yet, if the ancient Maya and WARPS theory are right, there have been 184,572 such events in the history of our planet – and – *it is still here*. At least 30 Hominid species must have emerged, evolved, and prospered through 273 such transformations – and *we're still here*. While all the human-like species that preceded ours are extinct, there is nothing to indicate their *instantaneous* extinguishing by a World Age Transformation – *changed, yes – extinguished, no*.

So how can a WARPS Transformation nondestructively optimize matter? The short answer appears to be that energy waves can be instantaneously trapped, stopped, and converted (phase transitioned) into optimizing information (i.e., converted into particle spin angles and states). Then, just as suddenly, that same information and be transformed back into energy and released to do work, or continue on.

Precedence

We know that energy waves can be phase transitioned into information, and that information can be turned back into energy again. That process was recently lab demonstrated by Lene Hau et al of Harvard University and reported in the December 4, 2009 issue of the Physical Review Letters.

Laura Sanders summarized the results of the Hau team experiment in a January 6, 2010 Science News article titled, "Trapped in a Cloud of Ultracold Atoms, Light Stayed Frozen for 1.5 seconds."[378] Sanders states that:

> A cloud of ultracold atoms can store a beam of yellow light for 1.5 seconds ... enough time for light to circle the Earth 10 times under normal conditions.
>
> ...
>
> As they fly in, the photons [energy waves] leave an imprint in a subset of the atoms. This imprint, stored in a quantum property known as spin [particle angle and state], contains the relevant information needed to reconstitute the light beam [back into energy waves]. But the imprint is fragile and deteriorates in milliseconds. The light's information is lost as

other atoms in the cloud interfere with the imprint.

Hau and colleagues overcame this problem by sequestering the matter imprint [information] from the rest of the atoms in the cloud. The team shone a pulse of laser light – which looks like yellow light from street lamps, Hau says – into a small cloud of sodium atoms. A three-microsecond pulse would produce a stretch of light about a kilometer long if unimpeded, but as the pulse entered the atom cloud, it began to compress. Like an accordion closing, the light folded up and crammed itself into a space 0.02 millimeters long.[379]

Figure 105 depicts the process reported by Hau.

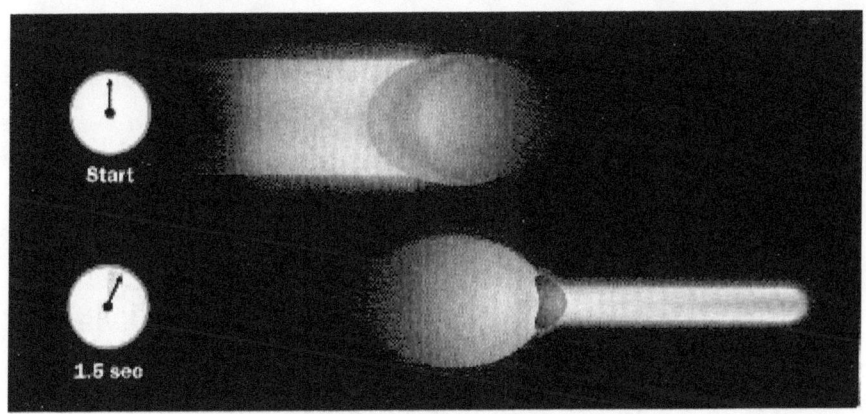

Figure 105 Energy Transformation to Information and Back to Energy
Source: Sanders, Laura, 2010, "Trapped in Cloud of Ultracold Atoms, Light Stayed Frozen for 1.5 seconds", Science News, Vol 177, No.2, January 6, 2010 (p.10)

Now imagine Gravitational Torque Wave energy undergoing that same sort of accordion fold-up into the spin states and angles of particles of dark energy, dark matter, and ordinary matter. Given such folding compression, one can imagine how GTW energy might be invisibly stored as information in atomic and sub-atomic spin states and angles over the course of a Maya World Age cycle.

There is, however, a much more natural and common example of such a process that occurs in bacteria and plants – *photosynthesis.* Quantum photosynthesis is an everyday instance of non-destructive energy transformation via phase transitions. Moreover, photosynthesis occurs at conventional ambient temperatures, as opposed to requiring outrageously super-cooled clouds of atoms. Unlike Hau's light-stopping lab experiment, Quantum photosynthesis results in close to 100 percent efficient transformation of wave energy into multi-particle spin states and angles, and it *optimizes* the allocation of light energy within plants and bacteria.

An April 12, 2007 article in Research News introduced the discovery this way.

> Through photosynthesis, green plants and cyanobacteria are able to transfer sunlight energy to [information in] molecular reaction centers for conversion into chemical energy with nearly 100-percent efficiency. Speed is the key – the transfer of the solar energy takes place almost *instantaneously* so little energy is wasted as heat. How photosynthesis achieves this near instantaneous energy transfer is a long-standing mystery that may have finally been solved.[380]

366

That study of quantum photosynthesis was led by researchers with the U.S. Department of Energy's Lawrence Berkeley National Laboratory and the University of California (UC Berkley). Results were detailed in the April 12, 2007 issue of the journal Nature in which Principal Investigator and Deputy Director of the Berkley Lab, Graham Fleming, reported:

> We have obtained the first direct evidence that remarkably long-lived wavelike electronic quantum coherence [the distance independent entanglement of the spin states of atomic and sub-atomic particles] plays an important part in energy transfer processes during photosynthesis. This wavelike characteristic can explain the extreme efficiency of the energy transfer because it enables the system to simultaneously sample all the potential energy pathways and choose the most efficient [*optimum*] one.[381]

Their journal article was titled, "Evidence for wavelike energy transfer through quantum coherence in photosynthetic systems." In it Fleming and his collaborators reported the detection of "quantum beating" signals, coherent electronic oscillations in both donor and acceptor molecules, generated by light-induced energy excitations [waves], like the ripples [folds] formed when stones are tossed into a pond."[382] Fleming explained:

> Electronic spectroscopy measurements made on a femtosecond (millionths of a billionth of a second) time-scale showed these oscillations meeting and interfering *constructively*, forming wavelike motions of

energy (superposition states [the assumption of all feasible spin states and angles simultaneously]) that can explore all potential energy pathways simultaneously and reversibly, meaning they can retreat from wrong pathways with no penalty [meaning instantly and with minimal loss of energy].[383]

Fleming et al go on to explain that their 2007 breakthrough in the description of quantum photosynthesis changed forever previously held views of the process and opened new doors in energy research and development.

> This finding contradicts the classical description of the photosynthetic energy transfer process as one in which excitation energy hops from light-capturing pigment molecules to reaction center molecules step-by-step down the molecular energy ladder.
> ...
> The photosynthetic technique for transferring energy from one molecular system to another should make any short-list of Mother Nature's spectacular accomplishments. If we can learn enough to emulate this process, we might be able to create artificial versions of photosynthesis that would help us effectively tap into the sun as a clean, efficient, sustainable and carbon-neutral source of energy.[384]

Generalization

The case of quantum photosynthesis is most instructive. In it, arriving light energy is stopped, trapped, converted into information, and optimally allocated to the spin states and angles of those molecules most in need of fuel in an instant (millionths of a billionth of a second) with minimal loss of the energy (i.e., optimal efficiency). Here, as theorized by WARPS, the array of ordinary matter atoms comprising each plant molecule receives incoming non-destructively captured and transformed energy information proportionate to its currently evolved atomic spin states and angles (needs). Then, those aggregate atom molecules apply the energy information gained by converting it back into energy (i.e. fuel). What is more, this entire process occurs at normal ambient temperatures in the absence of damaging thermodynamic effects (temperature and pressure changes). Figure 106 depicts the possible spin states and angles (m) of an atom which store the energy information and optimize the energy allocation.

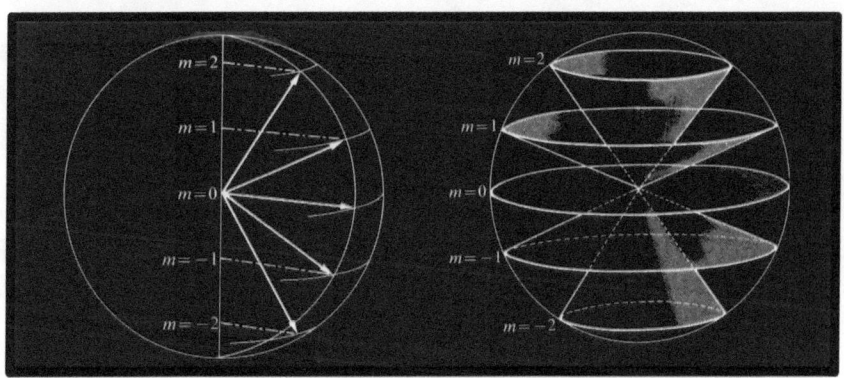

Figure 106 Spin States and Angles (m) of an Atom
Source: Atkins, P.W., 1978, Physical Chemistry, W.H. Freeman & Co, San Francisco, p.416

Biophysicist Rienk van Grondelle (2010) noted that researches had long asserted that:

At room temperature, the noisy environment would kill this kind of quantum interaction. But examining the light-harvesting systems of two species of photosynthetic algae, physical chemist Gregory Scholes of the University of Toronto and his colleagues observed that energy introduced to the system acted in a distinctly quantum manner, even at ambient temperatures."[385]

In 2009, Fleming and Engel once again demonstrated what they had first hypothesized in 2005, and initially demonstrated in 2006. Work solidly confirmed by Fleming and Engel's results in 2007 and 2009, and by Scholes et al in 2009 (reported in 2010). Scholes 2010 report put it this way:

While classical energy transfer theory predicts that the energy "hops, hops, hops" from one molecule to the next in a kind of 'random walk,' quantum theory predicts that energy flows through the system in a much more spread-out, directed fashion.

'Think about it as the energy moving [through the system] like a wave rather than like a ball bouncing from one molecule to another.

'Instead of traveling a single pathway – one molecule to the next – the wave-like energy can actually take three different pathways simultaneously.'

This wave-like motion provides the energy with a "memory" of where it's been that eliminates some of the randomness of how it moves through the cell. [It] can still follow many paths, [but] it will be certainly more directed' than the random walk of classic energy transfer. [386]

We return to the theory of World Age Resonant Phase Synchronization (WARPS). Only now we are armed with examples of instantaneous non-destructive energy phase transitions and transformation that optimize information to get work done. To generalize from the examples, we must extend our perspective to encompass particles of dark energy, dark matter, ordinary matter, and macro aggregations of ordinary matter. Further, we must substitute waves of compressive gravitational torque energy for the photonic light energy transformed by bacteria, plants, and lab lasers. Thus – with a fair bit of empirical galloping – we arrive at a plausible explanation of how the sudden release of the compressive energy of 2.05 GTW could conceivably result in 4,612.82 years worth of optimizing non-destructive evolution in under a minute, at every level of local energy and matter.

WARPS theory posits that 2.05 GTW slowly pass through and cumulatively compress every level and domain of our solar neighborhood. That cumulatively stored gravitational energy reaches the critical phase transition point at 2.05 GTW to end a World Age. Then, at the speed of light, a cascade of phase transitions propagates upward from the dark energy domain all the way to the level of macro ordinary matter. The spin states and angles of energy and material particles at each level non-destructively absorb that onrushing wave energy by converting it into changes in their spin states and angles that optimize those individual and collective spin states. All of this occurs with a minimum of thermodynamic change, and passing unabsorbed and unused wave energy continues on to the next level and domain.

In this nondestructive process, absorbed and used energy is instantaneously pre-allocated and applied, just as is done by plant molecules conducting quantum photosynthesis. The net result of the process is that the entire solar neighborhood is non-destructively synchronized with the Galactic Center and optimized once each World Age cycle. When the overall phase Transformation wave collapses to a single classic resolved state – at the start of a new World Age – the resulting change in each domain level is proportionate to the pre-Transformation spin states of the domain constituent particles. The result is the nondestructive World Age Resonant Phase Synchronization of energy and matter.

A Downside

Unfortunately, if WARPS is correct, there is both good and worrisome World Age news to report. The *good news* is that the World Age Transformation is almost certainly non-destructive. The bad news is that the ramp-up to the critical point of solar neighborhood Transformation could be seriously damaging. The available evidence strongly suggests that this World and its hosted humanity have survived many such Transformations. Unfortunately that same evidence suggests that serious stress and harm occurs in the run-up to the critical point of Transformation.

Ancient Mesoamericans associated each World Age Transformation with a symbol of destruction. This is what led to a great deal of later scholar confusion and sensationalist threats of wholesale destruction.

The Aztecs in particular associated earthquakes, fire, floods, wild animal attacks, and the like with both Epoch and Transformation dates. This was not because of the effects of the Transformation proper. Instead, it is the result of the fundamental definition and concept of change they shared with modern philosophers and scientists. Specifically, *change* by definition implies that something old is reduced or eliminated to give rise to something new.

The earlier bacteria and plant analogy can be used to point out the distinction between instantaneous World Age Transformation, and the potentially damaging effects of the build up of gravitational torque density compression.

Plant or cyanobacteria reap the gains of quantum photosynthesis daily, while at the same time suffering the adverse effects of local, regional, and global environmental variations. In earlier chapters, it became clear that the Earth's climate and geology are significantly influenced by accumulating GTW compression. The build up of that GTW compression over the course of a World Age reaches maximum just prior to a Transformation. Hence, one should expect – and the ancient Maya predicted increased rates of: geologic events (e.g. earthquakes and tsunami); meteorological events like extreme storms (e.g. hurricanes, tornadoes, floods); and global temperature extremes. To some extent, this book has reported that such geological and meteorological events have been increasing in frequency and severity.

How much damage will the pre-transformation GTW compression pressure do to the modern world and attendant civilization? No one knows. The last three times it happened *Homo sapiens* were hunter gatherers. The accepted fossil and genetic records indicate that already small populations of early humans became even more threateningly diminished just prior to associated Transformation dates. Those reductions were each further associated with stressful geologic and climatic changes.

Strangely, in some ways contemporary civilizations are more vulnerable to geologic and climatic variations than early man. For example, half the populations of developed countries live in fixed positions on the coasts. Others permanently dwell in areas that depend on a minimum of geologic and climatic variation. Modern power generation and distribution systems run continuously at maximum capacity. These systems are exceedingly vulnerable to solar flares, as are our space and earth-based information communications systems. Food tends to travel great distances to consumers, and rely on a minimum of change in the growing seasons. Yet, our true vulnerability may lie in our rapidly worsening state of global, regional, and local social disequilibrium.

The ancient Maya (and their surviving descendent shaman) predict that human beings will tend to become harmful to themselves and others as we approach the World Age end date. As related by Synthia and Colin Andrews (2008), the ancient Maya posit that:

> ... We're going through a difficult time leading into 2012. These difficulties are the result of our own disharmony of spirit. Governments and religions will let us down and lead us astray. Resources will be scarce and

people will be afraid. Natural disasters will increase through the period and the age will end with destruction by fire.[387]

The Andrews reiterate the ancient Maya prophecy of the *Chilam Balam* (referencing Maud Worcester Makemson's 1951, "The Book of the Jaguar Priest: A Translation of the Book of the Chilam Balm of Tizimin)."[388 & 389]

> The katuns [prophesies of the *Chilam Balam*] refer to this time as the final days of misfortune at the end of the Long Count Calendar. They predict earth changes in the form of storms, earthquakes, famine, and pestilence. In this time, people loose confidence in religion, priests, government, and officials."[390]

The prophecy of a loss of confidence in our leaders could well portend a secondary result of physical gravitational torque stress. It might be due to the accelerating rate of change, as opposed to just some mystical prophecy of doom and gloom. Chapter 2 quoted Everett M. Rogers as stating that, "*Disequilibrium* occurs when the rate of change is too rapid to permit the [human social] system to adjust."[391] Is the drastically increasing rate of innovation over the course of last century threatening to outstrip our existing social system's ability to absorb it and adjust to it? It appears so.

In March of 2010, the eminent British historian and author, Paul Johnson, described just such effects in an article in Forbes magazine titled, "The Sickness of the West."

Seldom in modern history has the lack of trust, now verging on contempt, been so deep, universal and comprehensive.

...

More devastating, in a sense, is the loss of trust in entire categories of people who once formed bastions of integrity at the heart of society. In Britain a half a century ago there were three categories of professionals who inspired general regard: bankers, scientists, and politicians.[392]

Johnson gets immediately specific about how our bankers and scientists are being divested of public trust.

Today the word 'banker' is a pejorative term. ... Bankers are associated with greed, recklessness and professional incompetence. It would be hard to think of another group that as fallen so fast and so far in public esteem. The same thing is happening with scientists.

Now, as the theory of man-made global warming unravels, scientists are suddenly and devastatingly revealed as fallible, mendacious, self-seeking, criminally secretive, furtively trying to hide their errors, debasing the system of peer review of scientific papers and conspiring to conceal the truth from once highly respected professional publications. The image of the scientist who puts the pursuit of truth before anything else has been shattered and replaced with a man on the make or quasi-religious enthusiast who wants to prove his case at any cost. Science is becoming the tool of campaigning warfare, in which truth is the first casualty.[393]

Johnson next proceeds to describe the great fall of British politicians in the eyes of the English people. Opinions that are increasingly reflected in political poles and media headlines throughout the free world. Views increasingly whispered between the constituents of oppressive nations as well.

> The fall of the scientists is nothing compared with the self-degradation of British politicians. I can remember a time when most people would regard it as an honor to shake hands with a Member of Parliament. Today the initials MP after a surname are a badge of shame.
>
> ...
>
> The House of Lords has been found to be no better. ... There's a general feeling that the entire system is rotten and that something drastic needs to be done, such as in 1653 when Oliver Cromwell dissolved the corrupt Long Parliament at the point of a sword.
>
> Indeed in Britain today it's a lamentable fact that, with the Church of England held in ridicule and contempt and lawyers seen as monsters of greed, the only group in society still treated with admiration and respect is the armed forces. Maybe that is a portent.[394]

The loss of trust reported by Paul Johnson for the British Isles appears to be no different in any country on any continent on this planet. One has only to flip on a news channel to gain immediate confirmation of it. The problem, however, extends far deeper than a loss of trust in our leaders.

The change driven population of the World is fractionalizing at the global, regional, local, small-group, and even interpersonal levels. In 2002 during a lecture in Santa Fe, New Mexico, a cultural anthropologist and practicing Maya Day Keeper Shaman, Carlos Barrios, put the implications of the Maya prophecy for the current time this way:

> Everyone thinks they are the most important, that their own understandings or their group's understandings are the key. There's a diversity of cultures and opinions, so there is competition, diffusion, and no single focus. ... They like the energy of the old declining Fourth World, the materialism. They do not want it to change. They do not want fusion. They want to stay at this level and are afraid of the next.
>
> The dark power of the declining Fourth World [the current World Age] cannot be destroyed or overpowered. It's too strong and clear for that. That is the wrong strategy. The dark can only be transformed when confronted with simplicity and open-heartedness. This is what leads to fusion, a key concept for the World of the Fifth Sun [The Transformed World after Friday December 21, 2012].[394]

The World Age Transformation prophesied by the ancient Maya and predicted by WARPS, posits an instantaneous nondestructive Transformation. Figure 107 plots the course of a phase transition indicative of those that may give rise to an optimizing Transformation.

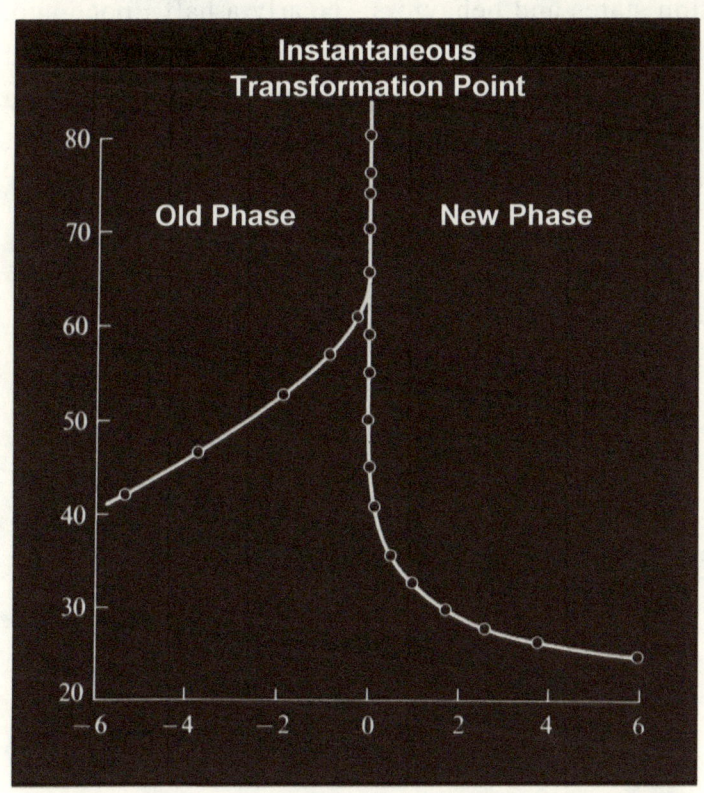

Figure 107 the Shape of an Example Phase Transition

Source: Source: Atkins, P.W., 1978, Physical Chemistry, W.H. Freeman & Co, San Francisco, p.187

Modern science has demonstrated many cases of such naturally occurring phenomena. Such energy and matter transformations exemplify how an optimizing set of cascading phase transitions might produce a World Age Transformation. However, the build up of compressive gravitational torque energy to the critical point can be seriously destructive. Already our solar system, planet, and people are exhibiting worrisome symptoms of such pre-transition states and behaviors. Nearly a half Epoch ago, the Maya predicted that decidedly dangerous states of man and nature would occur at the end of the current World Age. A set of predictions disturbingly in keeping with the results reported in the next two chapters.

17

WAVES OF EARTH COMPRESSION

Earth's Tummy Tuck Trend

E arth spins about its canted vertical axis, exhibiting a major bulge in its equatorial middle and stretching its poles out away from its center. This time varying relationship is called "oblateness."[395] "The oblateness of a body [like Earth] is calculated by subtracting its polar diameter from its equatorial diameter, and then dividing by the equatorial diameter."[396]

Figure 108 depicts the Earth as seen from Apollo 17. The figure shows a static view of Earth's dimensions, to include its axial tilt (called its Obliquity). If – as repeatedly demonstrated throughout this book – a World Age train of Gravitational Torque Waves has been at work on our World, the resulting space-time density compression – and hence Maya Degrees of Precession – should describe, explain, and predict a significant amount of the Earth's long term variation in oblateness.

Hypotheses H_{104} and H_{105} make that assertion a bit more specific; **Hypothesis H_{104}** posits that: the known linear rate of change in Earth's Oblateness is significantly correlated with the estimated rate of change in GTW imposed Density Compression. **Hypotheses Null$_{104}$** counters that there is no counting relationship between the rate of change in Earth's Oblateness and the estimated rate of GTW imposed Density Compression.

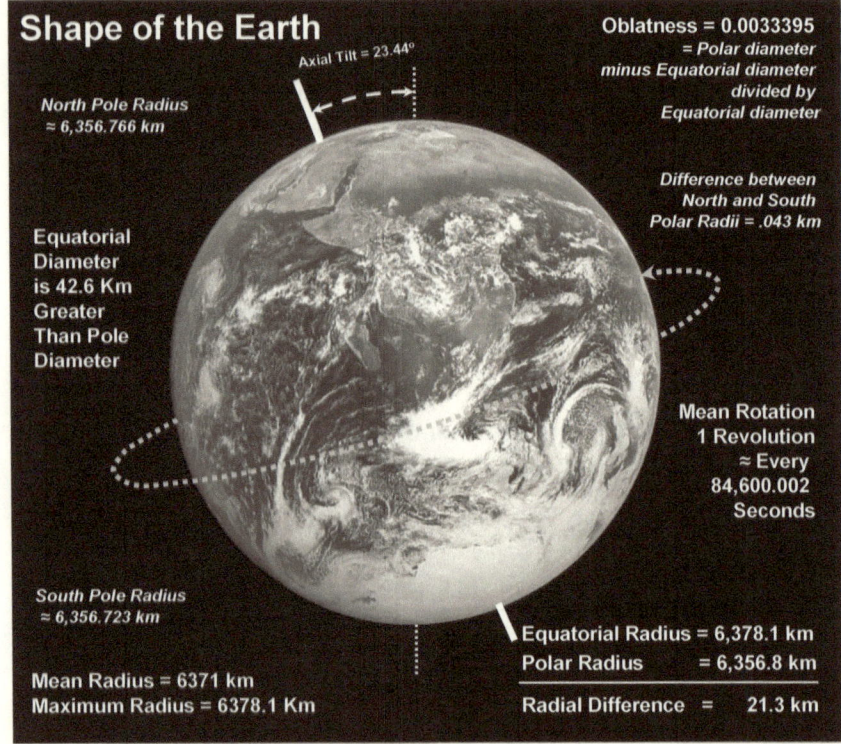

Figure 108 Shape and Dimensions of the Earth

Source: Http://en.wikipedia.org/wiki/File:The_Earth_ seen_from_Apollo_17.jpg, Revisited April 8, 2010

Hypothesis H_{105} alleges that: GTW imposed Density Compression describes, explains, and predicts a significant portion of the variation in the currently accepted linear rate of change in Earth's Oblateness. **Hypotheses Null$_{105}$** is the same as Hypotheses Null$_{104}$.

These two investigative assertions go well beyond the accepted view dominating Earth science and planetary geodetics. Prevailing scientists limit themselves to the Earth-local idea that:

> "Oblateness is the flattening of a planet or star at the poles, caused by its rotation; also known as ellipticity or polar flattening."[397]

That is a well established fact, as far as it goes, as is the related notion that the Earth's shifting mass causes the rotation rate – and hence its' Oblateness – to vary. But, there's considerably more to it.

Here we test the assertion that the majority of the observed and estimated variation in Earth's distribution of mass, and thus Oblateness, is caused by a *non*-Earth-bound force. The contention is nothing short of heretical. Of course, so were the earlier retained hypotheses that Galactic GTW Density Compression is the primary source of Solar System Precession and our planet's Obliquity, Eccentricity, Inclination, and climatic variations. As it turns out, Earth's Olateness is another case in point.

The linear trend in Earth's "Oblateness Coefficient," symbolized as "J_2," is widely estimated at "-2.75 x 10^{-11} per year," for the last 21,000 years of Earth history.[398] Different researchers report subtly different estimates of the long term linear rate of change in J_2.[399] All of these estimates of the rate of change in J_2 are, however, remarkably similar, linear, declining, and constant. Any of the oft reported estimates can be substituted one for another without seriously altering the results of the tests of Hypotheses H_{104} and H_{105}.

Figure 109 depicts the slow downward slope of the linear trend in the estimated rate of oblateness. That trend line was calculated using a constantly linear rate of "-2.75 x 10^{-11} per year," as estimated by Benjamin Fong Chao (2006) of the Space Geodesy Laboratory, NASA Goddard Space Flight Center.[400]

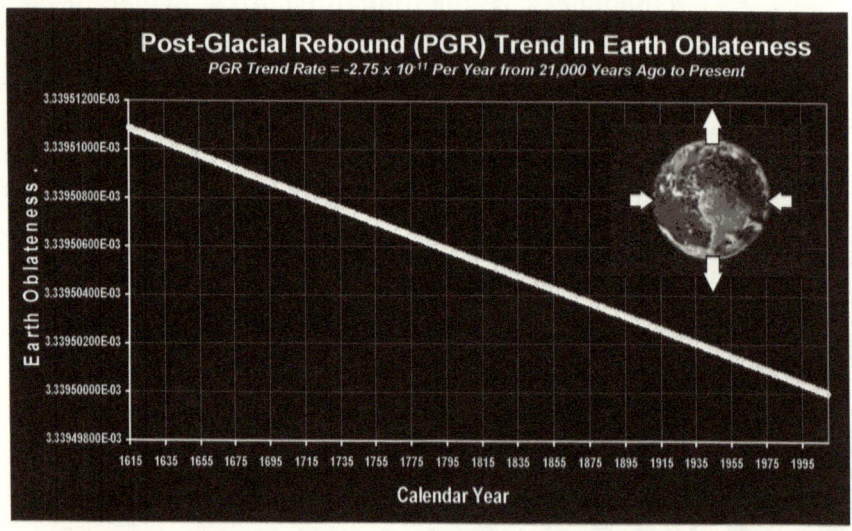

Figure 109 Linear Trend in the Estimated Oblateness Constant from 1615 to 2010

Source of PGR Secular Rate: Byron D. Tapley, 2004,
Http://www.thunderbolts.info/forum/phpBB3/viewtopic.php?p=25863&si
d=a4fed569b19c237d60f3416356e86f14,
Revisited April 8, 2010; Source of PGR Effect Earth Image:
http://www.gsfc.nasa.gov/topstory/20020801gravityfield.html, Revisited
March 25, 2010

The globe and arrows in the upper right-hand corner of Figure 109 symbolize the effects of this trend on Earth's equatorial and polar dimensions. Figure 109 is limited to the period from 1615 to 2010 AD. The plotted linear trend in Figure 109, however, is believed to have remained constant over the last 20,000 years of Earth history.[401] The following quotation states the widely accepted view of why this is the case.

> During the last glacial period, much of northern Europe, Asia, North America, Greenland and Antarctica were covered by ice sheets. The ice was as thick as three kilometers during the last glacial maximum about 20,000 years ago. The enormous weight of this ice caused the surface of the Earth's crust to deform and warp downward, forcing the fluid mantle material to flow away from the loaded region. At the end of the ice age when the glaciers retreated, the removal of the weight from the depressed land led to slow (and still ongoing) uplift or rebound of the land and the return flow of mantle material back under the de-glaciated area. Due to the extreme viscosity of the mantle, it will take many thousands of years for the land to reach an equilibrium level."[402]

The expanded view stated by J.M. Johansson in 2002, reflects recent changes in the accepted perspective of the Post Glacial Rebound (PGR).

> Recently, the term post-glacial rebound [PGR] is gradually being replaced by the term glacial isostatic adjustment. This is in recognition that the response of the Earth to glacial loading and unloading is not limited to the upward rebound movement, but also involves downward land

movement, horizontal crustal motion,[403 & 404] changes in global sea levels,[405] the Earth's gravity field,[406] induced earthquakes[407] and changes in the rotational motion."[408 & 409]

What all that amounts to is that Earth scientists still see the slow changes in Earth's Oblateness as the result of the planet local redistribution of the mass of the oblate globe. To give credit where credit is due, in the last three decades understanding of changes in Earth's Oblateness has improved tremendously. Figure 110 depicts the host of space based satellites used to regularly measure the Earth's geodetic dimensions and changing Oblateness. To quote NASA, the source of that illustration:

> "This graphic shows the constellation of satellites supported by the Satellite Laser Ranging (SLR) network. Data spanning 28 years from 8 satellites are used for measuring the large-scale mass movements on the Earth and global solid Earth dynamics."[410]

Data collected over the past 33 years have been combined with historical astronomical observations (much of it from the ancient civilizations of the last 2,000 years) theoretical assumptions, calculations, and simulations to produce longitudinal estimates of the rate of change in Earth's Oblateness Coefficient. According to a December 2006 report by NASA"

> "The Earth is inelastic and responds to changes by slowly recovering on time scales of tens of thousands of years, resulting in a

redistribution of mass toward the higher latitudes, effectively making the Earth more spherical. This phenomenon, called post-glacial rebound [PGR], still continues today, due to the slow response of the Earth's mantle."[411]

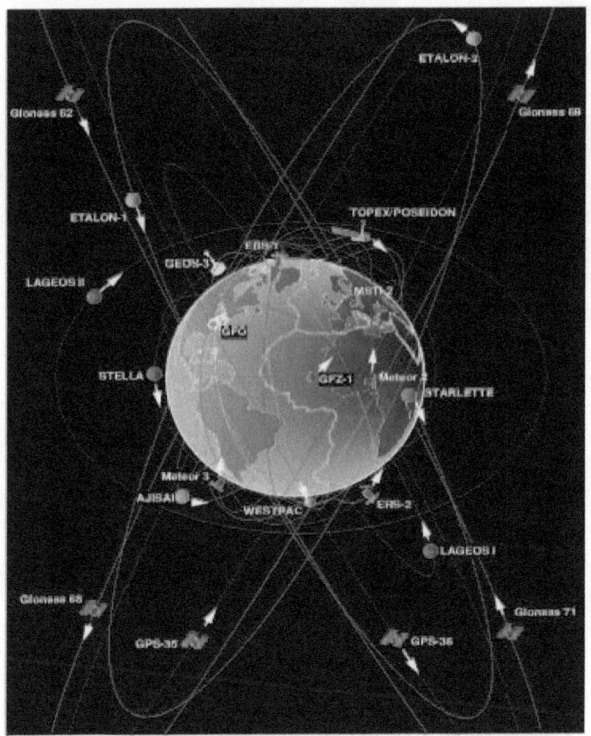

Figure 110 Satellite Laser Ranging Network
*Source: Http://www.nasa.gov/centers/goddard/
earthandsun/earthshape.html,Revisited April 11, 2010*

Earth's Long Term Waistline Reduction

Hypotheses H_{104}, was tested by calculating the correlation between the estimated rate of change in Galactic GTW Density Compression (1.16 x 10^{-05} per year, which equates to 0.014 degrees of Maya Solar System Precession per year) and the rate of change in Earth's Oblateness from 1615 to 2010 (n=395 years). Table 22 relates the resulting correlation coefficients.

Table 22 Correlation Between GTW Compression and Earth Oblateness.

Correlation Between PGR Earth Oblateness Degrees of Maya Precession & GTW Compression N = 393 Years			
	Oblateness	*Compression*	*Precession*
Oblateness	1.0000		
GTW Compression	-1.0000	1.0000	
Precession	-1.0000	1.0000	1.0000

It is apparent from Table 22 that for at least the last 395 years, the literature estimate of the declining rate of Earth's Oblateness Coefficient is perfectly correlated with the rate of change in GTW Density Compression, and Degrees of Maya Solar System Precession. The relations are significant at the .001 level of confidence and extend to a number of decimal places approaching infinity. Hypothesis H_{104} is retained and Hypothesis Null H_{104} rejected. As GTW Density Compression increases, Earth's Oblateness decreases. A decrease in Earth's Oblateness Coefficient is equated to a reduction in equatorial diameter and an increase in polar diameter.

For the last 20,000 years Earth's Oblateness has remained predominantly linear (described by a straight line). Hence, the perfect *negative* correlation between GTW Density Compression and Earth Oblateness holds for the last

5,125.36 years of Earth's history. At 5,125.36 years ago, the correlation changes sign (become *positive*), but remains 1.00 (perfect) for the period from 5,125.36 years ago to 10,250.7 years ago, the wavelength of one GTW. At 10,250.7 years ago, the relation turns *negative* again for another 5,125.36 years of history before turning positive again. Excepting for the described sign changes, the perfect correlation is applicable for a full 20,501.4 years of Earth's Oblateness. That duration equates to two Gravitational Torque wavelengths and the estimated duration of the PGR.

A regression analyses confirmed that the estimated rate of change in GTW Density Compression (and hence Degrees of Maya Precession) describes, explains, and predicts 100 percent of the variation in the linear estimate of the rate of change in Earth's Oblateness from 1615 to 2008. That predictive relation (Adjusted R^2 = 1.00) is significant well beyond the .001 level, as the F statistic approaches infinity. Hence, Hypothesis H_{105} is retained and Hypothesis Null$_{105}$ rejected. That perfect predictive relation between GTW Density Compression and Earth's linear Rate of Oblateness accurately predicts Earth's oblateness for the last 20,501.4 years of Earth history. Figure 111 depicts the regression analysis line fit plot resulting from the test of Hypothesis H_{105}.

It is here affirmed that the perfect correlation between the rate of change in GTW Density Compression and Earth Oblateness is *causal*. The density compression imposed by two passing GTW has applied compressive pressure to every level of energy and matter within the solar neighborhood. The resulting pressure and energy storage in our solar neighborhood over the past 20,501.4 years forced the slow redistribution of planetary mass. That redistribution has reduced Earth's equatorial diameter and increased its polar diameter. The Post Glacial Rebound (PGR) is thus an effect – as opposed to the cause – of the change in Earth's Oblateness Coefficient.

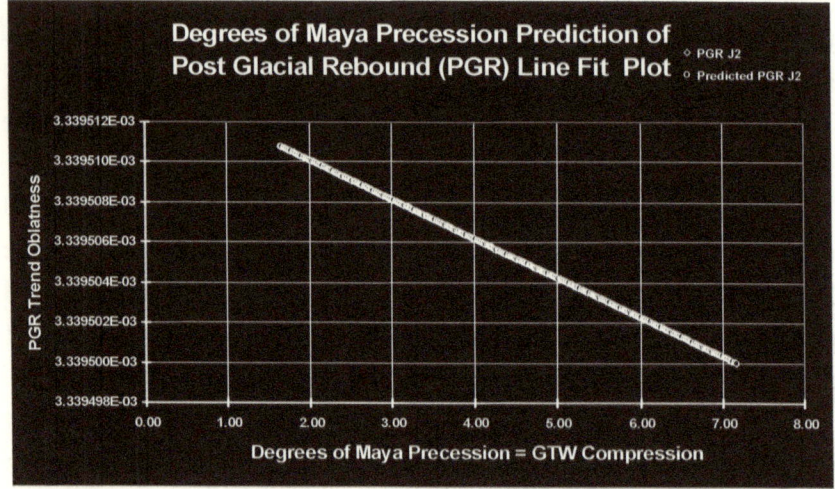

Figure 111 GTW Compression and Degrees of Maya Precession Prediction of Earth Oblateness

Predicting Earth's Waist Warps

Maya World Age theory and WARPS predict the cumulative storage of GTW energy in the constituent wave particles of space-time and ordinary matter. That energy storage is predicted to reach the critical Transformation point on Friday December 21, 2012. If this is indeed the case, Earth's Oblateness Coefficient (J_2) should be exhibiting signs of the energy storage build up to that critical point of Transformation. WARPS theory predicts that Earth's J_2 Oblateness should be exhibiting:

1. Significant variations in J_2 with periods *greater than* 20 years

2. Significant inter-annual variations in J_2 with periods *less than* 20 years

3. Significant short term J_2 *anomalies* (reversals of the longitudinal trend in oblateness).

Figure 112 indicates that over the past 33 years scientists have been observing all three of those types of changes in Earth's Oblateness.

WARPS **Hypothesis H_{106}** states that GTW Density Compression (and hence Maya Degrees of Precession) is significantly correlated with Earth's J_2 Oblateness variations having periods greater than 20 years from Earth's long term linear Oblateness trend. **Hypothesis Null$_{106}$** counters that there is no significant counting relationship between the rate of GTW Density Compression and the variations in J_2 Oblateness with periods greater than 20 years.

WARPS **Hypothesis H_{107}** states that GTW Density Compression describes, explains, and predicts a significant portion of variations from the linear trend in Earth's Oblateness that have occurred with periods greater than 20 years. **Hypothesis Null$_{107}$** counters that there is no significant counting relationship between GTW Density Compression and J_2 Oblateness variations from the longitudinal trend.

Figure 113 isolates the observed variations from Earth's longitudinal trend in J_2 Oblateness exhibiting periods of greater than 20 years.

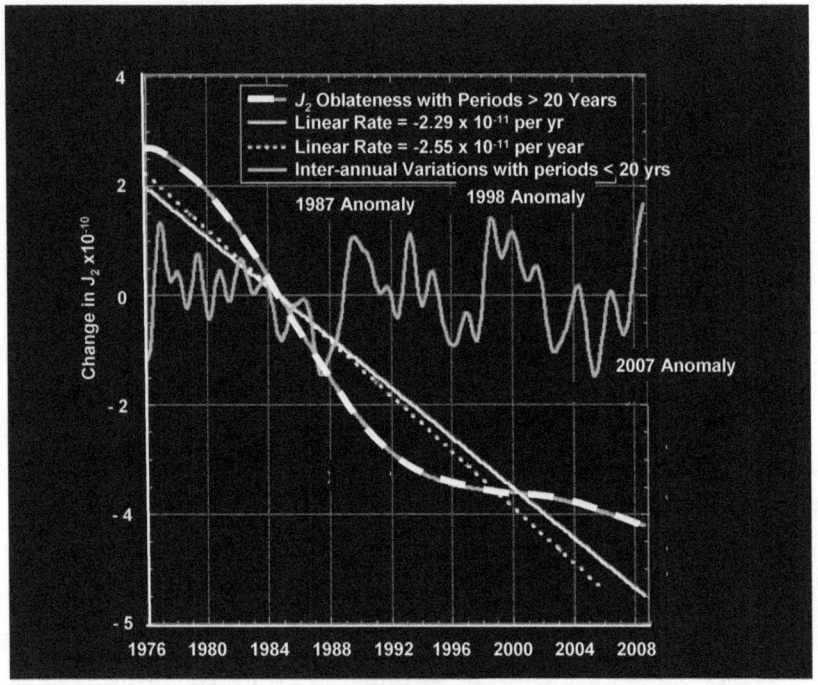

**Figure 112 WARPS Pre-Transformation
Predictions Now in Evidence**

Source: Minkang Cheng and Byron D. Tapley, 2009, A 33 Year Time History of the J2 Changes from SLR [Satellite Laser Ranging], ILRS Workshop, 2008, cheng@csr.utexas.edu

Analysis results confirm WARPS Hypothesis H_{106}. GTW Density Compression and Maya Degrees of Precession are correlated -0.9377 with variations with periods greater than 20 years from the longitudinal linear trend in J_2 Oblateness. Regression analysis corroborates Hypothesis H_{107}. GTW Compression and Maya Degrees of Precession respectively predict 87.53% of the observed variation in J_2 Oblateness exhibited over periods of greater than 20 years.

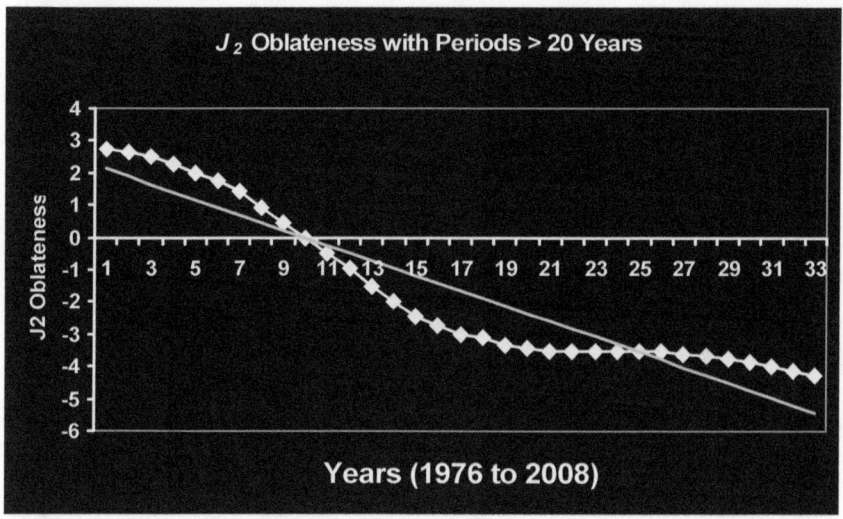

Figure 113: J_2 Oblateness Variations with Periods Greater Than 20 Years

Source: Minkang Cheng and Byron D. Tapley, 2009, A 33 Year Time History of the J2 Changes from SLR [Satellite Laser Ranging], ILRS Workshop, 2008, cheng@csr.utexas.edu

Figure 114 depicts the GTW Density Compression Line Fit plot prediction of J_2 Oblateness variations having periods greater than 20 years.

Hypothesis H_{106} and Hypothesis H_{107} are retained and Hypothesis $Null_{106}$ and $Null_{107}$ are rejected at the .001 level. The significance of the F statistic extends to 15 zeros after the decimal point. The sample size used, however, was limited to just 33 years (32 degrees of freedom). While the results appear valid and theoretically justified, the findings must be considered preliminary. Study findings must be replicated by independent researchers using larger samples of J_2 data points.

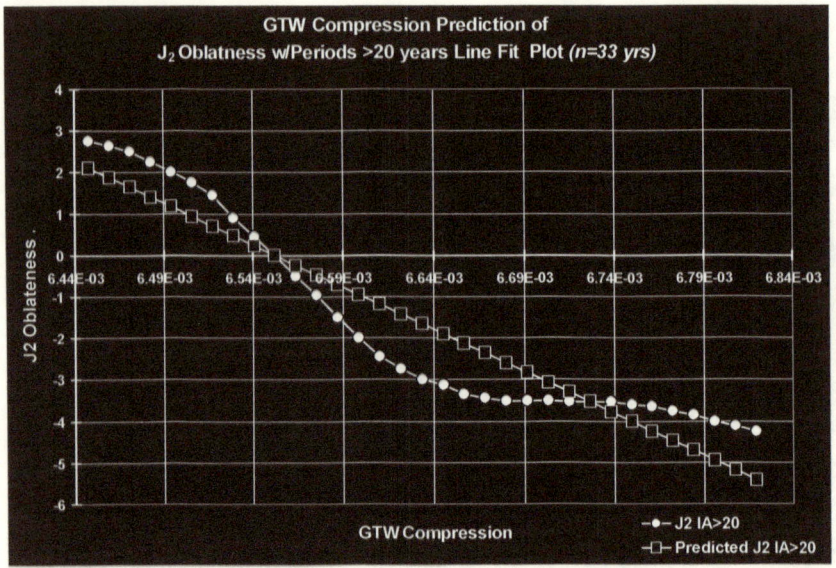

Figure 114 GTW Prediction of J_2 Oblateness Variations with Periods Greater than 20 Years

Predicting Earth's Waistline Anomalies

It was noticed that the shorter term J_2 Oblateness variation and anomalies that have been observed over the last 33 years occur in periods approximating one Maya Katun cycle. Each Maya Katun cycle equals 7,200 days, which equates to approximately 19.762 of our Gregorian calendar years. Similarly, the "Signal of 18.6 year tide" was described and "separated from other terms" of J_2 coeffecient data analysis reported by Minkang Cheng and Byron D. Tapley in 2009.[412] That year those researchers reported the results of a

33 Year Time History of the J_2 Changes from SLR (Satellite Laser Ranging) based studies. In their report they stated that:

> Analysis of satellite laser ranging (SLR) data indicates that the Earth's dynamic oblateness (J_2) has undergone significant variations during the past 33 years. The estimate of the secular decreasing rate is affected by the significant inter-annual variations [trend reversals] with amplitude as large as 0.5×10^{-10} and time scales of ~10 years. Two large inter-annual variations in J_2 are correlated with the strongest ENSO [El Niño-Southern Oscillation] events during the periods of 1986-1991 and 1996-2002. It appears a new cycle has started from 2007. [413]

Figure 115 depicts the raw and smoothed less than twenty year period J_2 Oblateness coefficient variations and anomalies cited by Cheng and Tapley. The longer term metrics of Gravitational Torque Wave Density Compression are nearly zero correlated (near totally unrelated) with the plotted raw J_2 Oblateness variations and anomalies depicted in Figure 115. Could a Maya Katun Cycle driven form of *K'ul* Response Effect, or *K'ul* Resonance, relate to the 5th order polynomial smoothed variations of Earth's Oblateness anomalies?

Hypothesis H$_{108}$ posits that the calculated Katun Cycle *K'ul* Response Effect is correlated with the less than 20 year period smoothed J_2 Oblateness variations containing anomalies. **Hypothesis Null$_{108}$** counters that the *K'ul* Response Effect measure is not related to the less than 20 year period J_2 variations containing anomalies.

Hypothesis H$_{109}$ asserts that the calculated Katun Cycle *K'ul* Response Effect describes, explains, and predicts a significant portion of the variation in the less than 20 year period smoothed J_2 Oblateness variations containing anomalies. **Hypothesis Null[109]** counters that the calculated Katun Cycle *K'ul* Response Effect measure is unrelated to the short term variations in J_2 containing the anomalies observed from 1976 to 2008.

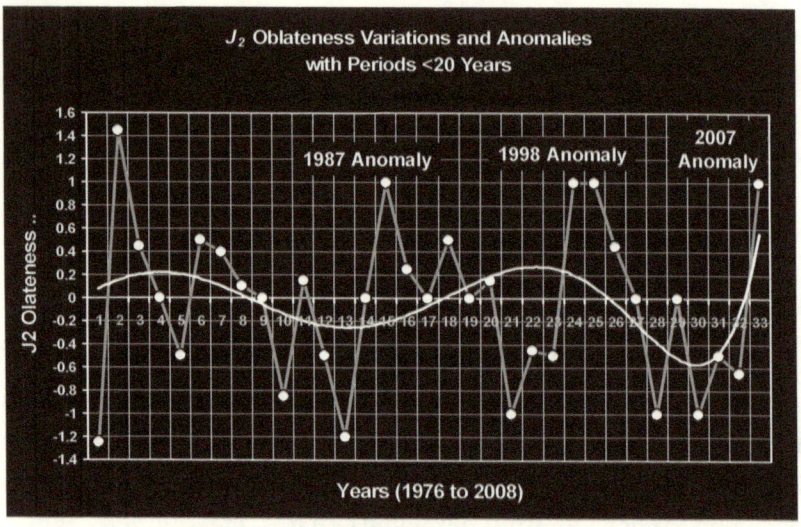

Figure 115 Oblateness Variations with Periods Less than 20 years and Anomalies

Source: Minkang Cheng and Byron D. Tapley, 2009, A 33 Year Time History of the J2 Changes from SLR [Satellite Laser Ranging], ILRS Workshop, 2008, cheng@csr.utexas.edu

The Katun Cycle *K'ul* Response Effect and Katun *K'ul* Resonance measures were calculated using the very same methods described earlier in this book for the calculation of the Baktun Cycle *K'ul* Response Effect and Baktun *K'ul* Resonance. The only change made was to *scale* the mean and standard deviations fed into the normal distribution generator to match the scaled shorter duration Katun Cycle. Instead of calculating the *K'ul*

Response Effect using a mean of 1.0 and a standard deviation of 0.3, a Katun scaled mean of .05 and a standard deviation of 0.15 were substituted.

Correlation Analyses revealed that the calculated Katun Cycle *K'ul* Response Effect measure is correlated 0.7347 with the smoothed J_2 Oblateness variations containing the observed anomalies. Katun *K'ul* Resonance is correlated 0.7241 with the smoothed J_2 Oblatness variations containing the anomalies. While both *K'ul* measures are strongly correlated with the smoothed trend in J_2 variations, *K'ul* Response Effect proved the slightly stronger predictor.

Regression analysis revealed that Katun *K'ul* Response Effect describes, explains, and predicts 54% (the Adjusted R^2 = 0.539759725) of the variation in the smoothed J_2 variations occurring in less than 20 year periods from 1976 to 2008 (n=33 years). The relationship is significant well below the .001 level (i.e. F = 0.0000013771) with 32 degrees of freedom.

The same cautionary notes stated earlier apply equally well to these findings. Those warnings are further amplified by the prediction of *smoothed J_2* variations, as opposed to the raw obliquity data.

Figure 116 plots the Katun *K'ul* Response Effect in relation to the smoothed and Raw J_2 Oblateness variations and anomalies. Given the results, Hypotheses Null[108] and Null[109] are tentatively rejected. Both *K'ul* measures are significantly related to the smoothed less than 20 year period J_2 Oblateness variation containing anomalies.

Caveats aside, it appears there maybe reason to believe that Maya World Age Transformation predictions may not be as much the stuff of fantasy science fiction as we might wish them to be. The ratio of this planet's equatorial to polar diameter has begun to oscillate, and in progressively smaller periods – just as predicted by the ancient Maya and implied by the WARPS Theory. Are there other indications of a build up to Transformation? The next chapter answers in the affirmative.

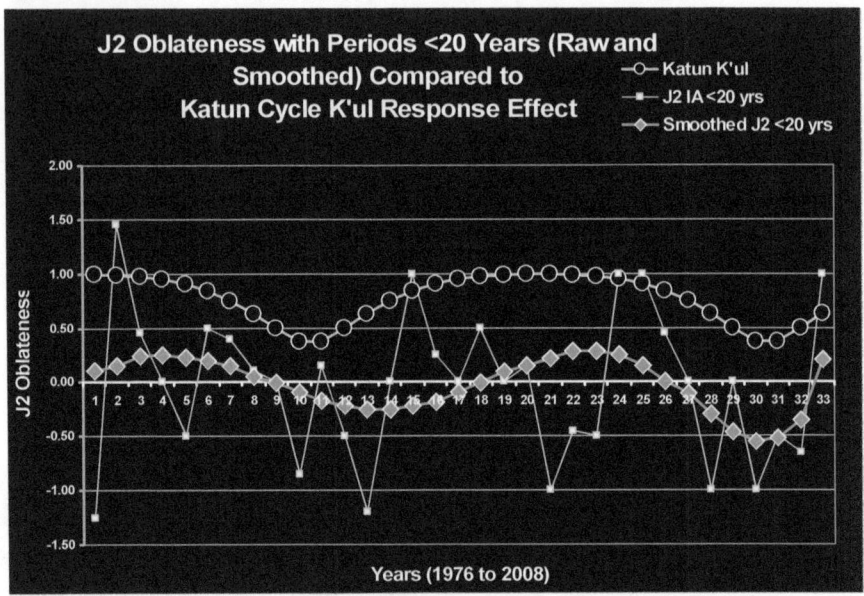

Figure 116 Relationship of Katun *K'ul* Response Effect to short term *J₂* Oblateness variations

18

WAVE TRAIN SIGNALS, SWITCHINGS, AND SIDINGS

The history of human change, global change, and our solar system's celestial motion count with the World Age cycles of the Maya Long Count calendar. Given all that, there ought to be still more indications that GTW are having their way with our solar system. This chapter describes three such *indications*. They are, the:

1. Accelerating rate of human genetic change

2. Accelerating rate of stressed populations and extinctions

3. Bombardments of the Earth's atmosphere by very high-energy particles emanating from a cosmic ray source atypically near to our solar system.

Make no mistake about it. The assertion that a solar system wide Transformation is about to occur on Friday December 21, 2012 is outrageously speculative. Still, something rare and extraordinary does appear to be looming on our near term horizon. If so, it is something that has happened:

- 184,572 times in the history of this planet

- 273 times since the first known Hominids appeared on Earth

- 11 times since the emergence of the first *Homo sapiens*

- 6 times since the literal birth of truly modern man

- 2 times since the origin of art and religion

- Once since our shift from hunter-gathers to an agrarian way of life.

This great audit trail of clocked events tends to strappingly endorse the cycles of the Maya Long Count Calendar, the ancient Maya prophecy of impending Transformation, and the theory of World Age Resonant Phase Synchronization (WARPS). We respectfully reserve judgment and urge you to do the same. The great pile of evidence for historic World Age Transformations may be large, but it is also largely circumstantial. Taking a hard-line legal perspective, there is no direct evidence from any living survivor of a prior Transformation. And yet, what follows is still more indirect evidence that suggests that another World Age Transformation will soon follow.

WARPS theory describes the energy build up to cascades of phase transitions resulting in a near instantaneous Transformation of the solar neighborhood. Three WARPS hypotheses emerge from that theoretic argument.

Hypothesis H_{110} argues that the rate of Human Genetic Change is accelerating proportionate to the increase in GTW pressure. **Null Hypothesis H_{110}** counters that the rate of Human Genetic Change has exhibited no observable and measurable increase for the last 50,000 years. **Alternate Hypothesis H_{110}** insists that any measured increase in the rate of Human Genetic Change is due solely to the increase in Human Population.

Hypothesis H_{111} asserts that the number of threatened species and the rate of species extinctions are increasing proportionate to the increase in GTW pressure. **Null Hypothesis H_{111}** counters that any increase in the number of threatened species and rate of species extinctions is independent of GTW pressure.

Hypothesis H_{112} alleges that evidence of the building GTW pressure to Transformation in the solar neighborhood is exhibiting as bombardments of the Earth's atmosphere by very high-energy particles emanating from sources atypically near to our solar system. **Null Hypothesis H_{112}** contends that there are no unexplained variations in the number and energy of particles hitting Earth's atmosphere. **Alternative Hypothesis H_{112}** argues that any observed increase in very high-energy particles hitting Earth's atmosphere is due to a nearby astrophysical object, such as a pulsar or micro-quasar.

Accelerating Rate of Human Genetic Change

A 2009 article by Kathleen McAuliffe cites the acclaimed Harvard Paleontologist Stephen Jay Gould as once stridently proclaiming, "There have been no biological changes. Everything we've called culture and civilization we've built with the same body and brain."[414] McAuliffe laments, "This view has become so entrenched that it is practically doctrine. ... So to suggest that humans have undergone an evolutionary makeover from Stone Age times to the present is nothing short of blasphemous."[415]

McAuliffe proceeds to cite a glaring problem with Gould's position. She states that a host of researchers continue to find "an abundance of recent adaptive mutations etched in the human genome."[416] What's more, "these mutations seem to be piling up faster and ever faster, like an avalanche."[417] McAuliffe's choice of the word *"avalanche"* is particularly noteworthy. At least one class of *avalanche* is considered a geophysical phase transition; one that results from the build up of excess energy to the critical point of transition in slab-like layers of matter.

McAuliffe argues that:

> Over the past 10,000 years [that period being approximately equal to one GTW wavelength of 10,250.7 years] data show [that] human evolution has occurred *a hundred times* more quickly than in any other period of our species' history.
>
> The new genetic adaptations, some 2,000 in total [and still counting], are not limited to the well-recognized differences among ethnic groups in superficial traits such as skin and eye

color. The mutations relate to the brain, the digestive system, life span, immunity to pathogens, sperm production, and bones – in short, virtually every aspect of our functioning.[418]

In a recent paper on human evolution, Henry Harpending of the University of Utah states, "It is likely that [the] human races are evolving away from each other. ... We are getting less alike, not merging into a single mixed humanity. ... We aren't the same as people even a thousand or two thousand years ago."[419] McAuliffe relates that Harpending's perspective is one long espoused by "historians," despite the entrenched views previously projected by aging paleontologists, archeologists, anthropologists, and geneticists. She suggests that these long clutched to views are changing, and relates: "Yet even skeptics now admit that some human traits, at least, are evolving rapidly, challenging yesterday's hallowed beliefs."[420]

John Hawks of the University of Wisconsin at Madison asserts, "You don't have to look hard to see that teeth are getting smaller, skull size is shrinking, and stature is getting smaller."[421] To this McAuliffe adds, "These overriding trends are similar in many parts of the world, but other changes, especially over the past 10,000 years, are distinct to specific ethnic groups."[422] Hawks concurs, stating:

> These variations are well known to forensic anthropologists. ... It beats me how leading biologists could look at the fossil record and

conclude that human evolution came to a standstill 50,000 years ago."[423]

Please note that the landmark date of 50,000 years ago is by convention rounded off to just 1,253.6 years short of two World Age Precession cycles ago (i.e., 25,626.8 x 2 = 51,253.6 years).

McAuliffe credits "stunning advances in sequencing and deciphering DNA in recent years," as the reason "scientists had begun uncovering, one by one, genes that boost evolutionary fitness. These variants, which emerged after the Stone Age [around 25,626.8 years ago, which equals one Maya World Age], seemed to help populations better combat infectious organisms, survive frigid temperatures, or otherwise adapt to local conditions. And they were popping up with surprising frequency."[424]

Thus, we lay Null Hypothesis H_{110} to rest. The rate of Human genetic change is indeed increasing, just as asserted by Hypothesis H_{110}. Practitioners of the science of genetics, however, are not rushing to cite the pressure of Gravitational Torque Waves as the source of this observed increase.

Population and the Rate of Genetic change

Stephen Hawks and Gregory Cochran wondered, "Why might evolution be picking up speed – what is fueling the trend."[425]   They were soon joined by West Coast Geneticists, Robert Moyzis of the University of California, and Eric Wang of Veracyte Inc, in San Francisco.

Moyzis and Wang had been leaders in the development of new computational methods for mining data sources like the International Haplotype Map. That map is a catalog of differences in DNA collected from Japanese, Han Chinese, Nigerians, and Northern Europeans. "They too suspected that the rate of human evolution was accelerating."[426] The four researchers came to a consensus. They suggested that the *source* of the accelerating rate of human evolution is the correspondent increase in rate of human population growth. Their argument, as paraphrased by McAuliffe, proceeds as follows:

> Ten thousand years ago [~ one GTW wavelength], there were fewer than 10 million people on Earth. That figure soared to 200 million by the time of the Roman Empire. Since around 1500 [the estimated date of the last GTW leading edge arrival in our solar system] the global population has been rising exponentially, with the total now surpassing 6.7 billion. Since mutations are the fodder on which natural selection acts, it stands to reason that evolution might happen more quickly as our numbers surge.[427]

There is of course a statistical correlation between modern population growth and the near term rate of human change. The idea that the increasing rate of genetic change exhibits some counting relationship with population change is understandable, but seriously misleading.

For anthropologists, biologists, and geneticists to interpret population growth as the *causal* source of evolution, is not only surprising, it's shocking. Here's why. Remember the "punctuated equilibrium theory of evolution" from earlier chapters? Those chapters demonstrated that human change and the *punctuations* of speciation, extinction, genetic mutation, bottlenecks, and migrations occur at regular fixed intervals. It did not, however, completely defeat the theory of punctuated equilibrium. Part of the basic logic of that theory remains sound. That part argues that bursts of genetic change occur when a population undergoes substantial stress, or threat of extinction, and that changes occur within one or more small suddenly isolated subpopulations. The reason is fundamental, and ardently proclaimed by Stephen Jay Gould.

> A new species [as well as most major genetic adaptations] can arise when a small segment of the ancestral population is isolated at the periphery of the ancestral range. Large, stable central populations exert a strong homogenizing influence. New and favorable mutations are *diluted* by the sheer bulk of the population through which they must spread. They may build slowly in frequency, but changing environments usually *cancel their selective value* long before they reach fixation.[428]

It may seem counterintuitive, but a large number of people concentrated in a geographic area suppress the effect of natural selection and genetic mutation. It is a fact even Charles Darwin was forced to recognize. It was, after all, a remote isolated island in the Pacific on which Darwin first found the evidence that led him to his famous theory. It was the isolation of small numbers of plants and animals that furnished the diversity-spawning means, motive, and opportunity that ultimately inspired Darwin.

So yes, there is currently a correlation between global population growth and the accelerating rate of human evolution. There is also a strong counting relation between the rate of global climate change and the diffusion of human innovation. As you may recall, however, those relationships are not directly *causal*.

In earlier chapters GTW Compression emerged as the causal source of both global climate change and human change. It is here argued that GTW compression is also the true *cause* of the correlation between human population growth and the increasing rate of human genetic change. If so, the simple counting relation between population growth and human genetic change is *spurious* (meaning not genuine, authentic, or truly causal). It is further suggested that the increasing rate of genetic change has emerged in spite of – rather than because of – the increase in population and decline of mortality rates.

As Henry Harpending said, "It is likely that human races are evolving away from each other. ... We are getting less alike, not merging into a single mixed humanity."[429] His point holds an important though oft overlooked fact. That is, there is more than one way for a group of individuals to become *isolated* from the greater population. A population can become culturally and even genetically fragmented, while remaining in geographic proximity.

For example, *Homo erectus*, Neanderthal man, and modern *Homo sapiens* shared proximate geography for at least 22,000 years. The fossil record plainly shows that each of these species continued to evolve independently of the others. There is an even more dramatic example. When Neanderthal man split off from our supposed shared common ancestor, the Neanderthal species underwent several hundred thousand years of significant genetic mutations. All the while, the adapting Neanderthal population remained extraordinarily small and widely dispersed. Essentially this same sort of population constriction and geographic dispersion was equally true for all the Hominid species of history, save our own. Our species alone achieved exponential population growth and reduced mortality. As depicted in Figure 117, however, that increased population and reduced mortality is exceedingly recent. In point of fact, humanity's population growth and mortality rates remained relatively flat until "about 1750" AD.[430]

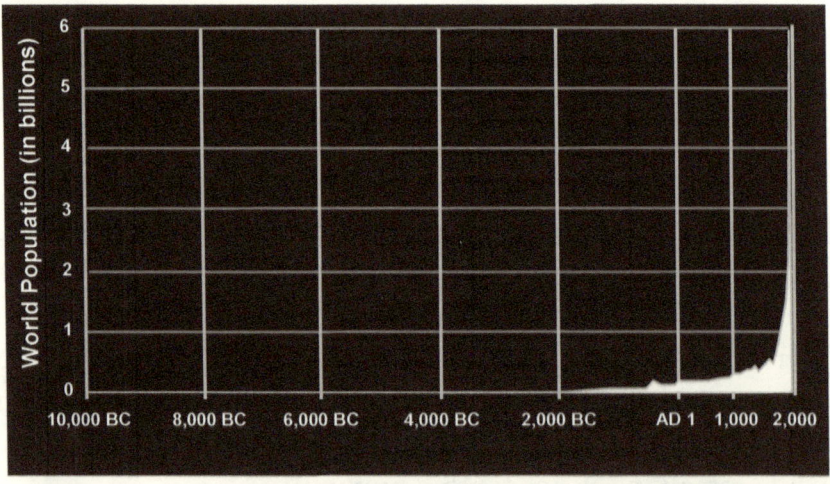

Figure 117 World Population from 10,000 BC to 2,000 AD *(Currently 6.8 billion)*
Source: http://en.wikipedia.org/wiki/File:Population_curve.svg,
Revisited January 15, 2010

More importantly, the great burst in human evolution that produced contemporary *Homo sapiens*" occurred in association with one of the most severe population bottlenecks in hominid history. It is estimated that during that particular genetic squeeze as few as 5,000 female *Homo sapiens* existed on this planet. That can hardly be considered a population growth rate driven genetic breakthrough.

There is little doubt that the researchers cited by McAuliffe are observing and measuring an accelerating rate of human genetic mutation. Equally valid and reliable, is the evidence that the rate of human genetic change has increased significantly in the last 10,000 years. Attributing those genetic changes to the increases in human population density limited to the last 259 years, however, is self nullifying. Thus, we must wave a confident goodbye to Alternative Hypothesis H_{110}. It is further suggested that genetic researchers have significantly underestimated the rate of increase in human genetic adaptations. The title of the next section alludes to the reason why.

Meaning is Usage; Not Possession

Biologists recently discovered a fact of life that came as a genuine shock to them. In 2008, Scean B. Carroll, Bejamin Prud'homme and Nicolas Gompel described the implications of that recent breakthrough in genetics:

> If we really want to understand what makes the human form different from that of other apes or what makes an elephant different from a mouse, for that matter, much of the information lies not in our respective genes and proteins but in an entirely different realm of our genomes that remains to be explored.
> ... Very different groups of animals can share most, if not all, the genes involved in body building and body patterning – contrary to

scientists' early expectations, it is mostly a matter of *how* and *when* those genes are used that shapes the different forms of the animal kingdom.[431]

Biologists had long "expected to find significant genetic differences among animals; reflecting their great diversity of forms. Instead, very dissimilar animals have turned out to have very similar genes. "Mutations in DNA 'switches' that control body shaping genes, rather than the genes themselves, have been a significant source of evolving differences among animals."[433] Carroll et al make it clear that as a result of this discovery, genetic science is undergoing a profound change in perspective.

For most of the last 40 years, or so, researchers have focused most of their attention on genes – the nucleotide sequences in DNA that encode the amino acid chains that form proteins. But to our surprise, it turned out that differences in appearance are deceiving: very different animals have very similar sets of genes. By following the trail of evolution, devices are being found within DNA – 'genetic switches' – that do not encode any proteins but regulate *when* and *where* genes are used. Changes in these switches are crucial to the evolution of anatomy and provide new insights into how the seemingly endless forms of the animal kingdom have evolved.[434]

What is further astounding is that these same geneticists now tell us that:

The approximate number of genes in each animal's genome (about 20,000 or so) and the relative positions of many genes have been fairly well maintained *over 100 million years of evolution*. That is to say there are no differences in gene number and location. But at first glance, nothing in these gene inventories shouts out 'mouse' or 'dog' or 'human.' When comparing mouse and human genomes, for example, biologists are able to identify a mouse counterpart for at least 99 percent of all our genes."

...

The DNA sequences of any two versions of a gene, as well as the proteins they encode, are generally alike to a degree that simply reflects the relative amount of time that has elapsed since the two species diverged from a common ancestor.[435]

But, the similarity in species doesn't stop there. "Body-building proteins are even more alike on average than other proteins ..."[436] "Animals as different as a mouse and an elephant are shaped by a common set of very similar, functionally indistinguishable body building proteins."[437]

It is no different for humans. Our "protein stretches of DNA make up only about 1.5 percent of our genome..."[438] For a long time scientists ignored the remaining non-coding DNA, "but some of those sequences participate in the very important task of regulating gene expression. And these regulatory sequences are key to evolution"[439] "The non-coding DNA play a critical part in directing *when* and *where* genes are expressed and are "components of 'genetic switches' that turn genes on or off at the right time and place in the body."[440] "Other components of the switch recognize those DNA sequences, often called 'enhancers.'"[441] It turns out that "every gene has at least one enhancer,"[442] and some genes have many. The difficulty has been that, "enhancers cannot be recognized solely on the basis of their DNA sequences and must be identified experimentally."[443]

In theory at least, "... mutations in enhancers would allow individual body traits to be [reversibly] selectively modified without changing genes or proteins themselves."[444] According to Carroll et al, "... in the past few years, direct evidence has emerged that this is frequently how the evolution of various body parts and patterns have occurred."[445]

Evolving Switch Settings

There are some serious misconceptions that emerge from incorrectly interpreting the effects of *natural* and *sexual selection*. Carroll et al cite two:

> Whereas we are naturally inclined to think that the presence of a feature in one species and its absence in another related species is the result of a gain by the first; that is often not the case. A flip side of evolution, the loss of features, is very common, though much underappreciated. The loss of body characteristics perhaps best illustrates why evolution of enhancers is the more likely path for the evolution of anatomy.
>
> ...
>
> Losses of features may or may not be beneficial for survival or greater reproductive

success, but some losses are adaptive because they facilitate some changes in lifestyle. Hind limbs, for example, have been lost many times in vertebrates – by snakes, lizards, whales, and manatees – and those losses are associated with adaptation to different habitats and means of locomotion.[446]

Researchers further note that the differences in pelvic morphology among certain species of fish have evolved repeatedly in just the 10,000 years since the last Ice Age.[447]

These points apply equally well to human beings. "Quite a few examples of evolutionary changes in regulatory sequences [those gene switches] that alter gene expression have been demonstrated for human traits as well."[448] Gregory A. Wray of Duke University cites an example of particular note:

> One of the most intriguing associations revealed thus far involves divergence in the great ape and human regulatory sequences controlling the *Prodynorphin* gene, which encodes a set of small opioid proteins produced by the brain and involved in perception, behavior, and memory. The human gene is more highly expressed in response to stimuli than is the chimpanzee version; and strong evidence suggests that the human regulatory sequence evolved under natural selection – that is, it was retained because it was advantageous.[449]

What is clear in all this is that evolutionary changes to anatomy are more likely to happen via changes to *gene enhancers* than to the genes themselves.[450] In this regard, nature is exceedingly clever. A full gene change may be irreversible and risk the survival of an entire species. A change in expression, however, is reversible, and significantly reduces the risk of extinction due to a premature adaptation to some temporal circumstance.

The evidence presented in this book suggests that significant changes in the synchronization and setting of *genetic expression and mutations* are associated with the World Age cycles of Precession. Hominid innovations, new species emergences, genetic bottlenecks, species migrations, and species extinctions all point to it. If the WARPS theory is correct, in addition to rapidly accelerating rates of genetic mutation, there ought to now be evidence of increasing stress in plant and animal populations over the last 20,000 years. Accordingly, **Hypothesis H$_{111}$** asserts that the number of threatened species and the rate of species extinctions should be increasing proportionate to the increase in GTW pressure. The next section addresses indications of exactly that.

Accelerating Rate of Stressed Populations

"We are in the midst of one of the most remarkable patterns of extinctions that has ever happened,"[451] says Christopher Austin, Herpetologist from Louisiana State University. "The current temperature spikes are destroying delicate ecosystems and threatening the survival of millions of species of plants and animals around the globe. 'We need to understand the balance between extinction and speciation ...'"[452]

Earlier chapters made the historic and current case for World Age Precession driving Gravitational Torque Wave

414

compression as the principal *cause* of Global Climate Change. Here we address the consequences and effects of those recent GTW imposed changes.

As reported by Christen Brownlee, back in 2006 Christina Halzpfel and William Bradshaw of the University of Oregon noted that:

> "Mosquitoes were one of the first organisms in which scientists observed genetic changes that might be attributed to global warming. Other scientists have more recently reported that the genetic makeup of organisms ranging from fruit flies to birds might be responding to climate trends."[453]

In the October 10, 2008 issue of Science, the international conservation monitoring organization (IUCN) reported (as they did in Barcelona at its World Conservation Congress) data from 1,700 experts who'd worked the five year review of the conservation status of known wild mammals. The result is the IUCN "Red List of Threatened Species, the main global scorecard for extinction risk."[454] Citing that report and the Red List, Susan Millus related that:

> Between a fifth and a generous third of the world's mammal species now face the threat of extinction, according to the first comprehensive review since 1996.
> Now 1,139 species rank in the most imperiled categories ... [455]

"'All in all, a major event,' said Don Wilson, a mammal curator at the Simthsonian Institution in Washington, D.C."[456] "836 species are data deficient, notes Jan Schipper of the IUCN,"[457] meaning that species experts could not assess their status. Millus quotes Shipper,

> If those mystery [data deficient] species are doing just fine, a situation Schipper considers improbable, then 21 percent of the currently known 5,487 mammal species face a serious threat of extinction. If all the little-known species turn out to be faltering too, that would mean 36 percent of known mammals are in trouble.[458]

"The IUCN sets quantitative criteria for tally's of the three worrisome categories of animals threatened with extinction: critically endangered, endangered, and vulnerable."[459] "The vulnerable species meet one of several thresholds, such as loosing more than a third of their population during either a decade or over three generations, whichever is longer."[460] Habitat loss or degradation ranks as the most widespread threat, grinding down the populations of some 40 percent of the species studied.[461]

"At present, 1,050 species in the United States and its neighboring waters are listed as endangered – at risk of extinction. Another 309 are listed as threatened, or likely to become endangered in the foreseeable future."[462] Verlyn Klinkenborg reports that in the US: "Since 1973, only 39 species have been removed from the endangered and threatened list. Nine of those went extinct, and 16 were removed when evidence emerged that the listed species was not, in fact, imperiled. Only 14 have recovered enough to be de-listed.   Meanwhile, listing is pending for nearly 300 official candidates ..."[463]

The World Age Resonant Phase Synchronization (WARPS) theory predicts that:

1. The compression effects of the first GTW in the current World Age Precession cycle produced significant, though relatively modest, changes in the:

    - Human rate of invention and innovation
    - Animal and plant genetic adaptation and mutation
    - Global Temperature and Long Lived Greenhouse Gases

2. The compression effects of the second GTW (a cumulative period of 20,501.4 years ending in 1500 AD) in the current World Age cycle produced significant changes in the:

    - Human rate of invention and innovation
    - Animal and plant populations under stress
    - All species genetic adaptation and mutation
    - Global Temperature and Long Lived Greenhouse Gases

3. The compression effects of the first 512.536 years of the third GTW in the current World Age

Precession cycle will send the stored energy of the DE, DM, EMF energy, OM $_{micro}$ and OM $_{Macro}$ systems to the critical point, resulting in a rapid cascade of resonant phase transitions, and the sudden abrupt transformation of the:

- Human species
- Animal and plant species
- Planetary Climate
- The state of the entire solar neighborhood

The first two numbered predictions of the WARPS theory are solidly in evidence. The third set of predictions, though increasingly in evidence, remain largely hypothetical.

The GTW density compression *effects* of current World Age suggest a definite buildup to something phenomenal. That buildup is exhibiting in the rates of: human invention and innovation, populations under stress, all species genetic adaptations and mutations, rising global temperature, and increases in the density of Long Lived Greenhouse Gases. All of the available evidence indicates that Null Hypothesis H_{111} is false. That null hypothesis insists that any increase in the number of threatened species and rate of species extinctions is independent of GTW pressure. Still it remains difficult to definitively demonstrate Hypothesis H_{111} true. The foundering definition of the term "species" and nature of current species data do not enable a demonstration that the number of stressed species and the rate of species extinctions are increasing proportionate to GTW pressure.

What about the state of the entire solar system? Are there indications of the buildup of energy to the critical point of Transformation near our solar neighborhood? The answer is an increasingly strong *maybe.*

GTW Cosmic Ray Forcing

Hypothesis H_{112} alleges that evidence of the building GTW pressure to Transformation in the solar neighborhood is now exhibiting as bombardments of the Earth's atmosphere by very high-energy particles occurring atypically near to our solar system. According to Susan Gaidos of Science News in her article of February 2009:

> There's an air of excitement in the astrophysics community, created by a surplus of particles from space invading Earth's atmosphere.
>
> Balloon flights high in the atmosphere over Antarctica detected electrons in numbers and energies much higher than what usually pours in from space, scientists on a project called ATIC [Advanced Thin Ionization Calorimeter] reported in November.
>
> About the same time, a separate report from Milagro, a ground based detector near Los Alamos, New Mexico, described two unexpected patches of high-energy protons in the sky. A review of seven years of Milagro data revealed an unusual distribution in the energies of these cosmic rays.
>
> Both experiments seem to show that the Earth is being bombarded with high-energy

cosmic rays from a mysterious, nearby source.[464]

These time dependent increasing amplitude high energy particle barrages are coming from areas behind our solar system that lie along our solar system's Galactic orbit. Could these localized particle showers be coming from areas that have already reached the WARPS predicted critical point of Transformation? A bit of desktop calculation and analysis strongly indicates that that argument explains the locations and the progressive increase in the very high energy cosmic ray bombardment observed over the last decade.

There are of course alternative scenarios. Susan Gaidos first qualifies the phenomena, before describing one such alternative interpretation:

> "Cosmic rays are actually subatomic particles, such as protons and electrons that slam into Earth's atmosphere with a wide variety of energies. About 90 percent are protons. The rest are helium nuclei, with a smattering of electrons.
>
> Billions of cosmic ray particles hit Earth's atmosphere every second; most come from the sun and are the low-energy variety. An energetic few, however, are believed to get an extra oomph because they are created by high energy cosmic objects and events ...[465]

Many of the cosmic ray particles are accelerated over a wide range of high energies from 600 to 800 billion electron volts (GeV). Those particles of primary interest here, however, have mean energies ranging from 1 to 4 Trillion electron volts (TeV), with some particles attaining energies of 100 TeV. The energy values equate to the particles rate of travel. Gaidos describes the observation problem and a surprising recent development.

> Pinning down a specific source that generates high-energy cosmic rays has proven difficult. Because the magnetic fields of the galaxy and Earth scramble the flight paths of these particles, scientists have not been able to trace their trajectories back to their sources. This random scrambling effect means a map of cosmic ray intensities should appear completely uniform throughout the sky.
> Or so scientists thought. In November [2008], Milagro researchers reported seeing [from April 1, 2000 to September 21, 2007 and continuing to the present] 'hot spots' of high energy cosmic ray protons in two distinct regions of the sky. It was the first time scientists could trace such protons back to a particular location.
> ...
> In 200 billion cosmic ray collisions recorded during a seven year period ending in April 2008, the scientists found two areas of the sky that appeared to have an excess number of high-energy protons in the background. The protons also appeared to have a higher average energy – up to 10,000 trillion electron volts – than the background.[466]

Figure 118 depicts the Milagro, NM detector measured cosmic ray bombardments traced back to sources at the tail-end of our solar neighborhood along the Sun's Galactic orbit.

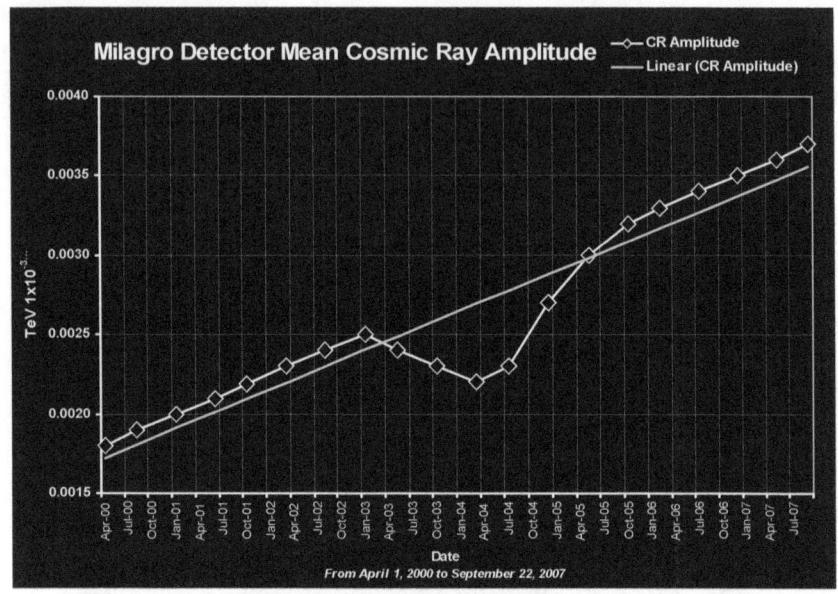

Figure 118 Localized Milagro Cosmic Ray Bombardment Data

Source: Sinnis, Gus, Cosmic-Ray Physics with the Milagro Gamma-Ray Observatory, Proc. Int. Workshop Advances in Cosmic Ray Science. J. Phys. Soc. Jpn. 78 (2009) Suppl. A, pp. 84-87, The Physical Society of Japan, 2009 The Physical Society of Japan, 2009, p.86, Http://www.jpsj.ipap.jp/link?JPSJS/78SA/84/pdf, Revisited May 28, 2010

We have hypothesized that localized high energy electrons and protons should result from locations experiencing the WARPS asserted 2.05 GTW driven critical point. That means that the mean amplitude of the high energy particles should be significantly correlated with GTW Compression for the period of Milagro detector cosmic ray data collection from April 1, 2000 to September 22, 2007. Moreover, GTW Compression (and hence Maya Degrees of Precession) should describe, explain, and predict the majority of variation in the amplitudes of those localized cosmic ray bombardments.

In point of fact, GTW Compression is correlated 0.9439 with the mean observed raw Cosmic Ray Amplitudes, which is significant well below the .001 level. Moreover, GTW is perfectly correlated (correlation equals 1.00) with the linear trend in those mean Cosmic Ray Amplitudes.

As depicted in Figure 119, GTW Compression describes, explains, and predicts 88.5 percent of the variation in the observed mean Cosmic Ray Amplitudes.

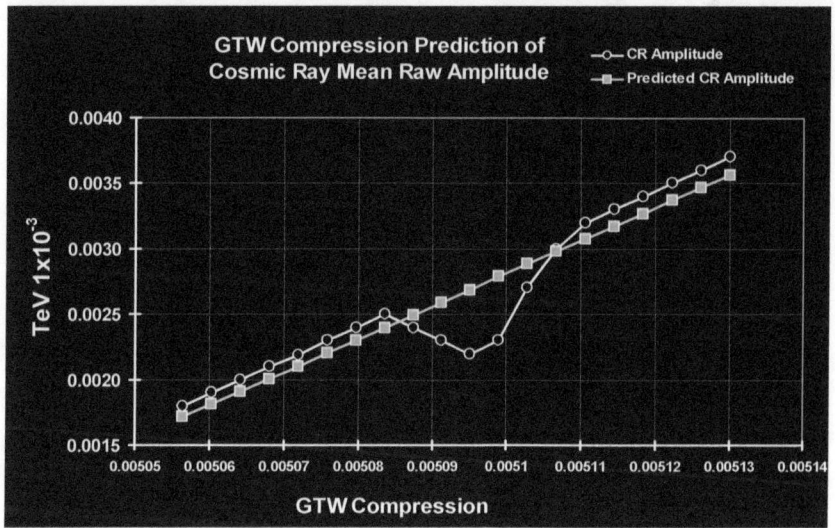

Figure 119 GTW Compression Prediction of Milagro Cosmic Ray Bombardment Data

Even more dramatic is the fact that GTW Compression describes, explains, and predicts 100 percent of the linear trend in those mean Cosmic Ray Amplitudes (Figure120). Both GTW Compression predictive relationships are significant at .001 level of confidence.

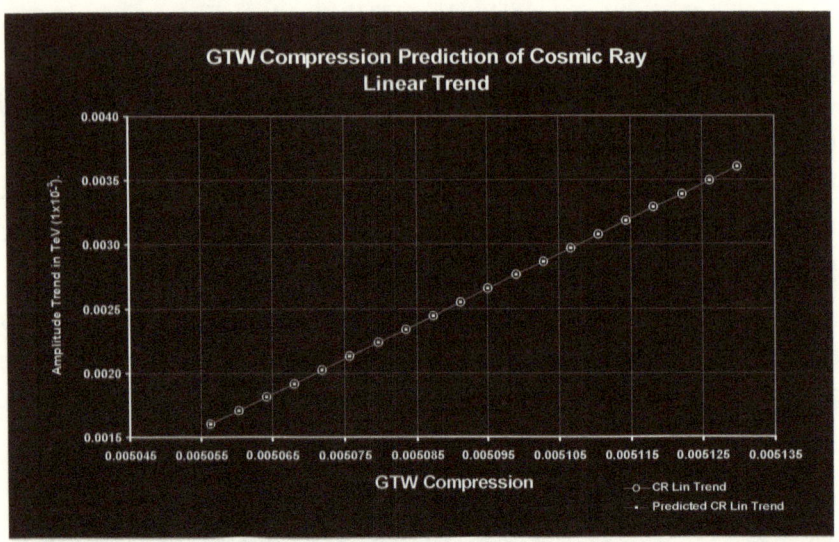

Figure 120 GTW Compression Perfect Prediction of the Linear Trend in Milagro Cosmic Ray Bombardment Data

Quite different explanations are being pursued by the astrophysics community to explain the source localized increasing amplitude trend in electron and proton spikes. One is that the cosmic rays are coming from "a nearby astrophysical object, such as a pulsar or micro-quasar ..."[467]; Null Hypothesis H_{112}.

The second explanation being offered up is that the high-energy particles, "were produced by dark matter" particle annihilations (two dark matter particles that meet, annihilate each other, and produce a spray of particle and antiparticle pairs).[468] The dark matter explanation is a highly subdued version of WARPS Hypothesis H_{112}. The WARPS hypothesis simply provides an explanation as to why the

424

posited dark matter particle annihilations are only now observable.

The nearby astrophysical object hypothesis has pronounced weaknesses. If objects like the named candidate – the pulsar "Geminga" – was the source of the high energy particles, why haven't our instruments seen them before? "While bumps in the cosmic ray electron spectrum have been measured before, they didn't cover the energy range seen in the ATIC experiment."[469] Second, if the high energy cosmic rays were coming from a nearby astrophysical object, scientists should be observing a much smoother bump in the energy distribution than is suggested by the data – and they aren't. Current data are of insufficient quantity and resolution to definitively rule out the astrophysical near object Null Hypothesis. Hence the unlikely notion lingers.

In support of the dark matter hypothesis, In 2008, "scientists reported hints of dark matter in measures taken by the Russian-European orbiting observatory known as PAMELA" [Payload for Antimatter Matter Exploration and Light Nuclei Astrophysics].[470] PAMELA researchers, "reported preliminary evidence for the detection of more positrons [the antimatter partner of electrons] from the Milky Way than could be accounted for by the standard model of particle physics."[471]

On December 17, 2008, Dan Hooper, Albert Stebbins, and Kathryn M. Zurek published their letter paper and proposal, *"The PAMELA and ATIC Excesses From a Nearby Clump of Neutralino Dark Matter."*[472] In that document, Hooper et al state that:

... A large nearby (within 1-2 kiloparsecs of the Solar System) clump of annihilating neutralinos [the still hypothetical Kaluza-Klein form of dark matter] can lead to a spectrum which is consistent with PAMELA and ATIC [measurements], while also producing an acceptable antiproton flux. Furthermore, the presence of a large dark matter clump can potentially accommodate the very large annihilation rate required to produce the PAMELA and ATIC [detected] signals.[473]

The authors analyzed the electron plus positron spectrum from a clump of 800 GeV neutralinos (dark matter) annihilating with the dark matter clump assumed to be stationary at alternative distances ranging from 0.1, 1, 2 or 4 kilo-parsecs from our Solar System. Distances equal to 0.1, 1, and 1.2 kilo-parsecs proved a good fit to both the PAMELA and ATIC data.[474] The scientists concluded that "the origin of the signal must lie within approximately 1 or 2 kilo-parsecs of the Solar System in order to generate the observed spectral shape," and these very large annihilation rates require "a very large or *dense* [as in compressed] clump of dark matter."[475] The scientists recommend:

To determine whether or not a nearby clump of annihilating neutralinos [dark matter] is in fact responsible for the ATIC and PAMELA signals, further data will be required. In addition to more data from PAMELA (at higher energies, and with greater exposure), ground based gamma ray telescopes such as HESS [High Energy Stereoscopic System] and VERITAS [the Very Energetic Radiation Imaging

Telescope Array System] should be capable of measuring the cosmic ray electron spectrum over the energy range of the ATIC feature with higher precision than is currently available.[476]

If they get their wish, Hooper et al will test their hypothesis that a solar system nearby *very dense* dark matter clump is the *source* of those PAMELA high energy positron and ATIC high energy electron signals. These researchers estimate the source of the measured cosmic ray anisotropy (meaning cosmic rays localized to sources at or about the tail-end of the solar neighborhood along the solar system galactic orbit) is *static* and at some 0.1 to 2 *kilo parsecs* behind us along the Galactic Orbit of the Solar System. A kilo-parsec is 1,000 parsecs, which equals 3,261.564 light years. That means that the researchers are asserting a static source of the observed cosmic ray anisotropy exists at 3,261.564 to 6,523.128 light years behind the solar system along its Galactic Orbit.

The WARPS Theory counters that the source of the cosmic ray anisotropy is the cumulative result of space-time density compression imposed by 2.05 GTW moving toward our solar system at the speed of light. The rationale is as follows. First, GTW Compression predicts 100 percent of the variations in the linear trend in the increasing mean amplitudes of source localized cosmic ray measurements. Second, the amplitude of the localized high energy cosmic ray measurements is *time dependent* (the nearer in time the measurement, the greater the amplitude observed). That means that current researcher estimates are off by a factor of 1,000. Instead of the posited static source 0.1 to 2 kilo parsecs distant, a rapidly approaching critical point of

Transformation explains the data. Figure 121 depicts that cosmic ray amplitude progression as a function of time and distance from our solar system. It is obvious that the data predict that the critical point of Transformation will intercept the center of our solar system on or about 2012.

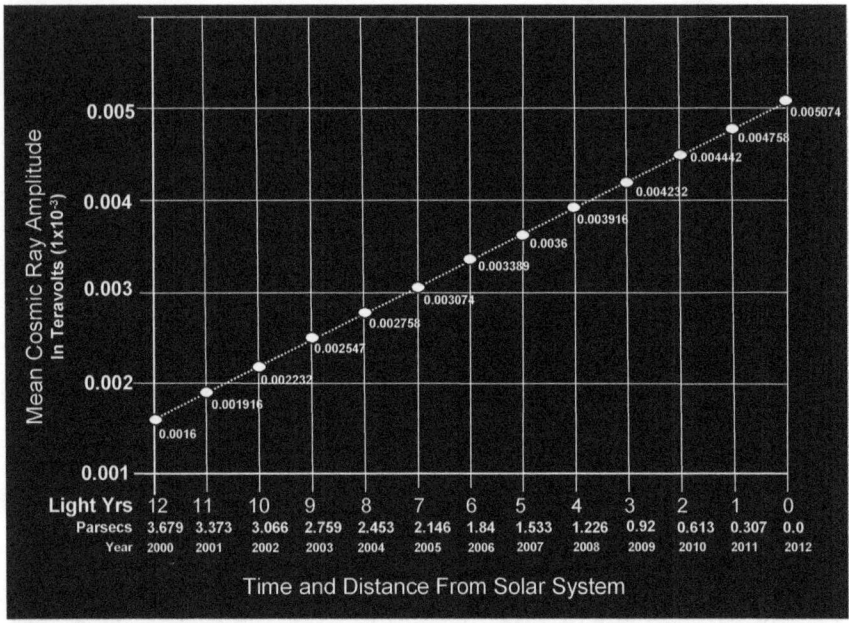

Figure 121 Cosmic Ray Amplitude as a Function of Critical Point Time and Distance from our Solar System

This particular line of high energy particle research may present the opportunity to look into the not so distant past and see our immediate future. WARPS hypothesis H_{112} could easily equal the time remaining to Transformation.

It is time to sum up, and the next chapter does exactly that.

428

19

CONCLUSIONS

Counting the Unknown Cause

Pre-Maya Mesoamerican sages rigorously measured the motions of the visible heavens. Succeeding generations of wise men continued that long standing tradition of data collection, quantitative innovation, and analysis. By 1200 BC the Maya priest elite had confirmed the existence of the repeating rhythms of Solar System Precession and validated their scaled nested sub-cycle predictors of it. Subsequent a day-keeper shaman faithfully refined the time scaled sub-cycles into the standardized Long Count calendar. The Maya then rigorously applied the Long Count to reliably describe the past, explain the present, and predict the future.

Around 200 BC, Maya shaman resolved long standing problems posed by competing theories. They did this by instituting a system of rule-enforced ceremonies that insured the calibration and maintenance of their precious Long Count. After that, the cyclic waveforms of Maya Precession and the derived *K'ul* response metric dominated Mesoamerican mathematics, religious mythology, pyramid architecture, and activities of daily living until the Spanish Conquest of 1519.

Just as Newton and Einstein mathematically described how gravity works without ever discovering its origin or what it is[477], so the ancient Maya described Solar System Precession and its subordinate cycles. All the Maya knew was that the source of those predictive rhythmic cycles had to be a gigantic wavelike force emanating regular as clockwork from the Galactic Center. A wavelike pressure they described as echoing the repeating wave-like shape of a wriggling serpent.

Measuring & Predicting Effects

The ancient Maya quantitatively estimated – as opposed to subjectively guessing at – the responses of people, places, and things to the unseen pressures metered out by their Long Count. Those calculated responses the Maya called *K'ul* are a superior predictor to the un-weighted cyclic values of the Long Count. It is conjectured that the Maya may well have discovered that the multiplication of Long Count cycle values times their quantitative *K'ul* response measure produces a 'K'ul Resonance' metric that yields a superior predictor of Earth bound events.

Revealing the Cause

Approximately 2,200 years after the Maya discovered the universality of their Long Count cycles, modern astrophysicists discovered Galactic Gravitational Torque Waves (GTW). GTW emanate from galactic center and propagate to well beyond the edges of galactic plane. Figure 122 depicts a head on view of the current GTW that entered our Solar system around 1500, just as predicted by the Maya Long Count 2,100 years ago.

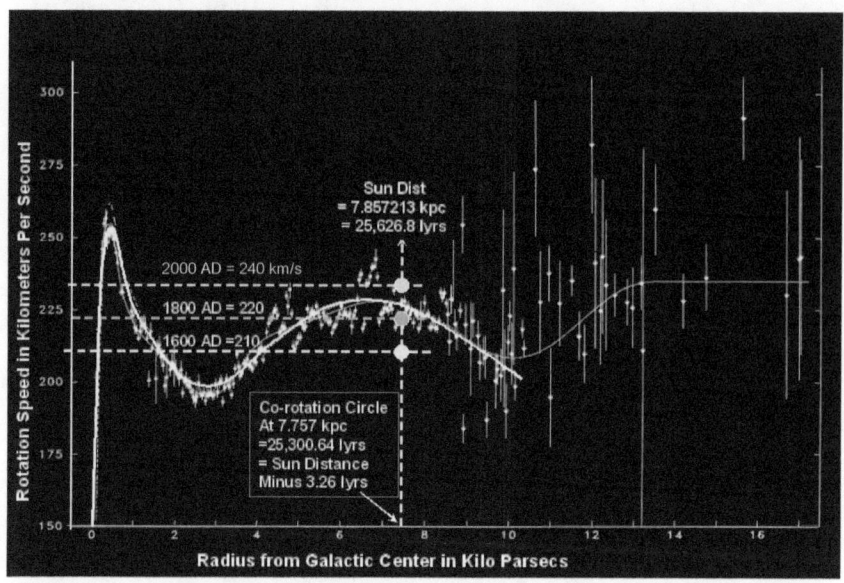

Figure 122 Pre 1995 Multi-Sky Survey Face-on Image of GTW
Source: Clemens (1985), Ap. J. 295, 422.)
http://web.njit.edu/~gary/321/Lecture19.html, Revisited January 28, 2008

Figure 122 is the result of multiple sky surveys, most done prior to 1995. Literally plotted in the figure is the "rotation curve"[478] of stars in the Galactic Plane of the Milky Way. The rotation curve shows the stars' orbital rotation distance from the Galactic Center (the x-axis) plotted relative to their Galactic orbital velocities (the y-axis). In the figure the orbital velocity of observed stars increases from 1600 AD to 2000. That change in stellar velocities exactly reproduces the negative image characteristic shape of a GTW coming straight at us in the Galactic Plane. The closer the GTW gets to us the faster the stars (like our Sun) orbit the Galaxy, and the higher the GTW curve becomes. For visual reference, the figure includes the Co-rotation radial distance from the Galactic Center. As a result of GTW synchronization effects, the stars inside the Co-rotation radius slow down, while stars (like our Sun) that reside outside the Co-rotation radius speed up.

The prior Figure 122 pictured GTW is compressing the very fabric of space-time, thereby concentrating objects at all scales. Cumulative compression results in the increased proximity and density of energy-mass objects. Increased proximity yields a correspondent increase in the probability of object interactions. Rather than imposing an absolute determinism on everything, GTW compression does the exact opposite. GTW compression increases the proximity of objects and multiplies the probabilities of object interactions. Hence, for human beings at least, that expansion of probabilities translates to a correspondent increase in the opportunities for the expression of free will.

Conjoining Cause and Effect

This book has shown that World Age cycles of Maya Solar System Precision equal the slow Galactic compression and synchronization effects of GTW. That GTW compression – and hence the Maya World Age Long Count – validly and reliably predict the motions of our Galaxy, Solar System, and planet. Those relations accurately predict past celestial events, explain the tendencies in currently observed motions, and predict most likely future motions. Moreover, GTW Compression and the Maya Long Count system account for the long term variations in Earth's climate and predict the cyclic variations in the diffusion of human evolution and innovation.

Drilling Deeper

Nearly four decades ago modern geologists discovered Gravitational and Microseism Hum modes embedded in the waves of Earth's gravitational field. They ascribed the origin of those modes to variations in Earth's atmosphere and ocean pressures. The greater originating source, however, remained a mystery. The most likely fundamental source of the variation in Earth's horizontal and vertical Hum modes is GTW imposed density compression.

The autonomic human sense of self is believed to be the emergent product of the 10 to 20 second neuronal waves indexed by the Default Mode Network of the brain. Evidence suggests that the Default Mode Network is the product of the vestibular sensing, decoding, and distribution of the waves of the Earth's Microseism Hum. It is concluded that:

1) Galactic GTW compression induces long term response variations in the Earth's Microseism Hum modes

2) The waves of the Microseism Hum are detected and decoded by the human vestibular system

3) The vestibular system neuronal outputs give rise to the waves of the Default Mode Network in the brain

4) The Default Mode Network gives rise to the human autonomic sense of self.

In healthy human beings, conscious cognitive processes ignore (by-pass) the autonomic sense of self. Conscious cognitive processing represents a very small percentage of total neuronal processing time and energy consumption in the human brain. The Default Mode Network, in contrast, predominates and persists in waking, sleeping, anesthetized, and even comatose states of existence.[479, 480, 481, 482, 483 & 484] This predominance and persistence of the autonomic sense of self appears to explain why GTW compression predicts the event dates and rates of human history, evolution, and innovation.

A Mind Warping Prediction

The ancient Maya constituted a *primitive* civilization by virtue of their era of existence. There was nothing the least bit crude about the scientific discipline and mathematics of their shaman. Their vast knowledge and understanding of our planet, solar system, and Galaxy led them to make an astounding prediction. Ancient Maya shaman predicted that on Friday December 21, 2012 our solar neighborhood will undergo an optimizing transformative cosmic makeover at every level of energy and matter.

The Maya predicted Transformation appears to be consistent with Rudolf Clausius' 1862 second law of thermodynamics. Clausius asserted that: *"The algebraic sum of all the transformations occurring in a cyclical process can only be positive, or, as an extreme [and hence unlikely] case, equal to nothing."*[485] This book described the scientific basis for the ancient Maya prophecy of Transformation. That objective basis is called the theory of World Age Resonant Phase Synchronization (WARPS).

The WARPS theory predicts that we will undergo 4,612.8240 years worth of motion and evolution at every level in less than a minute. The prediction is at once astounding and unimaginable. Yet, we are forced to acknowledge three facts. First, everything the Maya asserted about their Grand, Great, and subordinate Long Count Cycles has proven spot-on accurate. Second, the Maya assertion of prior transformations provides the only currently viable scientific explanation for the otherwise impossible rate of evolution of organic life on this planet. Third, the Maya prediction appears completely consistent with the fundamental discoveries and theoretic arguments of our best modern science. Hence, one cannot dismiss the possibility that the Maya were as right about an impending transformation of our solar neighborhood as they were about Precession. The Maya prophecy maybe as difficult to fathom as the implications of Einstein's General Theory of Relativity, but that does not make it false – just seriously scary.

Value Added

In 1903 Henri Poincaré complained,

"A very small cause which escapes our notice determines a considerable effect that we cannot fail to see, and then we say that the effect is due to chance."[486]

Given the conclusions previously stated, we are moved to rephrase Poincaré's indictment.

'A *very large* cause that has escaped modern notice determines considerable effects we cannot fail to see, but heretofore *falsely* insisted due to chance.'

That large cause is Galactic Gravitational Torque Wave density compression. Its' so-called chance effects are nothing less than the past, present, and future responses of our spiraled barred Galaxy, solar system, planet and fundamental physical selves to it.

Modern science and mathematics is dedicated to achieving and improving a probabilistic description of nature. One cannot help but conclude that the discovery of the effects of GTW compression driven Solar System change will be of significant use in the achievement that purpose. To paraphrase Deuteronomy 29:29, *the secret things belong to the universe, but those things that are revealed belong to us and our children, that we may work within its laws to contribute to it.*

19

HYPOTHESES SUMMARY

T his chapter states each of the hypotheses presented in the prior chapters of this book. Each hypothesis is preceded by its status (e.g., 'retained' or 'rejected'). Each group of hypotheses is listed under the chapter title heading in which it was introduced and addressed in this book. This chapter thus provides a comprehensive results summary of the 'Change; Described-Explained-Predicted' investigation.

Waves of Human Change

Retained: Maya inspired **Hypothesis H_1:** a significant portion of the variation in the literature reported Dates of Human Change occurring over the last 282,000 years, is described, explained, and predicted by Degrees of Maya Precession (e.g. World Age Cycles); *where one World Age Cycle equals 25,626.8 years and 360 degrees.*

Rejected: **Hypothesis Null$_1$:** there is no significant relation between the variation in reported dates of human change and Degrees of Maya Precession (e.g. Maya World Age Cycles).

Rejected: **Alternate Hypothesis,** *Extreme Gradualism*: Human change is continuously and equally distributed and probable over time (i.e. Human change exhibits as a flat rectangular distribution).

Rejected: **Alternate Hypothesis,** *Extreme Punctuated Equilibrium*: Significant Human changes are concentrated in discrete evolutionary jumps (i.e., exhibiting as time spaced spikes of dynamic change) when one or more small sub-populations becomes geographically separated from a greater population undergoing substantial stress or threat of extinction.

Rejected: **Alternate Hypothesis,** *Gradualism with Punctuated Equilibrium*: The rate of Human change continuously fluctuates from very fast to very slow with the static equilibrium condition representing nothing more than a period of ultra slow change.

Retained: **Hypothesis H$_2$:** the number of Human Changes per 5,125.36 year long Maya Epoch of 72 Degrees of Maya Precession, describe a distribution not significantly different from the continuity transformed Poisson distribution.

Rejected: **Hypothesis Null$_2$:** asserts that the frequency of Human Change is neither Poisson nor normally distributed across the 5,125.36 year long 72 degree partitions characterizing the five Epochs of Maya Precession defined World Ages.

Retained: **Hypothesis H$_3$:** the frequency of Human Change occurring within the current Epoch from 1500 AD to the present (positively signed and numbered Epoch 1) describes a distribution not significantly different from the continuity transformed Poisson and normal distributions.

Rejected: **Hypothesis Null$_3$:** the frequency distribution of Human Changes occurring within the current Epoch is significantly different from the Poisson and normal distributions.

Retained: **Hypothesis H$_4$:** a significant portion of the variation in the fossil record driven dates of Hominid species emergence (speciation) and extinction events is described, explained, and predicted by the range of dates comprising the first even numbered Epoch in each applicable Maya World Age.

Rejected: **Hypothesis Null$_4$:** Hominid speciation and extinction events exhibit no counting and predictive relation with Maya World Age Cycle first even numbered Epoch dates.

Retained: **Hypothesis H$_5$:** a significant portion of the variation in the fossil record driven dates of Hominid species genetic bottlenecks, mutations, and major migrations is described, explained, and predicted by the range of dates of the first even numbered Epoch in the applicable Maya World Ages.

Rejected: **Hypothesis Null₅:** the dates of Hominid genetic bottlenecks, mutations, and major migrations exhibit no counting relation with Maya World Age Cycle first even numbered Epoch dates.

Waves of Celestial Motion

Retained: **Hypothesis H₆:** for any given Maya World Age Precession cycle Epoch of 5,125.36 years, elapsed Maya Degrees of Precession describe, explain, and predict all (100 percent) of the variation in Earth's Obliquity, Inclination, and solar orbital Eccentricity.

Rejected: **Null Hypothesis Null₆:** Maya World Age Cycle Degrees of Precession are not significantly related to Earth Obliquity, Inclination, or Eccentricity.

Waves of Climatic Change

Retained: **Alternate Hypothesis:** Global Temperature describes, explains, and predicts a significant portion of the variation in the dates of Human Change.

Retained: **Hypothesis H₇:** from 250,000 to 100 years ago, Degrees of Maya Precession describe, explain, and predict a significant portion of the variation in Long-Lived

Greenhouse Gases (LLGGs), as represented by carbon dioxide (CO_2 in ppm) emissions extracted from Antarctic Ice Cores.

Rejected: **Hypothesis Null$_7$**: there is no statistically significant counting relationship between Degrees of Maya Precession and CO_2 emission levels from 250,000 to 100 years ago.

Retained: **Hypothesis H$_8$**: from 250,000 to 100 years ago Global CO_2 Levels are significantly correlated with Global Temperature Changes.

Rejected: **Hypothesis Null$_8$**: there is no significant counting relation between CO_2 levels and Global Temperature Change.

Retained: **Hypothesis H$_9$**: from 250,000 to 100 years ago, Maya Degrees of Precession describe, explain, and predict more of the variation in Global Temperature Change than that explained by Global CO_2 Levels.

Rejected: **Hypothesis Null$_9$**: Global CO_2 is a stronger predictor of Global Temperature Change than Maya Degrees of Precession.

Retained: **Hypothesis H$_{10}$**: Degrees of Maya Precession describe, explain, and predict a significant portion of the variation in Global Temperature Change from 1880 AD to 2007 AD.

Rejected: **Hypothesis Null$_{10}$:** there is no significant counting relationship between Maya Degrees of Precession and Global Temperature change from 1880 AD to 2007 AD.

Retained: **Hypothesis H$_{11}$:** Degrees of Maya Precession describe, explain, and predict a significant portion of the variation in long-lived atmospheric greenhouse gases (LLGGs), as represented by atmospheric carbon dioxide (CO_2) levels from 1880 AD to 2007 AD.

Rejected: **Hypothesis Null$_{11}$:** there is no significant counting relationship between Degrees of Maya Precession and CO_2 levels from 1880 to 2007.

Retained: **Hypothesis H$_{12}$:** Degrees of Maya Precession predict approximately the same amount of the variation in Global Temperature change as is predicted by CO_2 levels from 1880 AD to 2007 AD.

Rejected: **Hypothesis Null$_{12}$:** CO_2 levels predict substantially more of the variation in Global Temperature Change than is predicted by Degrees of Maya Precession over the time from 1880 to 2007.

Waves of Gravitational Torque

From 285,012 BC to 1500 AD

All Retained: **Hypothesis H$_{13}$:** for any given Maya World Age cycle Epoch of 5,125.36 years occurring between

285,012 BC and 1500 AD, Poisson distributed GTW Compression is significantly correlated with:

$H_{13.1}$: Solar System Precession and hence the Dates of Human Change

$H_{13.2}$: Earth Obliquity

$H_{13.3}$: Earth Inclination

$H_{13.4}$: Earth Eccentricity

$H_{13.5}$: Global Temperature Change

$H_{13.6}$: Long Lived Greenhouse Gas Levels as represented by CO_2

Rejected: **Null (and Alternate) Hypothesis 13 ($Null_{13}$):** The Center of the Galaxy is much too far away and much too weak a gravitational source to have any influence whatsoever on the geometric motions of the solar system, our weather, and ourselves.

From 1880 to 2007

All Retained: **Hypothesis H_{14}:** for the current even numbered Epoch from 1880 to 2007, Gravitational Torque Waves (GTW) emanating from the Galactic Center, imposed Poisson distributed space-time density changes (Compression strengths) that describe explain and predict a significant portion of the variation in:

$H_{14.1}$: Solar System Precession and the Dates of Human Change

$H_{14.2}$: Earth Obliquity

$H_{14.3}$: Earth Inclination

H $_{14.4}$: Earth Eccentricity

H $_{14.5}$: Global Temperature Change

H $_{14.6}$: Long Lived Greenhouse Gas Levels as represented by CO_2

Rejected: **Null (and Alternate) Hypothesis 14 (Null$_{14}$),** for the current even numbered Epoch from 1880 to 2007, Gravitational Torque Wave Compression is not significantly related to the cited measures of celestial motion, variation in Global Temperature, or LLGG levels.

Waves of Economic Change

All Retained: **Hypothesis H$_{15}$:** Maya Degrees of Precession and hence Gravitational Torque Wave (GTW) imposed Poisson distributed space-time density changes (Compression strengths), are significantly correlated (count with) the variation in the:

> **H$_{15.1}$:** US Per Capita Consumption (in Yr 2000 dollars) from 1889 to 2004
>
> **H$_{15.2}$:** Real US Gross National Product (GNP) from 1890 to 1974
>
> **H$_{15.3}$:** Global Hourly Wages (in English Pounds) from 1871 to 1994
>
> **H$_{15.4}$:** Real US Consumption (in Yr 2000 Dollars) from 1889 to 2004
>
> **H$_{15.5}$:** US Gross Domestic Product (GDP) from 1929 to 2007
>
> **H$_{15.6}$:** Money Supply from 1890 to 1974
>
> **H$_{15.7}$:** Velocity of Money from 1869 to 1970
>
> **H$_{15.8}$:** Consumer Price Index (CPI) from 1871 to 2007

H$_{15.9}$: Global Hourly Wages (in English Pounds) from 1500 to 1994

H$_{15.10}$: Dow Jones Industrial Average (DJIA) Annual Closing Price from 1928 to 2007

H $_{15.11}$: Common Stock Price from 1871 to 1970

H $_{15.12}$: Real Stock Price Index from 1871 to 2007

H $_{15.13}$: Stock Price Index from 1871 to 1994

H $_{15.14}$: Dow Jones Industrial Average (DJIA) Trading Volume from 1928 to 2007

Rejected: **Hypothesis Null$_{15}$:** the variation in the financial and trading behavior of human beings is fundamentally stochastic (predominantly random) and quantitatively unrelated to Degrees of Maya Precession and GTW imposed density changes.

All Retained: **Hypothesis H$_{16}$:** Gravitational Torque Wave (GTW) imposed Poisson distributed space-time density changes (Compression strengths), describe, explain, and predict a significant portion of the variation in:

H$_{16.1}$: US Per Capita Consumption (in Yr 2000 dollars) from 1889 to 2004

H$_{16.2}$: Real US Gross National Product (GNP) from 1890 to 1974

H$_{16.3}$: Global Hourly Wages (in English Pounds) from 1871 to 1994

H$_{16.4}$: Real US Consumption (in Yr 2000 Dollars) from 1889 to 2004

H$_{16.5}$: US Gross Domestic Product (GDP) from 1929 to 2007

H$_{16.6}$: Money Supply from 1890 to 1974

H$_{16.7}$: Velocity of Money from 1869 to 1970

H$_{16.8}$: Consumer Price Index (CPI) from 1871 to 2007

H$_{16.9}$: Global Hourly Wages (in English Pounds) from 1500 to 1994

H$_{16.10}$: Dow Jones Industrial Average (DJIA) Annual Closing Price from 1928 to 2007

H$_{16.11}$: Common Stock Price from 1871 to 1970

H$_{16.12}$: Real Stock Price Index from 1871 to 2007

H$_{16.13}$: Stock Price Index from 1871 to 1994

H$_{16.14}$: Dow Jones Industrial Average (DJIA) Trading Volume from 1928 to 2007

<u>Rejected</u>: **Hypothesis Null$_{15}$:** the variation in the financial and trading behavior of human beings is fundamentally stochastic (predominantly random) and quantitatively unrelated to GTW imposed density changes.

Wavelets of Synchronization

<u>Retained</u>: **Hypothesis H$_{17}$:** asserts that: *Response Effect K'ul* is significantly correlated with the smoothed Rate of Human Innovation for the current Maya *Baktun* Cycle.

<u>Rejected</u>: **Hypothesis Null$_{17}$:** counters that there is no significant relationship between Response Effect *K'ul* and the

smoothed rate of human innovation for the current Maya *Baktun* Cycle.

Retained: **Hypothesis H_{18}:** states that: *Response Effect K'ul* describes, explains, and predicts, more of the variation in the expected Rate of Human Innovation than GTW Compression as measured by the Maya *Baktun* Cycle units.

Rejected: **Hypothesis $Null_{18}$:** the Maya *Baktun* Cycle unit measure of GTW Compression is a better predictor of the rates of expected Human Innovation than Response Effect *K'ul*

Retained: **Hypothesis H_{19}:** *K'ul* x Sub-cycle Compression (*K'ul* Resonance) is significantly correlated with the expected Rates of Human Innovation for the current Maya *Baktun* Cycle.

Rejected: **Hypothesis $Null_{19}$:** there is no significant relationship between *K'ul* Resonance and the expected rates of human innovation during the current Maya *Baktun* Cycle.

Retained: **Hypothesis H_{20}:** *K'ul* Resonance describes, explains, and predicts, more of the variation in the expected Rates of Human Innovation than either sub-cycle indexed GTW Compression or Response Effect *K'ul* for the current Maya *Baktun* Cycle.

447

Rejected: **Hypothesis Null$_{20}$**: Response Effect *K'ul* is a better predictor of the expected rates of Human Innovation than *K'ul* Resonance.

Retained: **Hypothesis H$_{21}$**: *Response Effect K'ul* is a better predictor of the variation in the economic indicators listed in Table 17, than Sub-cycle indexed GTW Compression alone.

Rejected: **Hypothesis Null$_{21}$**: Sub-cycle indexed GTW Compression is the better predictor of the economic indicators in Table 17.

Retained: **Hypothesis H$_{22}$**: *K'ul Resonance* is a better predictor of the variation in the economic variables in Table 17, than *Response Effect K'ul*.

Rejected: **Hypothesis Null$_{21}$**: *Response Effect K'ul* is the superior predictor of the economic variables in Table 17.

Waves of Humming Earth

Retained: **Hypothesis H$_{23}$**: the human vestibular system is sensitive enough to detect and respond to the variations in the Gravitational Hum and Microseism Hum modes.

Rejected: **Hypotheses Null$_{23}$**: the human vestibular system is not sensitive enough to detect and respond to the variations in either the Gravitational Hum or the Microseism Hum modes.

Waves of Humming Identity

Retained: **Hypothesis H_{24}:** the neuronal signals of the human vestibular system are used by the brain to discriminate the internal human frame of reference (the self) from the external frame of reference, the non-self (i.e. the environment).

Rejected: **Hypotheses Null$_{24}$:** either no such discrimination takes place, or that the source of self and non-self discriminations in the brain is other than vestibular.

Retained: **Hypothesis H_{25}:** the autonomic human sense of self is the product of the Default Mode Network in the healthy adult brain.

Rejected: **Hypothesis Null$_{25}$:** discrimination of the autonomic self from the non-self (the environment), is not related to the cycles of the Default Mode Network in the brain.

Retained: **Hypothesis H_{26}:** the source of the Default Mode Network as evidenced by fMRI BOLD (Blood Oxygen Level–Dependent) Fluctuations is neuronal.

Rejected: **Hypothesis Null$_{26}$:** fMRI BOLD measurements of functional networks like the Default Mode Network are generated by artifacts such as aliasing of cardiac or respiratory pulsations in the low-frequency range, vasomotor oscillations, or effects of attention.

Rejected: **Hypothesis H$_{27}$**: the genetically controlled and externally cued rhythms of the suprachiasmatic nuclei source the persistent low frequency rhythms of the Default Mode Network.

Retained: **Hypothesis Null$_{27}$**: externally cued rhythms of the suprachiasmatic nuclei *do not source* the persistent low frequency rhythms of the Default Mode Network in the brain.

Rejected: **Hypothesis H$_{28}$**: Magnetoception (or magneto-reception, the ability to detect a magnetic field to perceive direction, altitude or location) is the source of the persistent low frequency rhythms of the Default Mode Network.

Retained: **Hypothesis Null$_{28}$**: Magnetoception *does not source* the persistent low frequency rhythms of the default mode network in the brain.

Waves of Human Influence

Hypotheses H$_{29}$ to H$_{100}$ state that $\mu_1 = \mu_2$, or in words, that the variances of the two compared samples are equal. **Hypotheses Null$_{29}$ to Null$_{100}$** counter that $\mu_1 \neq \mu_2$, or in words, that the variances of the two samples are *not* equal.

All Retained: **Hypotheses H$_{29}$ through H$_{40}$**, the resting Default Mode Network (DMN) waveform variances in *Wave Number* (k) equal the paired wave number variances of the Microseism

Hum (MSH) waveforms for the months from January through December.

Retained for the months of January through April (the assumed DMN period of data collection); Rejected for the months of May through December: **Hypotheses H_{41} through H_{52}** the DMN waveform variances in peak, zero crossing, and valley point *Arrival Times* equal the Arrival Time points of the paired MSH waveforms.

Retained for the month of February (the assumed month of DMN data collection); Rejected for the months of January, and March through December: **Hypotheses H_{53} through H_{64}** the DMN waveform variances in peak and valley point *Amplitudes* (in normalized units) equal the paired points of the MSH waveforms.

Retained for January (H_{65}), March through May (H_{67} to H_{69}), and July (H_{71}) through August (H_{72}); Rejected for the months of (February H_{66}, June H_{70}, and September H_{73} through December H_{76}): **Hypotheses H_{65} through H_{76}**, the DMN waveform variances in *Wavelength* (measured in seconds) equal the paired Wavelengths of the MSH waveforms.

Rejected for all twelve months (January through December): **Hypotheses H_{77} through H_{88}** the variances in DMN waveform *Velocities* (in degrees per second) equal the paired Velocity variances of the MSH waveforms.

Retained for the months of January (H_{89}), February (H_{90}), and April (H_{92}); Rejected for the months of March (H_{91}) and May through December (H_{93} through H_{100}: **Hypotheses H_{89} through H_{100}** variances in DMN waveform *Frequencies* (in vibrations per second) equal the paired Frequencies variances of the MSH waveforms.

Retained: **Hypothesis H_{101}** the *Velocity* (speed) of Default Mode Network waves is equal to or slower than the *Velocity* of paired Microseism Hum waves. Hypothesis H_{101} thus posits a cause and effect relationship between the MSH and DMN. The logic being that the speed of the *cause* must equal or exceed that of the *effected* wave.

Rejected: **Hypothesis Null$_{101}$:** the waves of the Microseism Hum are slower in Velocity that the paired waves of the Default Mode Network, precluding any causal relation between the two.

Tsunami Trains of Transition

None

World Age Resonant Phase Synchronization (WARPS)

Tentatively Retained: **WARPS Hypothesis H_{102}:** Our Solar System (and hence everything within it) experiences a hyper-jump (the functional equivalent of a sudden abrupt acceleration) of 4.6128224 light years and an equivalent amount of evolutionary change in under a minute, once every 21,013.9760 years. .

Tentatively Rejected: **Hypothesis Null$_{102}$:** the Solar system has not exhibited, nor will it exhibit, any solar neighborhood wide fixed-time-interval evolutionary transformations.

Tentatively Retained: **WARPS Hypothesis H_{103}:** Any residual evidence of the sudden (less than one minute long) compression and decompression of space-time imposed on the solar system by 2.05 GTW once every 21,013.9760 years, will appear to have occurred at an interval of 25,626.8 years; a chronological offset of 4,612.824 years (or 90 percent of one Maya Epoch) more than actual.

Tentatively Rejected: **Hypothesis Null$_{103}$** is the same as **Hypothesis Null$_{102}$.**

Waves of Earth Compression

Retained: **Hypothesis H_{104}:** the known linear rate of change in Earth's Oblateness is significantly correlated with the estimated rate of change in GTW imposed Density Compression.

Rejected: **Hypotheses Null₁₀₄**: there is no counting relationship between the rate of change in Earth's Oblateness and the estimated rate of GTW imposed Density Compression.

Retained: **Hypothesis H₁₀₅**: GTW imposed Density Compression describes, explains, and predicts a significant portion of the variation in the currently accepted linear rate of change in Earth's Oblateness.

Rejected: **Hypotheses Null₁₀₅**: there is no counting relationship between the rate of change in Earth's Oblateness, and the estimated rate of GTW imposed Density Compression.

Retained: **WARPS Hypothesis H₁₀₆**: GTW Density Compression (and hence Maya Degrees of Precession) is significantly correlated with Earth's J_2 Oblateness variations having periods greater than 20 years from Earth's long term linear Oblateness trend.

Rejected: **Hypothesis Null₁₀₆**: there is no significant counting relationship between the rate of GTW Density Compression and the variations in J_2 Oblateness with periods greater than 20 years.

Retained: **Hypothesis H₁₀₇**: GTW Density Compression describes, explains, and predicts a significant portion of variations from the linear trend in Earth's Oblateness that occurred with periods greater than 20 years.

Rejected: **Hypothesis Null$_{107}$**: There is no significant counting relationship between GTW Density Compression and J_2 Oblateness variations from the longitudinal trend.

Retained: **Hypothesis H$_{108}$**: Calculated Katun Cycle *K'ul* Response Effect is correlated with the less than 20 year period smoothed J_2 Oblateness variations containing anomalies.

Rejected: **Hypothesis Null$_{108}$**: the *K'ul* Response Effect measure is not related to the less than 20 year period J_2 variations containing anomalies.

Retained: **Hypothesis H$_{109}$**: the calculated Katun Cycle *K'ul* Response Effect describes, explains, and predicts a significant portion of the variation in the less than 20 year period smoothed J_2 Oblateness variations containing anomalies.

Rejected: **Hypothesis Null$_{109}$**: the calculated Katun Cycle *K'ul* Response Effect measure is unrelated to the short term variations in J_2 containing the anomalies observed from 1976 to 2008.

Wave Train Signals, Switches, and Sidings

Tentatively Retained: **Hypothesis H_{110}:** the rate of Human Genetic Change is accelerating proportionate to the increase in GTW pressure.

Rejected: **Hypothesis Null$_{110}$:** the rate of Human Genetic Change has exhibited no observable and measurable increase for the last 50,000 years.

Rejected: **Alternate Hypothesis H_{110}:** any measured increase in the rate of Human Genetic Change is due solely to the increase in Human Population.

Tentatively Retained – **Hypothesis H_{111}:** the number of stressed species and rates of species extinctions are increasing proportionate to the increase in GTW pressure.

Tentatively Rejected – **Hypothesis Null$_{111}$:** all increases in the number of stressed species and rates of species extinctions are independent of GTW pressure.

Tentatively Retained -- **Hypothesis H_{112}:** evidence of the building GTW pressure to Transformation in the solar neighborhood is exhibiting as bombardments of the Earth's atmosphere by very high-energy particles emanating from sources occurring atypically near to our solar system.

Rejected: **Hypothesis Null$_{112}$:** there are no unexplained variations in the number and energy of particles hitting Earth's atmosphere.

<u>Tentatively Retained</u> – **Alternative Hypothesis H$_{112}$:** any observed increase in observed very high-energy particles hitting earth's atmosphere is due to a nearby astrophysical object, such as a pulsar or micro-quasar.

END NOTES

1 INTRODUCTION

1. Greenspan, Alan, 2007-2008, *the Age of Turbulence*, Penguin Books, New York, NY, p.465

2 WAVES OF HUMAN CHANGE

2. The New Lexicon Webster's Dictionary of the English Language (Deluxe Encyclopedic Edition), 1990, Lexicon Publications Inc New York, NY, p.463

3. Zimmer, Carl, 2005, *Smithsonian Intimate Guide to Human Origins*, Madison Press Ltd., Toronto, Canada, p.23

4. Ibid, pp. 6-7

5. Ibid, p.12

6. Jenkins, John, Major, 2002, *Galactic Alignment*, Bear & Co., Rochester, Vermont p.17; Jenkins, John, Major, 1998, *Maya Cosmogensis 2012*, Bear & Co., Rochester, Vermont, p. XXXIII

7. Ibid, p. 113-114

8. Ibid, p.82

9. Barry Carolyn, 2007, *Rolling Back the Years; Radiocarbon Dating Gets A Remake*, Science News, December 1, 2007, Vol. 172, No. 22, p.344

10. Ibid, p.345

11. Rodgers, Everett M., 2003, *Diffusion of Innovations*, Free Press, New York, NY, pp.36-37

12. Ibid, p.470

13. Ibid, p.471

14. Ibid, p.281

15. Ibid.

16. Ibid, p.280
17. Rosenblum, Bruce and Kuttner, Fred, 2006, *Quantum Enigma; Physics Encounters Consciousness*, New York, NY, Oxford University Press, p.198

18. Santiago, Elena, F., Cooper, Vaughn S., and Lenski, Richard E., 1996, *Punctuated Evolution Caused by Selection of Rare Beneficial Mutations*, Science 21 June 1996: Vol. 272. no. 5269, pp. 1802 – 1804, abstract;
http://www.sciencemag.org/cgi/content/abstract/272/5269/1802, Revisited February 9, 2008

19. Ibid.

20. Mayr, Ernst, 1954, *Change of Genetic Environment and Evolution,* In J. Huxley, A. C. Hardy and E. B. Ford, *Evolution as a Process,* London: Allen and Unwin. pp. 157-180;
http://www.paleautonomy.com/evolution/equilibrium.html, Revisited February 19, 2010

21. Gould, Stephen, Jay, 1977, *Evolution's Erratic Pace*, Natural History 86 (May): 12-16;
http://www.paleautonomy.com/evolution/equilibrium.html, Revisited February 19, 2010

22. Gould, Stephen Jay, 1980, *the Panda's Thumb*. New York. W. W. Norton. p. 184;
http://www.paleautonomy.com/evolution/equilibrium.html, Revisited February 19, 2010

23. Ibid

3 WAVES OF WORLD AGE EPOCHS

24. Http://en.wikipedia.org/wiki/Poisson_distribution, Revisited May 9, 2008

25. Ibid.

26. Ibid.

27. Jenkins, John, Major, 1998, *Maya Cosmogensis 2012*, Bear & Co., Rochester, Vermont, p. XXXV

28. Ibid, p. 149

4 WAVES OF HUMAN EVOLUTION

29. Zimmer, Carl, 2005, *Smithsonian Intimate Guide to Human Origins*, Madison Press Ltd., Toronto, Canada, pp.6-7

30. Http://en.wikipedia.org/wiki/Homo_floresiensis, Revisited February 19, 2010; G.A. Lyras, M.D. Dermitzakis, A.A.E. Van der Geer, S.B. Van der Geer, J. De Vos, 2009,. *The Origin of Homo floresiensis and its Relation to Evolutionary Processes Under Isolation,* Anthropological Science 117(1), 33–43, April 2009.

5 WAVES OF CELESTIAL MOTION

31. Cruttenden, Walter, 2003, *Comparison of Precession Theories: An Argument for the Binary Model*, Binary Research Institute, Newport Beach,CA, p.1,
http://www.binaryresearchinstitute.org/bri/research/calculations/
precdata.shtml, revisited March 3, 2008

32. Rosen, Edward, 1978, *Copernicus, Nicholas. De revolutionibus orbium coelestium. (On the revolutions of the heavenly spheres)*, translation, Baltimore 1978

33. Cruttenden, Walter, 2003, *Comparison of Precession Theories: An Argument for the Binary Model*, Binary Research Institute, Newport Beach,CA, p.1& 5

34. Ibid.

35. Cruttenden, Walter, 2003, *Comparison of Precession Theories: An Argument for the Binary Model*, Binary Research Institute, Newport Beach,CA, p.1

36. Http://www.binaryresearchinstitute.org/bri/research/calculations/
precdata.shtml, Revisited March 3, 2008

37. Cruttenden, Walter, 2003, *Comparison of Precession Theories: An Argument for the Binary Model*, Binary Research Institute, Newport Beach,CA, p.11

38. Ibid, p.1-2

39. Ibid, p.2

40. Http://www.siriusresearchgroup.com/articles/precession.shtml, revisited March 3, 2008

41. Cruttenden, Walter, 2003, *Comparison of Precession Theories: An Argument for the Binary Model*, Binary Research Institute, Newport Beach,CA, p.10; http://www.binaryresearchinstitute.org/bri/research/ calculations/precdata.shtml, Revisited March 3, 2008

42. Ibid, p.2

43. Lin, Douglas N.G., 2008, *The Genesis of Planets*, Scientific American, Vol 298, No 5, May 2008, p.57

44. Cruttenden, Walter, 2003, *Comparison of Precession Theories: An Argument for the Binary Model*, Binary Research Institute, Newport Beach,CA, p.10, http://www.binaryresearchinstitute.org/bri/research/calculations/precdata .shtml, revisited March 3, 2008

45. Ibid, p.3

46. Sirius Research Group, 2008, *Precession; the Changing Cycle*, http://www.siriusresearchgroup.com/articles/precession.shtml, Revisited March 3, 2008

47. Cruttenden, Walter, 2003, *Comparison of Precession Theories: An Argument for the Binary Model*, Binary Research Institute, Newport Beach,CA, p.3, http://www.binaryresearchinstitute.org/bri/research/calculations/precdata .shtml, revisited March 3, 2008

48. Rand McNally, *New Concise Atlas of the Universe*, 1978, Mitchell Beaszley Publishers Limited, P.140

49. United States Navy, *the Tilt (Obliquity) of the Earth's Axis and Its Elliptical Orbit*, http://aa.usno.navy.mil/faq/docs/seasons_orbit.php, Revisited April 5, 2008

50. Castelvecchi, D., 2007, *Northern Exposure; the Inhospitable Side of the Galaxy*, Science News, April 21, 2007, Vol. 171, p.244

51. Ibid, p.244;

52. Frisch, Priscilla, *the Galactic Environment of the Sun: the heliosphere appears to protect the inner solar system from the vagaries of the interstellar medium*, http://www.americanscientist.org/template/AssetDetail/assetid/21173/page/1?&print=yes#20979, Revisited January 28, 2008

53. Bok, Bart and Priscilla, 1981, *the Milky Way*, Harvard University Press. 1981

54. Henbest, Nigel and Couper, Heather,1994, *the Guide to the Galaxy*, Cambridge University Press, 1994

55. *Kinematics and space distribution of disk stars*, http://www.astro.utu.fi/~cflynn/galdyn/lecture7.html, Revisited March 19, 2008

56. Castelvecchi, D., 2007, *Northern Exposure; the Inhospitable Side of the Galaxy*, Science News, April 21, 2007, Vol. 171, p.244

57. *Kinematics and space distribution of disk stars*,http://www.astro.utu.fi/~cflynn/galdyn/lecture7.html, Revisited March 19, 2008

58. Morton, Chris, and Thomas, Ceri, Louise, 1998, *The Mystery of the Crystal Skulls,* Bear & Company, Santa Fe, New Mexico, p.335

59. United States Navy, *the Tilt (Obliquity) of the Earth's Axis and Its Elliptical Orbit*, http://aa.usno.navy.mil/faq/docs/seasons_orbit.php, Revisited Saturday April 5, 2008

60. Http://en.wikipedia.org/wiki/Obliquity; Revisited April 26, 2008

61. Ibid.

62. Wikipedia, the free encyclopedia, http://en.wikipedia.org/wiki/Inclination, Revisited January 24, 2008

63. Http://en.wikipedia.org/wiki/Milenkovich_cycle, Revisited April 22, 2008

64. Http://astrobiology.ucla.edu/OTHER/SSO/, Revisited January 24, 2008

65. Http://en.wikipedia.org/wiki/Milenkovich_cycle, Revisited April 22, 2008

66. *Orbital Geometry and Radiative Forcing,* Http://www2.ocean.washington.edu/oc540/lec02-20/, Revisited March 3, 2008

6 WAVES OF CLIMATE CHANGE

67. Zimmer Carl, (2005) *Smithsonian Intimate Guide to Human Origins*, Smithsonian Books, Madison Press Books, Toronto, Canada, 2005, P.29

68. Paulo Artaxo (Brazil), Terje Berntsen (Norway), Richard Betts (UK), David W. Fahey (USA) et al, 2007, *Changes in Atmospheric Constituents and in Radiative Forcing, In: Climate Change 2007:* The Physical Science Basis. Contribution of Working Group I to the Fourth Assessment Report of the Intergovernmental Panel on Climate Change *[Solomon, S., D. Qin, M. Manning, Z. Chen, M. Marquis, K.B. Averyt, M.Tignor and H.L. Miller (eds.)].* Cambridge University Press, Cambridge, United Kingdom and New York, NY, USA, p. 131

69. Ftp://ftp.ncdc.noaa.gov/pub/data/paleo/icecore/antarctica/vostok/deutnat.txt, Revisited March 3, 2008

70. HTTP://WWW.SCIENCEMAG.ORG/CGI/CONTENT/FULL/317/5839/7 93, Revisited March 3, 2008

71. Jouzel, J., Masson-Delmotte, V., Cattani, O., and Dreyfus, G. et al, 2007, *Orbital and Millennial Antarctic Climate Variability over the Past 800,000 Years*, Science 10 August 2007:Vol. 317. no. 5839, pp. 793 – 796;

HTTP://WWW.SCIENCEMAG.ORG/CGI/CONTENT/FULL/317/5839/7
93, Revisited March 3, 2008

72. Http://earthobservatory.nasa.gov/Study/Paleoclimatology_Evidence/, revisited April 5, 2008

73. Jouzel, J., Masson-Delmotte, V., Cattani, O., and Dreyfus, G. et al, 2007, *Orbital and Millennial Antarctic Climate Variability over the Past 800,000 Years*, Science 10 August 2007:Vol. 317. no. 5839, pp. 793 – 796;
HTTP://WWW.SCIENCEMAG.ORG/CGI/CONTENT/FULL/317/5839/7
93, Revisited March 3, 2008

74. Ibid.

75. Http://en.wikipedia.org/wiki/Density, Revisited August 12, 2008

76. Jouzel, J., Masson-Delmotte, V., Cattani, O., and Dreyfus, G. et al, 2007, *Orbital and Millennial Antarctic Climate Variability over the Past 800,000 Years*, Science 10 August 2007:Vol. 317. no. 5839, pp. 793 – 796;
HTTP://WWW.SCIENCEMAG.ORG/CGI/CONTENT/FULL/317/5839/7
93, Revisited March 3, 2008

77. Ibid.

78. Http://www2.ocean.washington.edu/oc540/lec02-20/, Revisited March 3, 2008

7 WAVES OF GRAVITATIONAL TORQUE

79. Jenkins, John, Major, 2002, *Galactic Alignment*, Bear & Co., Rochester, Vermont p.17

80. Jenkins, John, Major, 1998, *Maya Cosmogenesis 2012*, Bear & Co., Rochester, Vermont p.86 & 283]

81.
Http://www.bibliotecapleyades.net/tzolkinmaya/esp_tzolkinmaya08.htma nd, Revisited August 21, 2012]

82. Jenkins, John, Major, 1998, *Maya Cosmogensis 2012*, Bear & Co., Rochester, Vermont, p. XXXV;

83.
Http://www.prophetsmanual.com/content/index.cfm?navID=15&itemID=15, Revisited August 21, 2008]

84. Jenkins, John, Major, 2002, *Galactic Alignment*, Bear & Co., Rochester, Vermont p.9

85. Ibid, p.9

86. Http://cddisa.gsfc.nasa.gov/vlbi_summary.html, Revisited August 15, 2008

87. Cruttenden, Walter, 2003, *Comparison of Precession Theories: An Argument for the Binary Model*, Binary Research Institute, Newport Beach, CA, p.5;
Http://www.binaryresearchinstitute.org/bri/research/calculations/precdat a.
shtml, revisited March 3, 2008

88. Combes, Françoise, October 2005, *Ripples in a Galactic Pond*, Scientific American, Vol.293, No. 4, p.47

89. Kaku, Michio, 2005, *"Parallel Worlds,"* Anchor Books, New York, 2005, P.35

90. Http://science.howstuffworks.com/fpte4.htm, revisited January 8, 2008

91. Serway, R. A., and Jewett, Jr. J. W. (2003), *Physics for Scientists and Engineers,* 6th Ed. Brooks Cole, ISBN 0-53440-842-7; http://en.wikipedia.org/wiki/Torque, Revisited February 19, 2010

92. Tipler, Paul (2004). *Physics for Scientists and Engineers: Mechanics, Oscillations and Waves, Thermodynamics* (5th-ed). W. H. Freeman, ISBN 0-7167-0809-4; http://en.wikipedia.org/wiki/Torque, Revisited February 19, 2010

93. Http://en.wikipedia.org/wiki/Wave, Revisited January 7, 2008

94. Ibid.

95. Http://en.wikipedia.org/wiki/Gravitational_wave, revisited January 7, 2008

96. Ibid.

97. Ibid.

98. Ibid.

99. Dehnen, Walter, *The Effect of the Outer Linblad Resonance of the Galactic Bar on the Local Stellar Velocity Distribution*, Theoretical Physics, Oxford OX1 3NP, UK

100. Max-Planck-Institut für Astronomie, Königstuhl 17, D-69117, http://www.iop.org/EJ/article/1538-3881/119/2/800/990421.text.html, Revisited January 4, 2004

101. Combes, Françoise, October 2005, *Ripples in a Galactic Pond*, Scientific American, Vol.293, No. 4, p.45

102. Bournaud, F., and Combes, F., June 17, 2002, *Gas Accretion on Spiral Galaxies: Bar Formation and Renewal*, Astronomy & Astrophysics, P.7

103. Drakos, Nikos, (Computer Based Learning Unit, University of Leeds, 1996, and Moore, Ross, Mathematics Department, Macquarie University, Sidney, 1999, http://cosserv3fau.edu/~cis/AST2002/Lectures/C15/Trans/Trans.html, Revisited November 4, 2007

104. Chapline, George, February 12, 1999, *A Natural Cosmological Model with $\Omega < 1$*, astro-ph/9902183, Lawrence Livermore National Laboratory, Livermore, CA

105. Chapline, George, December 17, 2004, *Dark Energy Stars,* Texas Conference on Relativistic Astrophysics, Stanford, Ca 12/12-12/17/04, p.1-4

106. Chapline, George, & Mazur, Pawel, Nov 17, 2005, *Tommy Gold Revisited: Why Does Not The Universe Rotate?* ArXiv: Astro-ph/0509230 v2, pp. 1-10

107. Chapline, George, 2006, *Blueprint for a Rotating Universe,*Lawrence Livermore National Laboratory, Livermore/0608389

108. Ott, Thomas, 2006, Max-Planck-Institut für extraterrestrische Physik, "The Galactic Centre," November 20, 2006 Http://www.mpe.mpg.de/ir/GC/index.php, Revisited November 2007

109. Http://en.wikipedia.org/wiki/SAX_J1808.4-3658, Revisited January 9, 2008

110. Http://www.astrophysicsspectator.com/topics/milkyway/ MilkyWayCenter.html, Revisited January 14, 2008

111. Http://science.nasa.gov/headlines/y2002/21feb_mwbh.htm, Revisited January 15, 2008

112. Http://math.ucr.edu/home/baez/physics/Relativity/BlackHoles/hawking.h tml, Revisited January 15, 2008

113. Dehnen, Walter, 1999, *the Pattern Speed of the Galactic Bar*, Astrophysical Journal Letters, 524:L35–L38, October 10, 1999 Http://www.journals.uchicago.edu/doi/abs/10.1086/312299, Revisited January 4, 2008

114. Lopez-Corredoira, M., Cabrera-Lavers A., Mahoney T. J., Hammersley P. L., Garzon F., Gonzalez-Fernandez C., 2007, *The Long Bar in the Milky Way; Corroboration of an old hypothesis*, Astrophysics (astro-ph), Astron.J. 133 (2007) 154-161 (arXiv:astro-ph/0606201v3), http://arxiv.org/abs/astro-ph/0606201, Revisited January 9, 2008

115. Http://www.news.wisc.edu/11405, Revisited January 9, 2007

116. Combes, Françoise, October 2005, *Ripples in a Galactic Pond*, Scientific American, Vol.293, No. 4, p.47

117. Ibid, p.44

118. Wilton S. Dias and J. R. D., 2005, *Direct Determination of the Spiral Pattern Rotation Speed of the* Galaxy, The Astrophysical Journal, volume 629, part 1, pages 825–831

119. Gonzalez, Guillermo, and Brownlee, Donald, 2001, *The Galactic Habitable Zone: Galactic Chemical Evolution*, Icarus 152, 185–200 (2001), doi:10.1006/icar,2001, 6617, p.1.@ Http://www.idealibrary.com, Revisited January 28, 2008]

120. Combes, Françoise, October 2005, *Ripples in a Galactic Pond*, Scientific American, Vol.293, No. 4, p.46

121. Ibid.

122. Gómez, Gilberto, 2006, *Errors in Kinematic Distances and Our Image of the Milky Way Galaxy*, The Astronomical Journal, 132:2376—2382, 2006 December p.1

123. Combes, Françoise, October 2005, *Ripples in a Galactic Pond*, Scientific American, Vol.293, No. 4, p.46

124. Ibid.

125. Http://physics.uoregon.edu/~jimbrau/astr123/Notes/Chapter23.html, revisited January 16, 2008

126. Zhang, Xiaolei and Buta, Ronald, J., 2007, *The Potential-Density Phase-Shift Method for Determining the Corotation Radii in Spiral and Barred Galaxies*, The Astronomical Journal, 133:2584-2606, June 2007, p.2596; Http://www.iop.org/EJ/ article/1538-3881/133/6/2584/205650.text.html, Revisited January 21, 2008

127. Http://adsabs.harvard.edu/abs/1999A&A...341...81M, Revisited January 30, 2008

128. Fresneau A., Vaughan, A.E., and Argyle, R.W., August 19, 2005, *Density Wave Streaming Motions in Stars Along the Sagittarius Spiral Arm*, The Astronomical Journal, 130:2701-2716; http://www.iop.org/EJ/article/1538-3881/130/6/2701/204489.text.html, Revisited January 17, 2008

129. Zhang, Xiaolei and Buta, Ronald, J., 2007, *The Potential-Density Phase-Shift Method for Determining the Corotation Radii in Spiral and Barred Galaxies*, The Astronomical Journal, 133:2584-2606, June 2007, p.2596; Http://www.iop.org/EJ/article/1538-3881/133/6/2584/205650.text.html, Revisited January 21, 2008

130. Zhang, Xiaolei, 2001, *Secular Evolution of Disks*, ASP Conference Series, Vol. 230, September 18, 2001, p.2

131. Ibid, p.3

132. Http://adsabs.harvard.edu/abs/2006astro.ph..6201L, Revisited May 2, 2008

133. Http://www.sciam.com/article.cfm?id=a-bar-at-the-heart-of-the, Revisited May 2, 2008

134. Http://en.wikipedia.org/wiki/Milky_Way, revisited May 3, 2008

135. Zhang, Xiaolei and Buta, Ronald, J., 2007, *The Potential-Density Phase-Shift Method for Determining the Corotation Radii in Spiral and Barred Galaxies*, The Astronomical Journal, 133:2584-2606, June 2007, p.2596; Http://www.iop.org/EJ/article/1538-3881/133/6/2584/205650.text.html, Revisited January 21, 2008

136. Http://www.daviddarling.info/encyclopedia/G/Galaxy.html, revisited January 22, 2008

137. Http://www.rave-survey.aip.de/rave/pages/project/ProjectDescription_2.jsp, Revisited May 3, 2008

138. Http://front.math.ucdavis.edu/0709.1926, Revisited May 3, 2008

139. Dehnen, Walter, 1999, *The Effect of the Outer Lindblad Resonance of the Galactic Bar on the Local Stellar Velocity Distribution*, The Astronomical Journal, 126: 2896-2909, 2003, December; http://www.iop.org/EJ/article/1538-3881/119/2/800/990421.text.html, revisited January 4, 2008

140. Korchagin, V.I., Girard, T.M., Borkova, T.V., Dinescu, D.I., and van Altena F., 2003, *Local Surface Density of the Galactic Disk from a Three Dimensional Stellar Velocity Sample*, The Astronomical Journal, 126:2896-2909, December 2003 p3 & p19; http://www.iop.org/EJ/article/1538-3881/126/6/2896/203205.text.html, Revisited January 22, 2008

141. Zhang, X., 2006, *The Potential Density Phase-Shift Method for Determining The Corotation Radii in Spiral Barred Galaxies*, The The Astronomical Journal, 133: 2584-2606, June 2007, p.2589

142. Korchagin, V.I., Girard, T.M., Borkova, T.V., Dinescu, D.I., and van Altena F., 2003, Local Surface Density of the Galactic Disk from a Three Dimensional Stellar Velocity Sample, The Astronomical Journal, 126:2896-2909, December 2003; http://www.iop.org/EJ/article/1538-3881/126/6/2896/203205.text.html, Revisited January 22, 2008], p.3.

143. Ibid, p.19

8 WAVES OF WORLD AGES

144. Kenyon, Kern, E., 1997, *Gravity Torques for Surface Waves*, Journal of Oceanography, Vol. 53, pp. 89 to 92. 1997

145. *Deleted.* Jouzel, J., Masson-Delmotte, V., Cattani, O., and Dreyfus, G. et al, 2007, *Orbital and Millennial Antarctic Climate Variability over the Past 800,000 Years*, Science 10 August 2007:Vol. 317. no. 5839, pp. 793 – 796;
HTTP://WWW.SCIENCEMAG.ORG/CGI/CONTENT/FULL/317/5839/793, REVISITED MARCH 3, 2008

146. *Deleted.* Zhang, Xiaolei, 2001, *Secular Evolution of Disks*, Astrophysics Journal, Vol. 230, arXiv:astro-ph/0109301v1 18 Sep 2001, p.3

147. *Deleted.* Ibid, p.2

148. *Deleted* Ibid.

149. *Deleted.* Cowen, Roy, *Black Hole Glowed Brightly*, Science News, Vol. 173, No. 16, May 10, 2008, p.12

150. *Deleted.* Zhang, Xiaolei, 2001, *Secular Evolution of Disks*, Astrophysics Journal, Vol. 230, arXiv:astro-ph/0109301v1 18 Sep 2001, p.2

151. Paulo Artaxo (Brazil), Terje Berntsen (Norway), Richard Betts (UK), David W. Fahey (USA) et al, 2007, *Changes in Atmospheric Constituents and in Radiative Forcing, In: Climate Change 2007:* The

Physical Science Basis. Contribution of Working Group I to the Fourth
Assessment Report of the Intergovernmental Panel on Climate Change
*[Solomon, S., D. Qin, M. Manning, Z. Chen, M. Marquis, K.B. Averyt,
M.Tignor and H.L. Miller (eds.)]*. Cambridge University Press,
Cambridge, United Kingdom and New York, NY, USA, p.133

152. Ibid.

153. Cowen, Ron, 2008, *Voyager 2 Reports from the Edge*, Science
News, Vol.174, No. 3, August 2, 2008, p.7

154. Ibid.

155. Ibid.

156. Ibid.

157. Gerlach, Arch, C. (1978) *The Rand McNally New Concise Atlas of
the Universe*, Rand McNally and Company, New York, NY, p.140

158. Http://en.wikipedia.org/wiki/Milky_Way, Revisited August 2, 2008

159. Http://www.universetoday.com/2008/05/06/comet-strikes-increase-
as-we-pass-through-the-galactic-plane/, Revisited August 2, 2008

9 WAVES OF ECONOMIC CHANGE

160. Groz, Marc M., *Forbes Guide to the Markets*, John Wiley & Sons,
New York, NY, 1999, p. 48

161. Ibid.

162. Forbes, Steve, *Keeping a Sense of Proportion*, Forbes magazine,
November 12, 2008, p.18

163. Ibid.

164. Greenspan, Alan, 2008, *The Age of Turbulence*, Penguin Books,
New York, NY, 2008 edition, p.490-491

165. Forbes, Steve, *Keeping a Sense of Proportion*, Forbes magazine,
November 12, 2008, p.18

166. Greenspan, Alan, 2008, *The Age of Turbulence*, Penguin Books, New York, NY, 2008 edition, p.488-489

167. Http://en.wikipedia.org/wiki/Stochastic, Revisited October 26, 2008

168. Http://en.wikipedia.org/wiki/Efficient_market_hypothesis, Revisited October 26, 2008

169. Ibid.

170. Http://en.wikipedia.org/wiki/Economics, Revisited November 7, 2008

171. Http://en.wikipedia.org/wiki/Consumption_(economics), Revisited November 7, 2008

172. Http://en.wikipedia.org/wiki/Gross_domestic_product,Revisited November 5, 2008

173. Ibid.

174. Forbes, Steve, February 16, 2009, *Hamilton Got It Right, Why Can't We?* Forbes Magazine, p.15

175. Http://en.wikipedia.org/wiki/Consumer_Price_Index, Revisited October 26, 2008

176. Forbes, Steve, February 16, 2009, *Hamilton Got It Right, Why Can't We?* Forbes Magizine, p.15

177. Jenkins, John Major, 1998, *Maya Cosmogenesis*, Bear & Co., Rochester, Vermont, p.211

10 WAVELETS OF SYNCHRONIZATION

178. Ibid.

179. Morton, Chris and Thomas, Ceri Louise, 1997, The Mystery of the Crystal Skulls, Bear and Co., Santa Fe, New Mexico, p.192Refs][[Jenkins, John Major, 1998, *Maya Cosmogenesis*, Bear & Co., Rochester, Vermont, p.147

180. Andrews, Synthia, and Andrews, Colin, 2008, The Complete Idiot's Guide to 2012; An Ancient Look at a Critical Time, Alpha Books, Penguin Group, New York, NY, p.19-20

181. Ibid, Page 72

182. Http://en.wikipedia.org/wiki/Day, Revisited November 25, 2008

183. Ibid and Resolution 1 of the 13th meeting of the CGPM (1967/68)

184. Castelvecchi, David, 2008, Half Live More or Less, Science News, Vol 174, No 11, November 22, 2008, p.21] [Jenkins J.H, Fischbach E., Buncher J.B., Gruenwald J. T., Krause D.E, and Matte J.J (2008), *Evidence for Correlations Between Nuclear Decay Rates and Earth-Sun Distance*, Astrophysics, arXiv:0808.3283v1, August 25, 2006

185. Http://en.wikipedia.org/wiki/Day, Revisited November 29, 2008

186. Http://en.wikipedia.org/wiki/Winter_Solstice, Revisited November 29, 2008

187. Jenkins, John Major, 2002, *Galactic Alignment*, Bear & Co., Rochester, Vermont, p.9

188. Ibid.

189. Ibid.

190. Jenkins, John Major, 1998, *Maya Cosmogenesis*, Bear & Co., Rochester, Vermont, p.211

191. Jenkins, John Major, 2002, *Galactic Alignment*, Bear & Co., Rochester, Vermont, p.9

192. Ibid.

193. Andrews, Synthia, and Andrews, Colin, 2008, *The Complete Idiot's Guide to 2012; An Ancient Look at a Critical Time*, Alpha Books, Penguin Group, New York, NY, Page 72

194. Http://weber.ucsd.edu/~anthclub/quetzalcoatl/olmecs.htm, Revisited February 12, 2009

195. Andrews, Synthia, and Andrews, Colin, 2008, *The Complete Idiot's Guide to 2012; An Ancient Look at a Critical Time*, Alpha Books, Penguin Group, New York, NY,, p. 168

11   WAVES OF HUMMING EARTH

196. http://en.wikipedia.org/wiki/Gravimeter, Revisited October 14, 2009

197. Ibid.

198. Ibid.

199. Virtanen, Heikki, 2006, *Studies of Earth Dynamics with the Superconducting Gravimeter*, Academic dissertation in Geophysics, University of Helsinki, Finland, Finish Geodetic Institute ISBN-13: 978-952-10-3057-4 (PDF) and ISBN-10: 952-10-3057-7 (PDF) p.10-11; Http://ethesis.helsinki.fi/julkaisut/mat/fysik/vk/virtanen/studieso.pdf, Revisited January 16, 2010

200. H. Virtanen, M. Bilker-Koivula, J. Mäkinen, M. Nordman, J. Virtanen, Finnish Geodetic Institute, DynaQlim/GGOS Workshop June 23-26, 2009, Espoo, Finland, p.3; http://dynaqlim.fgi.fi/texts/GGOS_DynaQlim2009/p_040.pdf, Revisited January 16, 2010

201 Virtanen, Heikki, 2006, *Studies of Earth Dynamics with the Superconducting Gravimeter*, Academic dissertation in Geophysics, University of Helsinki, Finland, Finish Geodetic Institute ISBN-13: 978-952-10-3057-4 (PDF) and ISBN-10: 952-10-3057-7 (PDF) p.10-11; Http://ethesis.helsinki.fi/julkaisut/mat/fysik/vk/virtanen/studieso.pdf, Revisited January 16, 2010

202 Flamstead, Sam, *GRACE in Space*, Discover Magazine, March 23, 2007, p.44

203. Tanimoto1, T. and Artru-Lambin, Juliette, 2006, *Interaction of Solid Earth, Atmosphere & Ionosphere,* Department of Earth Science, University of California, Santa Barbara, California, 93106, October 30, 2006, p.10-11

204. Ibid.

205. Virtanen, Heikki, 2006, *Studies of Earth Dynamics with the Superconducting Gravimeter*, Academic dissertation in Geophysics, University of Helsinki, Finland, Finish Geodetic Institute ISBN-13: 978-952-10-3057-4 (PDF) and ISBN-10: 952-10-3057-7 (PDF) p.10-11; Http://ethesis.helsinki.fi/julkaisut/mat/fysik/vk/virtanen/studieso.pdf, Revisited January 16, 2010; with reference to Peterson, J., 1993, *Observations and Modeling of Seismic Background Noise,* Open-File report 93-332, U.S.Department of Interior, Geological Survey, Albuquerque, New Mexico

206. Tanimoto1, T. and Artru-Lambin, Juliette, 2006, *Interaction of Solid Earth, Atmosphere & Ionosphere,* Department of Earth Science, University of California, Santa Barbara, California, 93106, October 30, 2006, p.53

207. Stehly, L., Campillo, M., and Shapiro, N.M, (2006), *A Study of the Seismic Noise from its Long Range Correlation Properties*, Draft: December 21, 2005, p.4; Stehly, L., M. Campillo, and N. M. Shapiro (2006), *A study of the seismic noise from its long-range correlation properties*, J. Geophys. Res., 111, B10306, doi:10.1029/2005JB004237. See also: http://www.agu.org/pubs/crossref/2006/2005JB004237.shtml, Revisited January 16, 2010

208. McNamara, D. E.; Aster, R.; Bromirski, P.; Hutt, C.; Gee, L., 2007, *Seismological Evidence for Increasing Oceanic Storm Intensity*, American Geophysical Union, Fall Meeting 2007, Abstract; http://adsabs.harvard.edu/abs/2007AGUFM.S11D..05M, Revisited November 5, 2009, Abstract

209. Gutenberg, B., (1951), *Observation and theory of microseisms*, American Meteorological Society as cited by; [Stehly, L., Campillo, M., and Shapiro, N.M, (2006), *A Study of the Seismic Noise from its Long Range Correlation Properties*, Draft: December 21, 2005, p.4;

210. Stehly, L., M. Campillo, and N. M. Shapiro (2006), *a study of the seismic noise from its long-range correlation properties*, J. Geophys. Res., 111, B10306, doi:10.1029/2005JB004237.

211. Http://www.agu.org/pubs/crossref/2006/2005JB004237.shtml, Revisited January 16, 2010

212. Ibid. p.4-6.

213 Http://adsabs.harvard.edu/abs/2008AGUFM.S41C1852K, Revisited October 28, 2009

214. Http://www.livescience.com/strangenews/080416-earth-hum.html, Revisited October 28, 2009

215. Http://www.abc.net.au/science/articles/2008/04/17/2219939.htm, Revisited October 28, 2009

216. Http://www.sott.net/articles/show/150040-Mystery-hum-puzzles-geologists, Revisited October 28, 2009

217. Stehly, L., Campillo, M., and Shapiro, N.M, 2006, *A Study of the Seismic Noise from its Long Range Correlation Properties, Draft: December 21, 2005*, p.22; Published in J. Geophys. Res., 111, B10306, doi:10.1029/2005JB004237

218. Http://www.agu.org/pubs/crossref/2006/2005JB004237.shtml, Revisited January 16, 2010

219. Cohen, Melvin, J. (UCLA), 1955, *the Function of the Receptors in the Statocyst of the Lobster Homarus Americanus*, Journal of Physiology. (1955) I30, pages 9-34, p.9

220. Day, B.L., and Fitzpatrick, R.C., 2005, Current Biology 2005 15:583, http://scienceweek.com/2005/sw050916-2.htm, Revisited October 29, 2009

221. Virtanen, Heikki, 2006, *Studies of Earth Dynamics with the Superconducting Gravimeter*, Academic dissertation in Geophysics, University of Helsinki, Finland, Finish Geodetic Institute ISBN-13: 978-952-10-3057-4 (PDF) and ISBN-10: 952-10-3057-7 (PDF) p.10-11; Http://ethesis.helsinki.fi/julkaisut/mat/fysik/vk/virtanen/studieso.pdf, Revisited January 16, 2010

222. Ibid, p.11

223. Kurrle, Dieter, and Widmer-Schnidrig, Rudolf, 2006, *Spatiotemporal features of the Earth's background oscillations observed in central Europe,* Geophysical Research Letters, VOL. 33, L24304, doi:10.1029/2006GL028429, 2006, p.1

224. Ibid.

225. Stehly, L., Campillo, M., and Shapiro, N.M, (2006), *A Study of the Seismic Noise from its Long Range Correlation Properties, Draft*: December 21, 2005, p.22; Http://www-lgit.obs.ujf-grenoble.fr/~campillo/PAPIERS/draft/stehly_origine_bruit.pdf, Revisited October 21, 2009, p.2

226. Kurrle, D.; Widmer-Schnidrig, R., 2008, *Discovery of the Torsional Hum of the Earth*, American Geophysical Union, Fall Meeting 2008, abstract #S41C-1852, December 2008, Abstract, [p.2] http://adsabs.harvard.edu/abs/2008AGUFM.S41C1852K, Revisited October 28, 2009

227. Brahic, Catherine, 2008, The New Scientist, Sat, 23 Feb 2008, http://www.sott.net/articles/show/150040-Mystery-hum-puzzles-geologists, Revisited October 28, 2009

228. Day, B.L., and Fitzpatrick, R.C., 2005, Current Biology 2005 15:583, http://scienceweek.com/2005/sw050916-2.htm, Revisited October 29, 200

229. Http://www.unmc.edu/physiology/Mann/mann9.html, Revisited October 14, 2009

230. Jones T. A., Jones S. M., and Hoffman L.F., 2008, *Resting Discharge Patterns of Macular Primary Afferents in Otoconia-Deficient Mice,* Journal of the Association for Research in Otolaryngology, Volume 9, Number 4 / December, 2008, Pages   490-505, Abstract p. 490, http://www.springerlink.com/content/6lg22537277627m4/, Revisited October 31, 2009

231. Ibid.

232. Ibid.

233. Davis, Xue, Peterson, and Grant, *Layer thickness and curvature effects on otoconial membrane deformation in the utricle of the red-ear slider turtle: Static and modal analysis,*" Journal of Vestibular Research, 2007; 17(4): 145–162, p. abstract

234. Ibid, p.1-2

235. Ibid, p.8

236. Ibid, p.30

237. Ibid, p.7

238. Http://www.ncbi.nlm.nih.gov/pmc /articles/PMC2112390/ Revisited October 26, 2009

12   WAVES OF HUMMING IDENTITY

239. Day, B.L., and Fitzpatrick, R.C., 2005, Current Biology 2005 15:583, Http://scienceweek.com/2005/sw050916-2.htm, Revisited October 29, 2009

240. Hhttp://espra.risc.cnrs.fr/Lopez_Blanke_2007_PROOFS2.pdf, Revisited February 9, 2010

241. Http://www.scholarpedia.org/article/Vestibular_system, Revisited October 6, 2009

242. Saey, Tina, H, Science News, *You Are Who You Are by Default*, Vol. 176, No. 2, July 18, 2009, page 16

243. Ibid.

244. Ibid, p.18

245. Ibid.

246. Ibid.

247. Ibid.

248. Ibid.

249. Ibid.

250. Oswalt, Angela, (MSW), 2009, *Early Childhood Emotional and Social Development; Identity and Self-Esteem,* [http://www.mhcdinfo.org/poc/view_doc.php?type=doc&id=12766&cn= 462, Revisited October 4, 2009

251. Saey, Tina, H, Science News, "You Are Who You Are by Default," Vol. 176, No. 2, July 18, 2009, page 19

252. Ibid, p.20

253. Ibid.

254. Ibid.

255. Ibid

256. Raichle, Marcus, E., 2010, *the Brain's Dark Energy*, Scientific American, March 2010, Volume 302, No. 3, p.47

257. Saey, Tina, H, Science News, "You Are Who You Are by Default," Vol. 176, No. 2, July 18, 2009, p.18-19

258. Ibid, p.19

259. Ibid.

260. Ibid.

261. Ibid.

262. Ibid.

263. Ibid.

264. Horovitz, Silvina, G.; Braun, Allen R. ; Carr, Walter, S. ; Picchioni, Dante; Balkin, Thomas, J. ; Fukunaga, Masaki; Duyn, Jeff, H., *Decoupling of the brain's default mode network during deep sleep,* Proceedings of the National Academy of Sciences of the United States of America, 2009, vol. 106, no27, abstract, pp. 11376-11381; Http://cat.inist.fr/?aModele=afficheN&cpsidt=21768249, Revisited October 4, 2009]

265. Http://futurity.org/health-medicine/depression-tied-to-over-active-brain-network/, Revisited October 4, 2009

266. Ibid.

267. Ibid.

268. Http://boingboing.net/2009/07/10/default-state-of-the.html, Revisited October 4, 2009

269.
Http://www.sciencedirect.com/science?_ob=ArticleURL&_udi=B6WNP-4RKMJ25-
1&_user=10&_rdoc=1&_fmt=&_orig=search&_sort=d&_docanchor=&view=c&_acct=C000050221&_version=1&_urlVersion=0&_userid=10&md5=f5f1ca0b0cc7f92148cdcfbc3cc9cb05, Revisited October 4, 2009

270. Http://www.ajnr.org/cgi/content/full/29/9/1722, Revisited October 4, 2009

271. Ibid.

272. Ibid.

273. Raichle, Marcus, E., 2010, *the Brain's Dark Energy*, Scientific American, March 2010, Volume 302, No. 3, p.47

274.
Http://thebrain.mcgill.ca/flash/d/d_11/d_11_cr/d_11_cr_hor/d_11_cr_hor.html, Revisited October 5, 2009

275. Ibid.

276. Http://en.wikipedia.org/wiki/Magnetoception, Revisited October 6, 2009

277. Baker, R R; J G Mather, J H Kennaugh (1983-01-06). *Magnetic bones in human sinuses*, Nature 301 (5895): 79–80. PMID 6823284

278. Baker, R., Robin, 1989, *Human navigation and magnetoreception,* Manchester University Press, 1989

279. Wiltschko, R; Wiltschko, W; 1995, *Magnetic orientation in animals*, Springer, June 1995, Page 73.8

280. Carrubba, S; C Frilot, A L Chesson, A A Marino (2007-01-05), 2006, *Evidence of a nonlinear human magnetic sense*, Neuroscience 144

(1): 356–67. doi:10.1016/j.neuroscience.2006.08.068. PMID 17069982.
Http://en.wikipedia.org/wiki/Magnetoception, Revisited October 6, 2009

281.
Http://www.sciencedirect.com/science?_ob=ArticleURL&_udi=B6T0F-
4M69JW3-2&_user=10&_rdoc=1&_fmt=&_orig=search&_sort=d&_
docanchor=&view=c&_acct=C000050221&_version=1&_urlVersion=0&

userid=10&md5=cd3cab6e97c3554b0b47b627bb08f511, Revisited
October 6, 2009

282. Ibid.

283. Ibid, Abstract.

284.
Http://en.wikipedia.org/wiki/File:EM_Spectrum_Properties_edit.svg,
Revisited October 7, 2009

285. Cohen, Melvin, J. (UCLA),1955, *The Function of the Receptors in
the Statocyst of the Lobster Homarus Americanus*, Journal of
Physiology. (1955) I30, page 9

286. Ibid.

287. Ibid, p.32

13   WAVES OF HUMAN INFLUENCE

288. Stehly, L., Campillo, M., and Shapiro, N.M, 2006, *A Study of the
Seismic Noise from its Long Range Correlation Properties*, December
21, 2005, p.22; Http://www-lgit.obs.ujf-
grenoble.fr/~campillo/PAPIERS/draft/stehly_
origine_bruit.pdf, Revisited October 21, 2009

289. Ibid.

290. Harrison, Ben J.; Pujol, Jesus; Lo´, Marina; pez-Sola`, et al, 2008,
*Consistency and functional specialization in the default mode brain
network*, PNAS (National Academy of Sciences), July 15, 2008 vol. 105
no. 28, 9781–9786, Http://www.pnas.orgcgidoi10.1073pnas.0711791105

291. McNamara, D. E.; Aster, R.; Bromirski, P.; Hutt, C.; Gee, L., 2007, *Seismological Evidence for Increasing Oceanic Storm Intensity*, American Geophysical Union, Fall Meeting 2007, abstract; Http://adsabs.harvard.edu/abs/2007AGUFM.S11D..05M, Revisited November 5, 2009

292. Sivaji, C., Nishizawa, O., Kitagawa, G., and Fukushima, Y., (2002) in, *A Physical-Model Study of the Statistics of Seismic Waveform Fluctuations in Random Heterogeneous Media*, Geophysics Journal International, 2002, 148, 575-595 p. 588

293. Saey, Tina, H, Science News, *You Are Who You Are by Default*, Vol. 176, No. 2, July 18, 2009, page 16

294. Http://www.scholarpedia.org/article/Vestibular_system, Revisited October 6, 2009

295. Cohen, Melvin, J. (UCLA), 1955, *the Function of the Receptors in the Statocyst of the Lobster Homarus Americanus*, Journal of Physiology. (1955) I30, pages 9-34, p.9

296. Markopoulou, Fotini, 1998, *The Internal Description of a Causal Set: What the universe looks like from the inside,* Center for Gravitational Physics and Geometry, Department of Physics, The Pennsylvania State University, arXiv:gr-qc/9811053v2, November 18, 1999, p.1-2

14   TSUNAMI TRAINS OF TRANSITION

297. Terence McKenna (1998), Forward to *Maya Cosmogenesis 2012*, Jenkins, John, Major, 1998, *Maya Cosmogensis 2012*, Bear & Co., Rochester, Vermont, p. XXVII

298. Jenkins, John, Major, 2002, *Galactic Alignment*, Bear & Co., Rochester, Vermont p.17; Jenkins, John, Major, 1998, *Maya Cosmogensis 2012*, Bear & Co., Rochester, Vermont, p. XXXIII

299. Jenkins, John, Major, 1998, *Maya Cosmogensis 2012*, Bear & Co., Rochester, Vermont, p. XXXV

300. Andrews, Synthia, and Andrews, Colin, 2008, The Complete Idiot's Guide to 2012, Alpha Books, the Penguin Group, New York, NY, p.92

301. Ibid, p.2

302. Ibid, p.91.

303. Jenkins, John, Major, 1998, *Maya Cosmogensis 2012*, Bear & Co., Rochester, Vermont, p. 149

304. Http://www.mayamysteryschool.com/pdf%20files/Carlos_Barrios.pdf, Revisited January 16, 2010

305. Terence McKenna (1998), Forward to *Maya Cosmogenesis 2012*, Jenkins, John, Major, 1998, *Maya Cosmogensis 2012*, Bear & Co., Rochester, Vermont, p. XXVII

306. Http://en.wikipedia.org/wiki/Pendulum, Revisited January 7, 2010

307. Http://en.wikipedia.org/wiki/Resonance, Revisited February 17, 2009

308. Http://www.nationmaster.com/encyclopedia/Phases-of-matter, Revisited February 8, 2009

309. Ibid.

310 http://en.wikipedia.org/wiki/Phase_transition, Revisited February 27, 2009

311. Http://en.wikipedia.org/wiki/Phase_%28matter%29, Revisited May 28, 2008

312. Http://www.nationmaster.com/encyclopedia/Phases-of-matter, Revisited February 8, 2009

313. Ibid.

314. Ibid.

315. Zang, Xiaolei and Buta Ronald J., 2007, *The Potential-Density Phase Shift Method for Determining the Corotation Radii in Spiral and*

Barred Galaxies, The Astronomical Journal, 133:2584-2606, June, 2007, p.2588

316. Bournaud, F. and Combes, 2002, *Gas Accretion on Spiral Galaxies: Bar Formation and Renewal*, arXiv:astro-ph/0206273v1, June 17, 2002, p.11

317. Fresneau, A., Vaughan, A.E., Argyle, R.W, 2005, *Density Wave Streaming Motions in Stars Along the Sagittarius Spiral Arm*, The Astronomical Journal, 130:2701-2716, http://www.iop.org/EJ/article/1538-3881/130/6/2701/204489.text.html, Revisited January 17, 2008, p.1

318. Korchagin, V.I., Girard, T.M., Borkova, T.V., ET AL, 2003, *Local Surface Density of the Galactic Disk From a Three Dimensional Stellar Velocity Sample*, The American Astronomical Journal, 126:2896-2909, December 2003

319. Griv, E., and Gedalin, M., 2004, *Changes in the Angular Momentum and Entropy Induced by Jeans-Unstable Density Waves in Stellar Disks and Flat Galaxies*, Astronomical Journal, 128:1965-1973, November 2004.

320. Http://www.nationmaster.com/encyclopedia/Phases-of-matter, Revisited February 8, 2009

321. Ibid.

322. Ibid.

323. Ibid.

324. Ibid.

325. Ibid.

326. Ibid.

327. Ibid.

328. Http://en.wikipedia.org/wiki/Phase_transition, Revisited February 27, 2009

329. Http://www.nationmaster.com/encyclopedia/Phases-of-matter, Revisited February 8, 2009

330. Ibid.

331. Ibid.

332. Ibid.

333. Ibid.

334. Http://en.wikipedia.org/wiki/Larmor_precession, Revisited February 18, 2009

335. Atkins, P.W., 1978, Physical Chemistry, W.H. Freemanp & Co, San Francisco, p.401

336. Ibid.

337. Http://en.wikipedia.org/wiki/Larmor_precession, Revisited February 18, 200.

338. Atkins, P.W., 1978, Physical Chemistry, W.H. Freemanp & Co, San Francisco, p.401

339. Nassim H., Hypson M., and Rauscher, A. E., 2008, Scale Unification – A Universal Scaling Law for Organized Matter, Proceedings of The Unified Theories Conference, in Cs Varga, I Dienes & RI, Amoroso (eds.) reproduced at, http://www.theresonanceproject.org/pdf/scalinglaw_paper.pdf, Revisited February 26, 2009

340. Ibid.

341. Van Hooydonk, G., February 17, 2005, *Equivalence of macro- (Van der waals) and micro- (quantum) phase transitions: consequences for BEC, antihydrogen and generic chiral symmetry breaking* Http://arxiv.org/ftp/physics/papers/0502/0502098.pdf, Revisited February 26, 2009

342. Parker, B, 1993, *Overcoming some of the problems*, pp.259-279; Http://en.wikipedia.org/wiki/Grand_unified_theory, Revisited February 27, 2009

343. Http://en.wikipedia.org/wiki/Grand_unified_theory, Revisited February 27, 2009

343. Http://en.wikipedia.org/wiki/Grand_unified_theory, Revisited February 27, 2009

344. Ibid.

345. Drewfs, Paul, R., 2007, *Waters of Creation and Reality*, Paul R. Drewfs, LuLu.com, 2007

15 WORLD AGE RESONANT PHASE SYNCHRONIZATION (WARPS)

346. Combes, Françoise, *Ripples in a Galactic Pond*, Scientific American, October 2005, Vol. 393, No. 4, p.46

347. Zhang, Xiaolei, 2001, *Secular Evolution of Disks*, ASP Conference Series, Vol. 230, 2001, arXiv: astro-ph/0109301v1 18 September 2001, p.1

348. Combes, Françoise, *Ripples in a Galactic Pond*, Scientific American, October 2005, Vol. 393, No. 4, p.46-47

349. Zhang, Xiaolei, 2001, *Secular Evolution of Disks*, ASP Conference Series, Vol. 230, 2001, arXiv: astro-ph/0109301v1 18 September 2001, p.1

350. Combes, Françoise, *Ripples in a Galactic Pond*, Scientific American, October 2005, Vol. 393, No. 4, p. 47

351. Http://en.wikipedia.org/wiki/Dark_energy, Revisited February 27, 2009

352. Http://en.wikipedia.org/wiki/Superposition_principle, Revisited January 16, 2010

353. Kane, Gordon, 2005, *The Mysteries of Mass*, Scientific American, Vol. 293 No. 1, July 2005, p. 44

354. Http://en.wikipedia.org/wiki/Dark_energy, Revisited February 27, 2009

355. Kane, Gordon, 2005, *the Mysteries of Mass*, Scientific American, Vol. 293 No. 1, July 2005, p. 44

356. Ibid, p.45.

357. Http://en.wikipedia.org/wiki/Electromagnetic_field, Revisited January 16, 2010

358 Kane, Gordon, 2005, *the Mysteries of Mass*, Scientific American, Vol. 293 No. 1, July 2005, p. 42

359. Randel, Lisa, 2005, *Warped Passages; Unraveling the Mysteries of the Universe's Hidden Dimensions*, Harper Perennial, New York, NY, p. 136

360. Ibid. p.79

361. Ibid.

362. Kane, Gordon, 2005, *The Mysteries of Mass*, Scientific American, Vol. 293 No. 1, July 2005, p. 44

363. Ibid, p.45.

364. Ibid.

365. Ibid.

366. Ibid, p.42.

367. Albert, David, Z., and Galchen, Rivka, 2009, *A Quantum Threat to Special Relativity*, Scientific American, Vol. 300, No. 9, March, 2009, p.34

368. Ibid.

369. Http://en.wikipedia.org/wiki/Warp_drive, Revisited March 11, 2009

370. Ibid.

371. Http://www.cosmosmagazine.com/news/2141/dark-energy-spacecraft-could-fly-faster-light, Revisited March 11, 2009

372. Http://en.wikipedia.org/wiki/Warp_drive, Revisited March 11, 2009

373. Http://www.cosmosmagazine.com/news/2141/dark-energy-spacecraft-could-fly-faster-light, Revisited March 11, 2009

374. Http://en.wikipedia.org/wiki/Panspermia, Revisited March 11, 2009.

375. Ibid

376. Hawking, Steven, 2001, *Universe in a Nutshell*, Bantam Books, New York, NY, p.83

377. Ibid.

16 WAVES OF CASCADING TRANSFORMATION

378. Sanders, Laura, 2010, *Trapped in Cloud of Ultracold Atoms, Light Stayed Frozen for 1.5 seconds*, Science News, Vol 177, No.2, January 6, 2010 (p.10)

379. Ibid.

380. Fleming, Graham and Gregory Engel, *Quantum Secrets of Photosynthesis Revealed*, Berkley Lab Research News, April 12, 2007, http://www.lbl.gov/Science-Articles/Archive/PBD-quantum-secrets.html, Revisited March 11, 2010]

381. Ibid.

382. Ibid.

383. Ibid.

384. Ibid.

385. Akst, Jef, *Quantum photosynthesis*, The Scientist.com, February 3, 2010, http://www.the-scientist.com/blog/display/57131/, Revisited March 11, 2010

386. Ibid.

387. Andrews, Synthia and Colin, 2008, *A Complete Idiot's Guide to 2012,* Alpha, Penquin Group, New York, NY, 2008, p.112

388. Ibid, p.114

389. Makemson, Maud, Worcester, 1951, *the Book of the Jaguar Preist, the Chilam Balam*, Henry Schuman, 1951

390. Andrews, Synthia and Colin, 2008, *A Complete Idiot's Guide to 2012,* Alpha, Penquin Group, New York, NY, 2008, p.114

391. Rodgers, Everett M., 2003, *Diffusion of Innovations*, Free Press, New York, NY, pp.471

392. Johnson, Paul, *The Sickness of the West*, Forbes magazine, March 15, 2010, p. 15

393. Ibid.

394. Http://www.mayamysteryschool.com/pdf%20files/ Carlos_Barrios.pdf, Revisited January 16, 2010

17   WAVES OF EARTH COMPRESSION

395. Http://www.encyclopedia.com/doc/1O80-oblateness.html, Revisited April 9, 2010

396. Ibid.

397. Ibid.

398. Byron D. Tapley, 2004, Http://www.thunderbolts.info/forum/phpBB3/ viewtopic.php?p=25863&sid=a4fed569b19c237d60f3416356e86f14, Revisited April 8, 2010

399. Jean O. Dickey, Steven L. Marcus, Olivier de Viron,2 Ichiro Fukumori, 2002, *Recent Earth Oblateness Variations: Unraveling Climate and Postglacial Rebound Effects,* Science 6 December 2002:

Vol. 298. no. 5600, pp. 1975 – 1977, DOI: 10.1126/science.1077777, Abstract, http://www.sciencemag.org/cgi/ content/abstract/298/5600/1975, Revisited April 9, 2010

400. Chao, Benjamin, Fong, 2006, *Earth's Oblateness and Its Temporal Variations*, C. R. Geoscience 338 (2006) 1123–1129; Available at www.sciencedirect.com, Revisited April 9, 2010

401. Hhttp://en.wikipedia.org/wiki/Post-glacial_rebound, Revisited April 10, 2010

402. Ibid.

403. Johansson, J.M.; et al. (2002). *Continuous GPS Measurements of Postglacial Adjustment in Fennoscandia, Geodetic results,* Journal of Geophysical Research 107: 2157. doi:10.1029/2001JB000400

404. Sella, G.F.; Stein, S., Dixon, T.H., Craymer, M., James, T.S., Mazzotti, S., Dokka, R.K. (2007), *Observation of Glacial Isostatic Adjustment in "Stable" North America with GPS,* Geophysical Research Letters 34: L02306, doi:10.1029/2006GL027081

405. Peltier, W.R., 1998, *Postglacial Variations in the Level of the Sea: Implications for Climate Dynamics and Solid-earth Geophysics,* Reviews of Geophysics 36: 603–689, doi:10.1029/98RG02638

406. Mitrovica, J.X.; W.R. Peltier, 1993, *Present-Day Secular Variations in Zonal Harmonics of the Earth's Geopotential,* Journal of Geophysical Research 98: 4509–4526. doi:10.1029/92JB02700

407. Wu, P.; P. Johnston, 2000, *Can Deglaciation Trigger Earthquakes in N. America?*, Geophysical Research Letters 27: 1323–1326, doi:10.1029/1999GL011070

408. Wu, P.; W.R.Peltier, 1984, *Pleistocene Deglaciation and the Earth's Rotation: A New Analysis*, Geophysical Journal of the Royal Astronomical Society 76: 753–792

409. Http://en.wikipedia.org/wiki/Post-glacial_rebound, Revisited April 10, 2010

410. Http://www.nasa.gov/centers/goddard/earthandsun/earthshape.html, Revisited April 10, 2010

491

411. Http://earthobservatory.nasa.gov/Newsroom/view.php?id =22829, Revisited March 25, 2010

412. Minkang Cheng and Byron D. Tapley, 2009, *A 33 Year Time History of the J2 Changes from SLR [Satellite Laser Ranging]*, ILRS Workshop, 2008, cheng@csr.utexas.edu, Revisited April 12, 2010

413. Ibid.

18   WAVE TRAIN SIGNALS, SWITCHINGS, AND SIDINGS

414. McAuliffe, K., 2009, *Are We Still Evolving?* Discover Magazine, March 2009, p. 51

415. Ibid.

416. Ibid

417. Ibid.

418. Ibid.

419. Ibid.

420. Ibid, p.52

421. Ibid.

422. Ibid.

423. Ibid.

424. Ibid.

425. Ibid.

426. Ibid, p.53

427. Ibid

428. Gould, Stephen Jay, 1980, "*The Episodic Nature of Evolutionary Change,*" *the Panda's Thumb.* New York. W. W. Norton. p. 182; Http://www.stephenjaygould.org/ctrl/gould_quotations.html, Revisited January 17, 2010

429. McAuliffe, K., 2009, *Are We Still Evolving?* Discover Magazine, March 2009, p. 51

430. Http://www.deathreference.com/Nu-Pu/Population-Growth.html, Revisited January 15, 2010

431. Carroll, S. B., Prud'homme, B., and Gompel, 2008, *Regulating Evolution*, Scientific American, Vol. 298, No.5, May 2008, p.67

432. Ibid, p.61

433. Ibid.

434. Ibid.

435. Ibid.

436. Ibid, p.62

437. Ibid.

438. Ibid.

439. Ibid.

440. Ibid.

441. Ibid.

442. Ibid.

443. Ibid.

444. Ibid.

445. Ibid.

446. [Ibid, p.64-65]

447. Ibid, p. 65

448. Ibid. p.66

449. Ibid. p.67

450. Ehrenberg, Rachel, 2008, *Gene Activity Makes the Difference in Development of Human Qualities*, Science News, Vol. 174, No.7, September 27, 2008, p.13

451. Marsa, Linda, *Galapagos Next*, Discover Magazine, March 2009, p.48

452. Ibid.

453. Brownlee, Christen, *Inherit the Warmer Wind*, Science News, Vol. 170, No. 23, December 2, 2006, p. 362

454. Millus, Susan, *Mammals encounter tougher times New Assessment of Species Shows*, Science News, Vol.174 No.10, November 8, 2008, p.16

455. Ibid.

456. Ibid.

457. Ibid.

458. Ibid.

459. Ibid.

460. Ibid.

461. Ibid.

462. Klinkenborg, Verlyn, 2009, *Countdown to Extinction*, National Geographic, January 2009, p.92

463. Ibid, p.93

464. Gaidos, Susan, 2009, *Cosmic Mystery*, Science News, February 28, 2009, Vol. 175, No. 5, P.17

465. Ibid, p.17

466. Ibid, p.17-18

467. Ibid, p.18

468. Ibid.

469. Ibid.

470. Ibid.

471. Cowen, Ron, 2008, *PAMELA Spots the Dark Stuff Maybe*, Science News, Vol. 174, No. 7, September 27, 2008, p.8

472. Hooper, Dan; Stebbins Albert; and. Zurek, Kathryn M, 2008, *The PAMELA and ATIC Excesses From a Nearby Clump of Neutralino Dark Matter*, FERMILAB-PUB-08-566-A, Fermi National Accelerator Laboratory, Batavia, IL, (*arXiv:0812.3202v1[hep-ph]*), December 17, 2008, *Abstract, p.1*

473. Ibid.

474. Ibid. p.3

475. Ibid.

476. Ibid, p.4

19   CONCLUSIONS

477. Verlinde, Erik, 2010, *On the Origin of Gravity and the Laws of Newton,* arXiv: 1001.0785v1 hep-th, 6 Jan 2010, p.3

478. Clemens (1985), Ap. J. 295, 422.)
http://web.njit.edu/~gary/321/Lecture19.html, Revisited January 28, 2008

479. Lucina Q. Uddin 1, A.M. Clare Kelly 1, Bharat B. Biswal 2, F. Xavier Castellanos 1 *, Michael P. Milham, (2007) *Functional connectivity of default mode network components: Correlation, anticorrelation, and causality,* National Institute of Mental Health (NIMH); Grant Number: 5T32MH067763, 5R21MH066393, abstract

480. Malia F. Mason,1* Michael I. Norton,2 John D. Van Horn,1 Daniel M. Wegner,3 Scott T. Grafton,1 C. Neil Macrae4, *Wandering Minds: The Default Network and Stimulus-Independent Thought*, Science 19 January 2007:
Vol. 315. no. 5810, pp. 393 – 395, DOI: 10.1126/science.1131295, Abstract

481. Abler, Birgit; Hofer, Christian; Viviani, Roberto, 2008, *Brain ImagingHabitual emotion regulation strategies and baseline brain perfusion* NeuroReport: 8 January 2008 - Volume 19 - Issue 1 - pp 21-24
doi: 10.1097/WNR.0b013e3282f3adeb, Abstract
[http://journals.lww.com/neuroreport/Abstract/2008/01080/Habitual_em otion_regulation_strategies_and.4.aspx, Revisited January 25, 2008

482. Kelly CAM, Uddin LQ, Biswal BB, Castellanos FX, Milham MP. 2008.
Competition between functional brain networks mediates behavioral Variability, Neuroimage, 39:527—537;

483. A. Di Martino1,2, A. Scheres3, D.S. Margulies1, A.M.C. Kelly1, L.Q. Uddin1, Z. Shehzad1, B. Biswal4, J.R. Walters5, F.X. Castellanos1,6 and M.P. Milham1, *Functional Connectivity of Human Striatum: A Resting State fMRI Study*, Cerebral Cortex, December 2008;18:2735—2747 doi:10.1093/cercor/bhn041, Advance Access publication April 9, 2008; (see pdf for Kelly ref page number)
http://cercor.oxfordjournals.org/cgi/reprint/18/12/2735, Revisited January 25, 2010

484. Http://cercor.oxfordjournals.org/cgi/reprint/18/12/2735, Revisited January 25, 2010

485. Http://en.wikipedia.org/wiki/Entropy_(classical_ thermodynamics), Revisited January 1, 2010

486. Bais, F. Alexander and Farmer, J. Doyne, 2007, *The Physics of Information*, arXiv:0708.2837v2 [physics.class-ph] 18 Dec 2007, Http://arxiv.org/PS_cache

/arxiv/pdf/0708/0708.2837v2.pdf, Revisited January 18, 2010, p.15

19 HYPOTHESES

None

TABLE 8 ENDNOTES FOR ESTIMATES OF DISTANCE FROM SUN TO GALACTIC CENTER

[1] Torra, J.; Fern´andez, D.; and Figueras, F., Departament d'Astronomia i Meteorologia, Universitat de Barcelona, Av. Diagonal 647, E-08028 Barcelona, Spain and F. Comer´on, European Southern Observatory, Karl-Schwarzschild-Strasse 2, D-85748 Garching, bei M"unchen, Germany.

[2] Jenkins, John Major, 2002, "Galactic *Alignment; the Transformation of Consciousness,* Bear & Company, Rochester, Vermont, 2002, Page 9

[3] Nikos Drakos, Computer Based Learning Unit, University of Leeds. 1993, 1994, 1995, 1996 and Ross Moore, Mathematics Department, Macquarie University, Sydney 1997, 1998, 1999, http://cosserv3.fau.edu/~cis/AST2002/Lectures/C15/Trans/Trans.html, Revisited November 14, 2007; & Fresneau A., Vaughan, A.E., and Argyle, R.W., August 19, 2005, *Density Wave Streaming Motions in Stars Along the Sagittarius Spiral Arm*, The Astronomical Journal, 130:2701-2716; http://www.iop.org/EJ/article/1538-3881/130/6/2701/204489.text.html, Revisited January 17, 2008, p.1

[4] Http://www.en.wikipedia.org/wiki/Milky_Way, Revisited November 14, 2007

[5] M. Colleen Gino, 2001, "The Distance to the Galactic Center," http://www.astrophys-assist.com/educate/distance/distance_gc.htm, Revisited November 14, 2007.

[6] JustUs & Associates, Paul O. Hewit (1996) Http://www.horary.com/hhcrl/galact.html, Revisited November 14, 2007

[7] Reid 1993, Ann. Rev. Astron. Astrophys., 31, 345; Reid et al 1999, ApJ, 524, 816) *"The Distance to the Center of the Milky Way,"* Http://www.cfa.harvard.edu/~reid/trigpar.html, Revisited November 14, 2007

[8] NASA, University of Massachusetts, D. Wang et al, 2002, *"X-Ray Mosaic of Galactic Center; Chandra Takes In the Bright Lights, Big City Of The Milky Way,"* Reference, Q.D. Wang et al. Nature, 415, 148, 2002

[9] McNamara, D.H.; Madsen, J.B; Barnes, J; and Ericksen F.B, 2000, *"The Distance to the Galactic Center."* PASP 112, Feb 2000, p. 202

[10] Paczyn'ski, B. and Stanek, K. Z., 1998, *GALACTOCENTRIC DISTANCE WITH THE OGLE AND HIPPARCOS RED CLUMP STARS,* The Astrophysical Journal, 494:L219–L222, 1998 February 20, p.493 http://www.iop.org/EJ/article/1538-4357/494/2/L219/975607.web.pdf?request-id=8a2ec0e1-e3c4-4821-8dc4-c8d8455c7049, Revisited January 30, 2010; Http://cdsweb.cern.ch/record/331714/files/9708080.pdf , Revisited January 30, 2010, Alcock, Ch. et al., 1997, Astro-Physics Journal, submitted (astro-ph/9706292)

[11] F. Eisenhauer, R. Genzel, T. Alexander, R. Abuter, T. Paumard, T.Ott, A.Gilbert, S.Gillessen, M.Horrobin, S.Trippe, H.Bonnet, C.Dumas, N.Hubin, A.Kaufer, M.Kissler-Patig, G.Monnet, S.Stroebele, T.Szeifert, A.Eckart, R.Schoedel, S.Zucker, *"SINFONI in the Galactic Center: young stars and IR flares in the central light month,"* Astrophys.J. 628 (2005) 246-259; http://arxiv.org/abs/astro-ph/0502129, Revisited November 14, 2007

[12] Http://www.mpe.mpg.de/ir/GC/index.php, Revisited November 14, 2007 Thomas Ott, Infrared and Submillimeter Astronomy Group at MPE, Max-Planck-Institut für extraterrestrische Physik, November 20, 2006

TABLE 8 PRECESSION ESTIMATE ENDNOTES

[1] Http:///www.wikipedia.org/wiki/Precession, Revisited November 14, 2007

[2] Jenkins, John Major, 2002, "Galactic *Alignment; the Transformation of Consciousness,* Bear & Company, Rochester, Vermont, 2002, Page 9

[3] Rand McNally, 1978, *New Concise Atlas of the Universe*, Mitchell Beazley Publishers Ltd, 1978, page 183

[4] Http://www.crystalinks.com/precession.html, Revisited November 14, 2007

[5] Dr. David P. Stern, stargaze http://www.phy6.org/stargaze/Sprecess.htm (Last updated: September17, 2004), Revisited November 14, 2007

[6] Beatty, J. K.; Petersen, C. C.; and Chaikin, A. (Eds.), 1990, *the New Solar System*, 4th ed. Cambridge, England: Cambridge University Press, 1990, p. 105, http://scienceworld.wolfram.com/physics/PrecessionoftheEquinoxes.htm l Revisited November 14, 2007

[7] Http://www.revealer.com/platonic.htm, Revisited November 14, 2007, Andrew Raymond, "The Platonic Year. Secrets of the Sphinx," Chapter 2

[8] Http://ancientegypt.hypermart.net/royalarch/index.htm, Revisited November 14, 2007

[9] Http://csep10.phys.utk.edu/astr161/lect/time/precession.html, Revisited November 14, 2007

[10] Dr Shepherd Simpson, Astrological Historian, "Precession of the Equinoxes" http://www.geocities.com/astrologyages/precessionequinoxes.htm, Revisited November 14, 2007

[11] The Columbia Electronic Encyclopedia, 6th ed. Copyright © 2007, Columbia University Press,

http://www.infoplease.com/ce6/sci/A0840032.html, Revisited November 14, 2007

[12] J. C. Evans, Physics & Astronomy Department, George Mason University Maintained by J. C. Evans; jevans@gmu.edu http://physics.gmu.edu/~jevans/astr103/CourseNotes/cyclicPhenomena_ precessionEquinoxs.html Latest Modification: July 22, 2002, Revisited November 14, 2007

SUBJECT INDEX

A

A Physical-Model Study of the Statistics of the Seismic Waveform, 301
Accelerating Rate of Human Genetic Change, 402
Accelerating Rate of Stressed Populations and Extinctions, 414
Advanced Thin Ionization Calorimeter (ATIC), 419
Aliasing of Attention Effects. 282
Aliasing of Cardiac Pulsations, 282
Aliasing of Respiratory Pulsations, 282
Aliasing of Vasomotor Oscillations, 282
Alzheimer's Disease and the DMN, 278
Amyotrophic Lateral Sclerosis and the DMN, 281
An abundance of Recent Adaptive Mutations; Human Genome, 402
An Ancient Clue; the Galactic Center, 129
Angular Momentum; Galactic, Solar System, Co-Rotation &, 96,153
Antarctic Ice Cores, 106-119
Antarctic Vostok Ice Core Deuterium Data, 106
Astronomical Units (AU), 173
Atomic Age, 54
Attention-Deficit/Hyperactivity Disorder and the DMN, 281
Autism, 281

B

Baktun Cycle; Maya, 222,226,231,232
Binary Star Systems, 81
Blood Oxygen Level Dependence (BOLD), 274
Body-Centric & Geo-Object-Centric Frames of Reference, 272
Bombardment of Earth's Atmosphere by High Energy Particles, 419
Bureau of Weights and Measures (BIPM), 223

C

Calcium Carbonate Crystals, 265
California MLAC and PHL Seismograph Station Sites, 319
Carbon Dating Problem, 20
Carbon Dioxide (CO_2), 170
Cascade of Phase Transitions within the Systems, 363
Celestial Motion; Precession, Obliquity, Inclination, & Eccentricity, 5, 92,96, 99,102, 122,163,165, 168
Celestial Sphere, 174

CHAMP (CHAllenging Mini-satellite Payload), 250
Child Development; Default Mode Network &, 276-278
Chronological Offset of Transformation Period; 4,613.3 years, 354
Circumstellar Habitable Zone (Galactic), 147
Collapse into the Analytic Phase of a New World Age, 353
Collapse the Entangled Wave State, 353
Common Stock Price,193, 242
Communication Channels; Definition, 25
Composite Stock Price Index, 190
Compression and Resonance Distributions, 240
Computer Age, 54
Conclusions, 429
Consistency of DMN Wave Cycles in Adults, 278
Consumer Price Index (CPI), 212
Co-Rotation (Corotation) Circle; Galactic, 146-147
Co-Rotation (Corotation) Circle; Galactic Sign Change Effect, 151
Cosmic Rays; Definition, 419-420
Current Epoch Motion, Temperature, and CO_2, 120
Cyanobacteria, 373
Cycles of Re-invention, 23

D

Dark Energy Star, 142
Dark Energy, 345
Dark Matter (DM), 345
Dates of Human Change, 19
Dawn of the Fifth Sun, the Great Alignment, 130
Day; Maya, Long Count, 222
Day; Modern, Earth Rotations, 222
December 21, 2012, Friday, 231,232,245,323,325,336,390,400,420,426,434
December 22, 2012; Saturday, 358
Default Mode Network (DMN), 272
Default Mode Network Map, 274
Default Mode Network Waves, 275
Degrees of Maya Precession, 35
Denis Poisson, 51
Depression and the DMN, 281
Digital Revolution, 54
Disequilibrium, 27
Distribution of Expected and Observed Human Change Events, 49
Divination; Prediction, Maya, 220
Dow Jones Industrial Average (DJIA), 190

Dust Lanes: Galactic, 344
Dynamic Equilibrium, 27

E

Earth surface gravity, 248
Eccentricity; Earth, 96,99,102, 122,163,165, 168
Economic History Lessons, 178
Electroencephalograph Recording of the DMN, 281
Electromagnetic Field (EMF), 346
Electronic Quantum Coherence, 367
Electronic Spectroscopy, 367
Electro-Strong Force, 341
End of space and time& the Time of the Great Transformation, 130
Endofluid; Vestibular Organs, 263
Energies and Wave Functions of a Particle, 338
Energy and Surface Wave Distribution Characteristics, 50
Energy Mass Density of the Universe, 347
Entangled with the Galactic Center, 351
Entangled; Collective Resonant Synchronization, 371
Epoch Boundaries, 47
Equilibrium; Social; Innovation &, 27
Equivalence of Macro- and Microscopic Phase Transitions, 341
European Project for Ice Coring in Antarctica (EPICA) Deep Ice Core Data, 106
Evolutionary Transformation, 354
Evolving Switch Settings, 412
External Frame of Reference (the non-self), 271-273, 320
Extinct Animal & Plant Species Date Support, 44
Extreme Gradualism, 10
Extreme Punctuated Equilibrium, 10

F

Femtosecond, 367
FERMI Gamma Ray Space Telescope (formerly GLAST), 425
Flip Side of Evolution; Loss of Features, 412
fMRI BOLD (Functional Magnetic Resonance Imaging-Blood Oxygen Level Dependent), 274
fMRI Functional Magnetic Resonance Imaging, 273
Forensic Dating Techniques, 44
Fossil Dating; Error of Measurement Limits, 20
Fossil Record, 9

Fossilization, 16
Fossils; Accepted, 14
Frequency of Human Change, 10

G

Galactic Bulge Location and Dimensions, 152-153
Galactic Center (Center of the Galaxy), 147
Galactic Co-Rotation Radius, 146-147, 343
Galactic Habitable Zone, 147
Galactic Measurement Distortions; astronomical, 148
Galactic Rotation Curve; Plotting Radii by Velocity of Rotation, 431
Galactic Stellar Bar Dimension Approximate Equalities, 152-153
Galactic Stellar Bar; Rotation, Dimensions, and Pitch Angle, 146-155
Genetic & Population Bottlenecks, 17
Genetic Expression and Mutations, 409
Genetic Markers, 15
Genetic Mutations, 409
Genetic Switches; Gene Enhancers, 410
Geologic Strata Depth and Soil Analysis Dating, 44
Global Climate Data; 800,000 years of, 106
Global Hourly Wages, 189
Global Land Ocean Temperature, 126
Global Network of Superconducting Gravitometer (SG) stations, 251
Global Temperature, 126
Global Warming, 125
GRACE (Gravity Recovery and Climate Experiment), 250
Gradualism with Punctuated Equilibrium, 33
Grand Cycle; Maya Precession, 221
Grand Unification theory (GUT), 341
Graviatational Torque Waves; Description and Explanation, 139
Gravitational Field, 247
Gravitational Hum (Modes); Earth, 252
Gravitational Modes; Wave Bands; Frequency Spectra, 231
Gravitational Rhythms that drive Galactic Change, 360
Gravitometer; instrument, 248
Gravity Waves; Description, 137
Gravity; Operational Definition, 134
Great Cycle; Maya Precession, 221
Gregorian calendar, 13
GTW Amplitude Estimates; Wave Height, 131, 145, 153-158
GTW Compression and Earth Gravitational & Microseism Hum Modes, 254
GTW Compression Strengths, 162-167

GTW Compression; Density Estimates, 162-167
GTW Cosmic Ray Forcing, 419
GTW Galactic Orbital Speed Control Effects, 99
GTW Geometry, 157
GTW Imposed Density Changes, 157
GTW Pitch Angle Estimates, 156
GTW Poisson distribution; Profile of a GTW, 158
GTW Propagation and the Sun's Galactic Orbit, 148
GTW Stretching and Squeezing of Space-Time Density, 161
GTW; Conjoining Cause and Effect, 432

H

Heat Capacity, 333
Heliospheric Current-Sheet about the Sun Out to the Orbit of Jupiter, 291
HESS; High Energy Stereoscopic System, 424
High Energy Particle Bombardment, 419
Historic Long Lived Greenhouse Gas Emissions, 114, 123, 162, 170
Hominid Species Emergence and Extinction; Accepted Dates, 62
Hominid Species Emergences (speciation), 33-59
Hominid Species Extinctions, 33-59
Hominid Species; Missing; Estimates of, 65-66
Hominids; *Hominidae,* 14
Homo, 14
Homo sapiens, 14
How GTW Compression Stimulates Human Physiology, 297-318
Human Autonomic Sense of Self, 270, 272
Human Change in the Current Period, 57
Human Change Summary, 69
Human External Frame of Reference (the non-self; environment), 271-273, 320
Human Frames of Reference (Autonomic Self), 271-273, 320
Human Genome, 9
Human Population; Growth Rate, 404
Human Races Evolving Away from Each Other, 403
Human Sensing of Earth's Hum, 258
Human Vestibular System, 264,438
Human; what is a, 14
Hyper-Jump; Distance & Evolution, 354
Hypotheses Summary, 437
Hypotheses; Alternative, 4
Hypotheses; Investigative, 4
Hypotheses; Null, 4

I

Ice Ages; 100 in 2.5 Million Years, 96, 108
Inclination Angle; Earth; Orbital, 94, 99,102, 122,163,165, 168
Increasing opportunities for the expression of free will, 432
Indications GTW is Wedging Its Way Into Our Solar System, 399-419
Industrial Revolution, 54
Inflation, Economic, 196
Inflection Points; Maya; Epoch, 42, 47
Inner versus Outer Galaxy, 147
Innovation; Definition & Characteristics of, 25
Innovativeness & Innovation Adopter Categories, 29
Internal Panel on Climate Change (IPCC), 105
International Conservation Monitoring Organization (IUCN), 415
International Haplotype Map, 405
Introduction, 1
IUCN sets quantitative criteria; Red List, 416

J-K

Katun Cycle; Maya, 205, 226, 227, 228, 319, 394, 395, 396, 397, 398
Kin Cycle; Maya, 205, 226, 227, 228, 319
K'ul (as Quantitative Response Effect), 220, 222, 229, 246
K'ul (as Subjective Association), 220
K'ul Improved Economic Predictions, 229-246
K'ul Resonance (Calculated), 220, 222, 229-246
K'ul Resonance Prediction of Human Innovation, 242
K'ul Response Effect Index Distribution, 240
Kul; Maya, 220

L

Larmor Precession; Atomic Physics, 337
Life-Friendly Environment Impossibly Short; Earth, 357-358
Light harvesting systems, 364
Long-Lived Greenhouse Gases (LLGGs), 114, 123, 162, 170
Lunisolar Model of the Precession of the Equinoxes, 76

M

Macro level Ordinary Matter (OM macro), 345
Magnetoception (or Magneto-Reception), 288

Market Volatility, 182
Maya and Toltec Great Year of Precession, 58
Maya Asserted change Intervals, 11
Maya Day-Keeper Shaman, 324
Maya Epoch (Even, Odd, and 1st Even in World Age), 47
Maya Long Count Calendar, 129
Maya Long Count Calendar; End Date, 129
Maya Prophecy of Transformation, 323, 360, 375, 378, 400,433
Maya Solar System Precession, 81
Maya Sub-cycles, 205, 226, 227, 228, 319, 394, 395, 396, 397, 389
Maya Theory of Human Evolution; Definition of, 11
Maya Theory of Human Evolution; Test of, 35
Maya World Age Cycle, 205, 226, 227, 228, 319, 394, 395, 396, 397, 398
Measuring Earth's Gravity, 247
Medial Prefrontal Cortex, 274
Metastable, forms of energy and matter, 334
Methane (CH_4), 170
Micro level Ordinary Matter (OM Micro), 346
Microseism Hum MSH Mode; Gravitational Free Oscillations, 252-257
Milagro Ground Based Detector; Los Alamos, 399
Milky Way Galaxy; Remapping, 141
Money Supply; Economic, 185

N

NASA's Very Long Baseline Interferometry (VBLI) Group, 130
Natural and Sexual Selection, 11, 405
Natural Selection, 11, 405
Neutralino Dark Matter Hypothesis, 423
New Fire Ceremony; Mesoamerican, 19
New Low Noise Model (NLNM); Gravitational Modes, 253
Nitrous Oxide (N_2O), 170
Non-Analytic, 325
Non-destructive captured and transformation of energy information, 363
Non-destructive Transformation, 363
Normal Probability Distribution, 47, 52, 53, 232
Number of Prior Transformations; Estimated, 356, 363

O

Obliquity; Earth, 92,99,102, 122,163,165, 168
Olmec, Pyramid of the Sun, 241
One Hundred Million Years of Evolution; Genes &, 411

Optimum Energy & Information Distribution, 351
Otoconia Membrane, 265
Otolith Organs; Utricle and Saccule, Inner Ear, 259

P

Paleoclimate Evidence, 106
PAMELA; Payload for Antimatter Matter Exploration & Light Nuclei
Astrophysics, 423
Paradox of a Continuous Cycle that Ends in a Discontinuous Point, 326
Paradoxical Views of Man's Climate Interactions, 105
Particle Energy Absorption and Radiation; Half Wavelengths, 338
Per Capita US Consumption, 183
Period Year Transformation, 36
Periods; Historical, Archeological, & Current, 36
PET; Positron Emission Tomography, 273
Phase Boundaries; Energy and Matter Phase Transitions, 331-342
Phase Diagram, 354
Phase Transitions; Thermodynamic and Non-Thermodynamic, 331-342
Phase Transitions; Universal, Non-Analytic, & Largely Matters of Scale, 331-
332
Photosynthesis, Quantum, 366
Planck distance; 10^{-35} centimeters, 340
Poisson Distribution, 51,157
Population and the Rate of Genetic change, 404
Post Traumatic Stress Disorder (PTSD) and the DMN, 281
Posterior Cingulate Cortex (Precuneus), 272
Pre 1995 Multi-Sky Survey Face-on Image of GTW, 429
Pre and Post Transformation States, 372-380
Precession Periods; Literature Estimates of, 146-147
Precession, Degrees of Maya Solar System Precession, 35
Prediction of Bottlenecks, Mutation, and Migrations, 66-68
Prediction of CO_2 Levels, 120
Prediction of Global Temperature, 120
Prediction of Hominid Speciation and Extinction Events, 65
Prediction of the Dates of Human Change, 41
Pre-industrial Age, 54
Principal of Universality; Physics, 336
Prior and Pending World Age Transformations, 356
Prodynorphin Gene, 413
Pyramid of the Sun, 241

Q-R

Quantum Photosynthesis, 364
Radiometric & Accelerated Mass Spectroscopy (AMS) Dating, 44
Rate of Human Innovation, 231, 233-246
Raw Years Ago, 36
Real Stock Price Index, 195
Real US Consumption, 185
Real US Gross National Product (GNP), 184
Rectangular Distribution, 10
Red List of Threatened Species, ICUN, 416
Relation of Maya Epochs to Earth Motion Geometry, 99
Relation; Degrees of Precession with Obliquity, Inclination, & Eccentricity, 99
Renaissance, 54
Renormalization at the Plank distance, 340
Resonance Frequency; Physics, 329
Resonant Synchronization of Energy Emission, Absorption, & Exchange, 348
Resonant Synchronization; Dates, Maya, 218

S

Saccule, Inner Earth, 259
Sagittarius A*, 142
Sagittarius A*; Rotation Rate and Emissions, 142-143
Scale Unification - A Universal Scaling Law for Organized Matter, 339
Schizophrenia and the DMN, 281
Seasonal Variations in Microseism Hum, 257
Second Law of Thermodynamics, 434
Second; Time Unit, Determination, 223
Sefer Yetzirah ; the Book of Creation or Formation, 342
Seismic Background Energy Noise, 254
Seismic Noise from its Long Range Correlation Properties, 257
Seismographs; Instrument, 254
Semicircular Canals of the Inner Ear, 259
Sensitivity of the Vestibular System, 261
Sexual Selection, 11
Shape of a Phase Transition and Transformation, 379
Sidereal Year, 79
Simultaneity; Parallel Emergence of Innovations, 23
Social Diffusion of Innovation and Change; Process, 24
Social System; Definition, 26
Solar (Tropical) Year, 79
Solar System Galactic Orbital Speed, 91
Solar System Precession, 80

Solar System Z-Axis Vertical Gain, 88
Solar Wind; Bubble, 173
Solutions Involving torque and Coriolis Effects, 340
Source of the Default Mode Net Rhythm, 275
Sourcing and Forcing GTW Wave Trains, 161
Space Age, 54
Space-Time Compression and Decompression, 353
Space-Time Density Changes, 345
Species Perfecting Optimization; Transformation &, 358
Speed of Light, 6, 132
Spin Horizon' as a result of Space-Time Torque, 340
Stable and Unstable Transitions and Outcomes; Physics, 334
States, spin and angles of an atom, 337
Static Equilibrium, 27
Statistical Significance Level, 4
Statocyst, 258-263
Stereocilia Hair Cells, 269
Stochastic Mechanism; economics, 181
Stock Price Index, 195
Stopping and Releasing Photonic Energy, 364
Stress of the run-up to the Transformation critical point, 372
Sun Distance to the Galactic Center; Literature Estimates of, 149
Superconducting Gravimeter (SG), 248
Super-Massive Black Hole, 142
Suprachiasmatic Nuclei and Pineal Gland of the Brain, 285

T

Testing for World Age Epochs, 47
Testing Gravitational Torque Wave Economic Effects, 195
Testing Gravitational Torque Waves, 165
The *Katuns* (prophesies of the *Chilam Balam*), 375
The Maya World View, 323
Tikal's Pyramid of the Great Jaguar, 241
Torque Force; Definition, 135
Tourette Syndrome and the DMN, 281
Transformation Precedence, 331
Transformation Waveform Collapse; History &, 359
Transformation; Maya Prophecy-Prediction, 57, 59, 130
Transforming Cosmic Makeover, 434
Transition & Transformation Interval, 351
Tsunami Trains of Transition, 323
Tun Cycle; Maya, 205, 226, 227, 228, 319

Two-hundred billion cosmic ray collisions, 422

U

Uinal Cycle; Maya, 205, 226, 227, 228, 319
Unified Theory of Celestial Motion, 74
United Nations (UN) Intergovernmental Panel on Climate Change (IPCC), 105
Universal Equation of State for Any Phase Transition in Any System, 341
Universal Scaling Law for All Organized Matter, 339
US Consumer Price Index (CPI), 190
US Gross Domestic Product (GDP), 186
US Money Supply, 186
US National Debt Outstanding, 193

V

Vacuum Dark Energy (DE), 142
Velocity of Money, 187
VERITAS; the Very Energetic Radiation Imaging Telescope Array System, 424
Vestibular Hair Cell Bundles; Stereocilia Hairs, 269
Vestibular Inputs and Distribution, 259-269
Vestibular Neurons, 259-269
Vestibular Nuclei, 259-269
Vestibular Resolution of Human Head & Body Motion, 264
Vestibular Sensory Apparatus, 259
Voyager; 1 and 2, 171

W

Warp Drive, 353
WARPS Predictions, 354-361
WARPS; Falsifiable Theory, 358
Wave Amplitudes (γ) (Wave Height), 134, 300
Wavelength, 136, 300
Wave Arrival Times (Peak, Zero Crossing, & Valley), 300
Wave Frequencies, 300
Wave Number (k), 300
Wave Train Signals Switches and Sidings, 399
Wave Velocities (Speed), 300
Wave Warp Drive, 353
Wavelengths (λ), 136, 300
Wavelengths; Epoch Pairs, 48

Wavelets of Synchronization, 217
Waves of Cascading Transformation, 363
Waves of Celestial Motion, 73
Waves of Climatic Change, 105
Waves of Economic Change, 177
Waves of Gravitational Torque, 129
Waves of Human Change, 61
Waves of Human Evolution, 61
Waves of Human Influence, 297
Waves of Humming Earth, 247
Waves of Humming Identity, 271
Waves of World Age Epochs, 47
Waves of World Ages, 161
Waves; Definition and Representation, 136, 300
Weather Born Change Illusion, 110
Winter Solstice, 224
Winter Solstice of Friday December 21, 2012, 18, 58, 235, 250
World Age Circle; Cycle, 12
World Age Resonant Phase Synchronization (WARPS), 343
World Age Transformation, 323
World Age Wave Force Criteria, 132
World Population from 10,000 BC to 2,000 AD, 408

XYZ

PAUL R. DREWFS
CHANGE;
DESCRIBED-EXPLAINED-PREDICTED

Educated at the Portland Art Museum School, the Nova Scotia College of Art and Design, Florida State University, and Memphis State University, Paul R. Drewfs holds degrees in both the arts and sciences. With 38 years experience as a practicing scientist, he worked for 22 of those years as a Research Scientist for Science Applications International Corporation (SAIC). Today he is an independent research scientist, author, and sculptor living and working in New Mexico.

www.ingramcontent.com/pod-product-compliance
Lightning Source LLC
Chambersburg PA
CBHW020719180526
45163CB00001B/27